Organic Pollutants: Updated Reviews

Organic Pollutants: Updated Reviews

Edited by **Bruce Horak**

New York

Published by Callisto Reference,
106 Park Avenue, Suite 200,
New York, NY 10016, USA
www.callistoreference.com

.

Organic Pollutants: Updated Reviews
Edited by Bruce Horak

© 2015 Callisto Reference

International Standard Book Number: 978-1-63239-497-2 (Hardback)

Printed in the United States of America.

Contents

Preface   VII

Methods of Decontaminating
the Environment from Organic Pollutants   1

Chapter 1   Fundamental Mechanistic Studies of the Photo-Fenton
Reaction for the Degradation of Organic Pollutants   3
Amilcar Machulek Jr., Frank H. Quina, Fabio Gozzi,
Volnir O. Silva, Leidi C. Friedrich and José E.F. Moraes

Chapter 2   Fenton´s Process for the Treatment
of Mixed Waste Chemicals   25
Cláudia Telles Benatti and Célia Regina Granhen Tavares

Chapter 3   Photocatalytic Degradation
of Organic Pollutants: Mechanisms and Kinetics   49
Malik Mohibbul Haque, Detlef Bahnemann
and Mohammad Muneer

Chapter 4   Study on Sono-Photocatalytic Degradation of POPs:
A Case Study Hydrating Polyacrylamide in Wastewater   83
Fanxiu Li

Chapter 5   Chemical Degradation of Chlorinated
Organic Pollutants for In Situ Remediation
and Evaluation of Natural Attenuation   101
Junko Hara

Chapter 6   Research on Pressure Swing Adsorption
of Resin for Treating Gas Containing Toluene   121
Ruixia Wei and Shuguo Zhao

Chapter 7   **Electrochemical Incineration of Organic**
             Pollutants for Wastewater Treatment:
             Past, Present and Prospect   137
             Songsak Klamklang, Hugues Vergnes,
             Kejvalee Pruksathorn and Somsak Damronglerd

Chapter 8   **Organic Pollutants Treatment**
             from Air Using Electron Beam
             Generated Nonthermal Plasma – Overview   155
             Yongxia Sun and A. G. Chmielewski

Chapter 9   **Vapor Phase Hydrogen Peroxide –**
             Method for Decontamination of Surfaces
             and Working Areas from Organic Pollutants   179
             Petr Kačer, Jiří Švrček, Kamila Syslová, Jiří Václavík,
             Dušan Pavlík, Jaroslav Červený and Marek Kuzma

Chapter 10   **Alternative Treatment of Recalcitrant**
             Organic Contaminants by a Combination
             of Biosorption, Biological Oxidation
             and Advanced Oxidation Technologies   211
             Roberto Candal, Marta Litter, Lucas Guz, Elsa López Loveira,
             Alejandro Senn and Gustavo Curutchet

             Permissions

             List of Contributors

Preface

Even a decade after the Stockholm Convention on Persistent Organic Pollutants (POPs) was signed, a large variety of organic chemicals still pose ecological hazards of top most priority. The expansion of information base on organic pollutants (OPs), environmental fate and effects, as well as the cleansing methods, is accompanied by a rise in importance of certain pollution sources, associated with possible generation of new hazards for humans and nature. This book specifies these details, particularly in the light of Organic Pollutants risk assessment and analyzing methods of decontaminating our environment from Organic Pollutants. Providing an analytical and environmental update, this text can specifically be important for engineers and experts.

The information contained in this book is the result of intensive hard work done by researchers in this field. All due efforts have been made to make this book serve as a complete guiding source for students and researchers. The topics in this book have been comprehensively explained to help readers understand the growing trends in the field.

I would like to thank the entire group of writers who made sincere efforts in this book and my family who supported me in my efforts of working on this book. I take this opportunity to thank all those who have been a guiding force throughout my life.

Editor

Methods of Decontaminating the Environment from Organic Pollutants

Fundamental Mechanistic Studies of the Photo-Fenton Reaction for the Degradation of Organic Pollutants

Amilcar Machulek Jr.[1], Frank H. Quina[2], Fabio Gozzi[1],
Volnir O. Silva[2], Leidi C. Friedrich[2] and José E. F. Moraes[3]
[1]Universidade Federal de Mato Grosso do Sul, Departamento de Química – UFMS,
[2]Universidade de São Paulo, Instituto de Química and NAP-PhotoTech – USP,
[3]Universidade Federal de São Paulo, Escola Paulista de Engenharia – UNIFESP,
Brazil

1. Introduction

Very few regions of the planet possess abundant fresh water and access to adequate fresh water resources can be expected to worsen as a result of population growth and industrial demands for water. Liquid effluents containing toxic substances are generated by a variety of chemistry-related industrial processes, as well as by a number of common household or agricultural applications. The inadequate management of these residues can cause contamination of the soil and of subterranean and surface water sources.

In general, the recovery of industrial effluents containing low levels of organic substances by conventional treatments is not economically viable. Thus, for example, removal of the pollutant by adsorption onto active charcoal, while often efficient, requires subsequent recovery or incineration of the charcoal and merely transfers the pollutant from one phase to another (Matthews, 1992). Substances that are biocides or that are non-biodegradable represent a particular threat to the environment and prevent the use of conventional biological treatments. Social and legal demands for environmental safety increasingly require that effluents discharged into the environment have minimal impact on human health, natural resources and the biosphere. These demands have fueled increasing research into the development of new, more effective and economically viable methods for pollution control and prevention. When applied to the degradation of pollutants, these reactions are usually grouped together under the designations of Advanced Oxidation Processes (AOP) or Advanced Oxidation Technologies (AOT).

2. The principal advanced oxidation processes (AOP)

Advanced Oxidation Processes typically employ chemical oxidizing agents in the presence of an appropriate catalyst and/or ultraviolet light (Legrini, et al., 1993; Sonntag, 2008; Matilainen & Sillanpää, 2010) to oxidize or degrade the pollutant of interest. AOPs have been employed for the degradation of a variety of organic pollutants, such as aliphatic and aromatic hydrocarbons, halocarbons, phenols, ethers, ketones, etc. Examples of the major types of AOPs that have been proposed in the literature include (oxidant [catalyst, when present]/light): H_2O_2/UV (Gryglik et al., 2010; Ho & Bolton, 1998); O_3/UV (Esplugas et al., 1994; Machulek et al., 2009a); O_3-H_2O_2/UV (Yue, 1993); [TiO_2]/UV (Gaya & Abdullah, 2008; Henderson, 2011; Jenks, 2005; Matthews, 1992); Fe(III)/[TiO_2]/UV-Vis (Domínguez et al., 1998); direct photolysis of water with vacuum UV (Gonzalez et al., 2004); Fenton reaction or H_2O_2-Fe(II) (Dao & Laat, 2011; Haddou et al., 2010; Kwon et al., 1999; Pignatello et al., 2006; Pontes et al., 2010); and the photo-Fenton reaction or H_2O_2 [Fe(II)/Fe(III)]/UV (Benitez et al., 2011; Huston & Pignatello, 1999; Kim & Vogelpohl, 1998; Kiwi et al., 1994; Machulek et al., 2007; Martyanov et al., 1997; Nichela et al., 2010; Pignatello et al., 2006; Ruppert et al., 1993).

In most AOP, the objective is to use systems that produce the hydroxyl radical, $HO^{\bullet}$, or another species of similar reactivity such as the sulfate radical anion ($SO_4^{\bullet-}$). The hydroxyl radical is one of the most reactive species known in aqueous solution, surpassed only by fluorine atoms, and reacts with the majority of organic substances with little or no selectivity and at rates often approaching the diffusion-controlled limit (unit reaction efficiency per encounter). The principal modes of reaction of $HO^{\bullet}$ with organic compounds include hydrogen abstraction from aliphatic carbon, addition to double bonds and aromatic rings, and electron transfer (Bauer & Fallmann, 1997). These reactions generate organic radicals as transient intermediates, which then undergo further reactions, eventually resulting in final products (P_{oxid}) corresponding to the net oxidative degradation of the starting molecule (R). Overall, a typical AOP process can be can be generically represented as:

$$HO^{\bullet} + R \rightarrow \rightarrow \rightarrow \rightarrow \rightarrow P_{oxid} \tag{1}$$

Evidently, however, this representation belies the underlying complexity of the intermediate steps that occur on the pathway(s) from R to P_{oxid}.

From an environmental standpoint, one must be sure that the degradation of the initial pollutant does not produce intermediate products that are as toxic as or more toxic than the initial pollutant that one wishes to degrade. Thus, it is absolutely essential that the progress of the degradation and the final degradation products be adequately characterized. Since the degradation of organic pollutants by the hydroxyl radical is typically stepwise, even a relatively simple molecule like 4-chlorophenol can give rise to a plethora of intermediates, with from six to two carbon atoms, in various states oxidation (Li et al., 1999a, 1999b; Jenks, 2005) before converging to the final mineralization products, CO_2, water and HCl. Identification and quantification of all of the intermediates and determination of the kinetics and mechanisms of the individual reactions represent formidable tasks. For actual industrial effluents or wastewaters, which are often complex mixtures of pollutants, the intricacy of the degradation reactions can be enormous and one or more of the components or intermediates may be resistant to degradation and accumulate in the system. Not surprisingly, basic research into degradation pathways and mechanisms is still in its infancy for most of the AOP.

Photochemical and photocatalytic processes have enormous potential for becoming viable alternatives to conventional chemical AOP for the treatment of polluted waters and effluents. Currently available photochemical technology permits the conversion of organic pollutants having a wide range of chemical structures into substances that are less toxic and/or more readily biodegradable. In favorable cases, they can cause total decomposition of the organic constituents of the pollutant, generally referred to as "total mineralization" (complete oxidation to carbon dioxide and water, plus inorganic salts of all heteroatoms other than oxygen). Light of wavelengths in the range of 250-400 nm, corresponding to the ultraviolet (UV) region of the spectrum, is most commonly used in photochemical degradation processes (Braslavsky et al., 2011). Since ultraviolet light is a natural component of solar radiation, the sun provides a low-cost, environmentally friendly, renewable source of ultraviolet photons in photochemical processes. Thus, the use of solar photochemical reactors is an extremely interesting, cost-effective option for treatment of effluents in many of the tropical and sub-tropical regions of the planet. In areas with marginal or inadequate solar radiation intensity, conventional photochemical reactors fitted with ultraviolet lamps or hybrid UV lamp/solar photoreactors can be employed. In addition to light, the common AOPs use low to moderate concentrations of environmentally compatible chemical reagents and are capable, in favorable cases, of complete mineralization of the organic constituents of aqueous effluents. Although the initial or primary photochemical steps of the reactions employed to treat effluents can be conceptually rather simple, a large number of subsequent quite complex chemical steps are often involved in the overall degradation process.

Of the AOPs that have been proposed thus far, the Fenton reaction and the light-accelerated Fenton reaction, commonly known as the photo-Fenton reaction, appear to be the most promising for practical industrial applications (Pignatello et al., 2006). In the remainder of this chapter, we shall focus our attention on fundamental mechanistic details of the Fenton and photo-Fenton reactions, information that is essential for the adequate design and control of photo-Fenton processes for the degradation of organic pollutants.

3. The Fenton reaction

Over a century ago, Fenton (Fenton, 1894) demonstrated that a mixture of H_2O_2 and Fe(II) in acidic medium had very powerful oxidizing properties. Although the precise mechanism of this reaction, now known as the Fenton reaction, is still the subject of some discussion (Bossmann et al., 1998; Pignatello et al., 1999, 2006), it is generally assumed to be an important chemical source of hydroxyl radicals. The classical mechanism is a simple redox reaction in which Fe(II) is oxidized to Fe(III) and H_2O_2 is reduced to hydroxide ion and the hydroxyl radical:

$$Fe^{2+}_{aq} + H_2O_2 \rightarrow Fe^{3+}_{aq} + HO^{\bullet} + OH^- \qquad (2)$$

For the degradation of organic molecules, the optimum pH for the Fenton reaction is typically in the range of pH 3-4 and the optimum mass ratio of catalyst (as iron) to hydrogen peroxide is 1.5 (Bigda, 1995).

In the conventional Fenton reaction, carried out in the absence of light, the ferric ion produced in reaction (2) can be reduced back to ferrous ion by a second molecule of hydrogen peroxide:

$$Fe^{3+}_{aq} + H_2O_2 + H_2O \rightarrow Fe^{2+}_{aq} + H_3O^+ + HO_2^{\bullet-} \qquad (3)$$

However, this thermal reduction (reaction 3) is much slower than the initial step (reaction 2). Thus, although chemically very efficient for the removal of organic pollutants, the Fenton reaction slows down appreciably after the initial conversion of Fe(II) to Fe(III) and may require the addition of relatively large amounts of Fe(II) in order to degrade the pollutant of interest. Another important limitation of the Fenton reaction is the formation of recalcitrant intermediates that inhibit the complete mineralization. Particularly noteworthy is the formation of oxalic acid, $H_2C_2O_4$, a poisonous and persistent oxidation product of many degradation reactions. Since Fenton reactions are typically run at an initial pH of about 3 and oxalic acid is a relatively strong acid (with a first pK_a of 1.4), the accumulation of oxalic acid causes further acidification of the reaction mixture (to ca. pH 2) as the reaction proceeds. In addition, Fe(III) is very efficiently chelated by the oxalate anion. This prevents the reduction of Fe(III) back to Fe(II) and hence the complete mineralization of the organic matter. A still unresolved question in the mechanism of the Fenton degradation of organic material is the relative importance of other potential oxidizing species besides the hydroxyl radical. The stoichiometry of Fenton degradation reactions is complex and, in addition to Fe^{2+}/Fe^{3+} and hydrogen peroxide, can involve the participation of the hydroperoxyl radical, $HOO^\bullet$, iron(IV) or ferryl, FeO^{2+}, dissolved molecular oxygen, organic hydroperoxides, and other intermediates formed during the degradation. Despite these potential limitations, the conventional Fenton reaction has been widely used for the treatment of effluents (Benitez et al., 1999; Pignatello et al., 2006).

Any reaction or process that enhances the rate of conversion of Fe(III) back to Fe(II) will in principle accelerate the rate of the Fenton reaction. In the electro-Fenton reaction, this is accomplished electrochemically. Chemically, the Fenton reaction can be efficiently catalyzed by certain types of organic molecules, especially benzoquinones or dihydroxybenzene (DHB) derivatives. The catalytic influence of DHBs on the Fenton reaction was originally reported by Hamilton and coworkers (Hamilton et al., 1966a, 1966b). Because DHBs such as catechol or hydroquinone and their analogs are common initial intermediates in the degradation of aromatic molecules, their presence in the reaction medium can result in efficient DHB-catalyzed redox cycling of Fe^{3+} back o Fe^{2+} (Chen & Pignatello, 1997; Nogueira et al., 2005). Indeed, in several cases, it has been shown that the addition of catechol or catechol derivatives can enhance the rate and the overall mineralization efficiency of Fenton reactions (Aguiar et al., 2007; Zanta et al., 2010). In addition, DHB-catalyzed redox cycling of iron may be important in fungal degradation of lignin and several possible mechanisms have been suggested (Aguiar et al., 2007). The third method of accelerating the Fenton reaction is via irradiation with ultraviolet light, generally known as the photo-assisted Fenton or photo-Fenton reaction, which will be the primary focus of the remainder of this chapter.

4. The Photo-Fenton reaction

4.1 General features of the Photo-Fenton reaction

About two decades ago, it was found that the irradiation of Fenton reaction systems with UV/Visible light strongly accelerated the rate of degradation of a variety of pollutants (Huston & Pignatello, 1999; Ruppert et al., 1993). This behavior upon irradiation is due principally to the photochemical reduction of Fe(III) back to Fe(II), for which the net reaction can be written as:

$$Fe^{3+}_{aq} + H_2O \xrightarrow{h\upsilon} Fe^{2+}_{aq} + HO^\bullet + H^+ \tag{4}$$

More detailed studies of the pH dependence of the photo-Fenton reaction have shown that the optimum pH range is ca. pH 3. The reason for this pH dependence becomes clear when one examines the speciation of Fe(III) as a function of pH (Figure 1) and the absorption spectra of the relevant Fe(III) species (Figure 2). At pH < 2, the dominant species is hexaquoiron(III), $Fe(H_2O)_6^{3+}$ [or simply Fe^{3+} for convenience], which absorbs weakly in the ultraviolet above 300 nm. At pH > 3, freshly prepared solutions of Fe(III) are supersaturated with respect to formation of colloidal iron hydroxide, $Fe(OH)_3$ and prone to precipitation of hydrated iron oxides upon standing for a prolonged period. At pH 3, however, the predominant Fe(III) species present in aqueous solution is $Fe(H_2O)_5(OH)^{2+}$ [or simply $Fe(OH)^{2+}$], which absorbs throughout much of the ultraviolet spectral region (Martyanov et al., 1997).

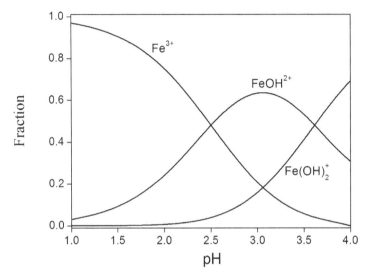

Fig. 1. Speciation of 0.5 mM Fe(III) between pH 1 and 4 at 25°C and an ionic strength of 0.1. Mole fractions of each species were calculated with the public domain program Hydra (Puigdomenech, 2010), employing the equilibrium constants for complexation supplied with the program (excluding insoluble iron species).

Studies of the photochemistry of $Fe(OH)^{2+}$ have shown (Pozdnyakov et al., 2000) that $Fe(OH)^{2+}$ undergoes a relatively efficient photoreaction to produce Fe(II) and the hydroxyl radical:

$$Fe(OH)^{2+} \xrightarrow{\;h\upsilon\;} Fe^{2+} + HO^{\bullet} \qquad (5)$$

Thus, irradiation of the Fenton reaction not only regenerates Fe(II), the crucial catalytic species in the Fenton reaction (reaction 2), but also produces an additional hydroxyl radical, the species responsible for the degradation of organic material. As a consequence of these two effects, the photo-Fenton process is faster than the conventional thermal Fenton process. Moreover, since Fe(II) is regenerated by light with decomposition of water (equations 4-5) rather than H_2O_2 (reaction 3), the photo-Fenton process consumes less H_2O_2 and requires only catalytic amounts of Fe(II).

The photo-Fenton reaction has several operational and environmental advantages. The classes of organic compounds that are susceptible to photodegradation via the Fenton reaction are rather well known (Bigda, 1995). The photo-Fenton process produces no new pollutants and requires only small quantities of iron salt. At the end of the reaction, if necessary, the residual Fe(III) can be precipitated as iron hydroxide by increasing the pH. Any residual hydrogen peroxide that is not consumed in the process will spontaneously decompose into water and molecular oxygen and is thus a "clean" reagent in itself. These features make homogeneous photo-Fenton based AOPs the leading candidate for cost-efficient, environmental friendly treatment of industrial effluents on a small to moderate scale (Pignatello et al., 2006). An early example of an industrial-scale application of the photo-Fenton process was the decontamination of 500 L batches of an industrial effluent containing 2,4-dimethylaniline in a photochemical reactor fitted with a 10 kW medium pressure mercury lamp (Oliveros et al., 1997).

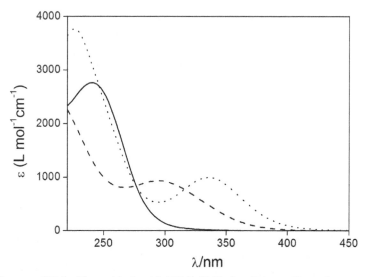

(a) Solid curve: pH 1.5 with perchloric acid, 90% $Fe(H_2O)_6{}^{3+}$ or Fe^{3+}, see Figure 1;
(b) Dashed Curve: pH 2.5 with perchloric acid, ca. 50:50 $FeOH^{2+}$:Fe^{3+}, see Figure 1;
(c) Dotted Curve: pH 1.5 plus added 500 mmol L^{-1} NaCl, 30% $FeCl^{2+}$ and 65% $FeCl_2{}^{+}$.

Fig. 2. Absorption spectra of 0.43 mM Fe(III) perchlorate under three typical conditions, expressed as the apparent extinction coefficient, ε, defined as the observed absorbance divided by the total Fe(III) concentration.

4.2 Inhibition of the Photo-Fenton reaction by chloride ion

Our interest in the fundamental mechanistic aspects of the photo-Fenton reaction arose from the necessity to optimize the degradation of organic material in effluents or wastewaters containing high concentrations of chloride ion. With such effluents, such as petroleum wastewaters from offshore marine environments or residues from organochloride pesticide production, the photo-Fenton reaction is strongly inhibited. Thus, although the photo-Fenton reaction results in essentially complete mineralization of phenol in the absence of chloride ion (Figure 3, curve e), in the presence of the chloride ion the mineralization process (indicated by the reduction in the amount of total organic carbon, TOC) stops after only partial decomposition of the organic material (Figure 3, curve c).

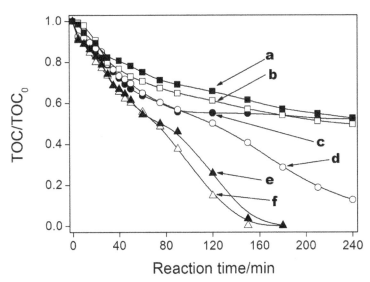

Fig. 3. Degradation of 12 mM phenol in the presence of 200 mM H_2O_2 and 0.5 mM Fe^{2+}. Fenton oxidation at $pH_{initial}$ 3.0 in the presence (curve a, ■) and absence (curve b, □) of 0.5 M NaCl. Photo-Fenton degradation (400 W medium pressure Hg vapor lamp – incident photon intensities were 1.2×10^{-4} Ein/s) at $pH_{initial}$ 3.0 with no pH control in the absence (curve e, ▲) or presence (curve c, ●) of 0.5 M NaCl; or with pH 3.0 maintained throughout the reaction in the absence (curve f, △) or presence of 0.5 M NaCl (curve d, ○). For details, see (Machulek et al., 2007).

In the presence of chloride ion, the extent of mineralization in the photo-Fenton reaction (Figure 3, curve c) is comparable to that observed in the analogous thermal Fenton reaction carried out in the dark in the presence or absence of chloride ion (Figure 3, curves a and b). Although we (Moraes et al., 2004a, 2004b) and others (De Laat & Le, 2006; Kiwi et al., 2000; Maciel et al., 2004; Pignatello, 1992) had ascribed this inhibition to the preferential formation of the less-reactive $Cl_2^{\bullet-}$ radical anion instead of the desired hydroxyl radical, optimization of photo-Fenton reactions required a fuller understanding of the mechanistic details of the inhibition. For this purpose, we (Machulek et al., 2006) used nanosecond laser flash photolysis to investigate the influence of added chloride ion on the photocatalytic step that converts Fe(III) back to Fe(II) (equation 5), deliberately omitting H_2O_2 from the reaction mixture to prevent the thermal Fenton reaction. Although direct spectroscopic detection of the hydroxyl radical has proved elusive (Marin et al., 2011), the $Cl_2^{\bullet-}$ radical anion, which absorbs at 340 nm (ε_{340nm} = 8000 $M^{-1}cm^{-1}$), can be readily detected upon excitation of aqueous solutions of iron(III) at acidic pH in the presence of added sodium chloride at 355 nm with the third harmonic of a Nd-YAG laser. Differential absorption spectra (Figure 4) and kinetic traces (insert, Figure 4) showed fast formation of the $Cl_2^{\bullet-}$ radical anion, within the lifetime of the laser pulse (5 ns), and its subsequent decay via mixed first and second order kinetics. The net decrease in absorption at longer times, relative to that prior to the laser pulse, reflects the conversion of Fe(III) to Fe(II), which does not absorb in this spectral region.

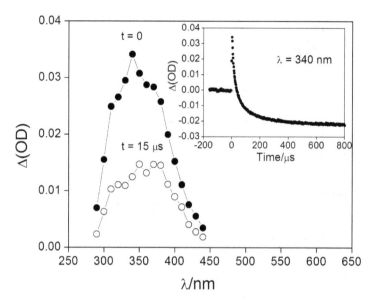

Fig. 4. Laser flash photolysis at 355 nm of 0.50 mM Fe(III) in the presence of 0.5 M NaCl at pH 1. Transient absorption spectra of $Cl_2^{\bullet-}$ immediately ($\bullet$) and 15 μs (○) after the laser pulse. The insert shows the kinetics of $Cl_2^{\bullet-}$ disappearance monitored at λ= 340 nm.

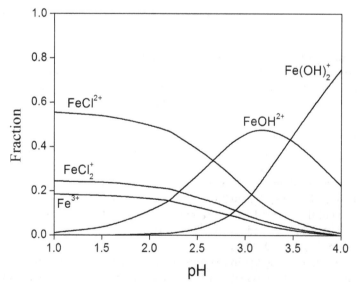

Fig. 5. Speciation of 0.5 mM Fe(III) between pH 1 and 4 at 25°C and an ionic strength of 0.2 in the presence of 0.1 M NaCl. Mole fractions of each species were calculated with the public domain program Hydra (Puigdomenech, 2010), employing the equilibrium constants for complexation supplied with the program (excluding insoluble iron species).

Laser flash photolysis data for the production and decay of $Cl_2^{\bullet-}$ were then obtained over a wide range of pH and concentration of Fe(III) and chloride ion and these data used to develop an explicit mechanistic model for the initial photoinduced processes involved in the photo-Fenton reaction in the presence of chloride ion (Machulek et al., 2006). As indicated in Figure 5, in the presence of chloride ion, the dominant species present at pH < 2.5 is no longer Fe^{3+}, but rather $FeCl^{2+}$ and $FeCl_2^{+}$ (coordinated waters being omitted).

Moreover, $FeCl^{2+}$ not only absorbs further out in the ultraviolet than $Fe(OH)^{2+}$ (Figure 2), but it also undergoes photolysis (equation 6) with a quantum yield higher than that of $Fe(OH)^{2+}$ (equation 5).

$$Fe(Cl)^{2+} \xrightarrow{\;h\upsilon\;} Fe^{2+} + Cl^{\bullet} \tag{6}$$

The photoproduced chlorine atoms rapidly react with chloride ions to form $Cl_2^{\bullet-}$ (equation 7), the species actually detected in the laser flash photolysis experiments:

$$Cl^{\bullet} + Cl^{-} \rightarrow Cl_2^{\bullet-} \tag{7}$$

and whose second order decay component is due to the disproportionation reaction:

$$Cl_2^{\bullet-} + Cl_2^{\bullet-} \rightarrow 2\,Cl^{-} + Cl_2 \tag{8}$$

The model employed to fit the experimentally observed decay curves for $Cl_2^{\bullet-}$ took into account the initial photochemical formation of the hydroxyl radical from $Fe(OH)^{2+}$ (equation 5) and of chlorine atoms from $FeCl^{2+}$ (equation 6) and the subsequent reactions of these via the set of elementary steps outlined in Table 1 (Machulek et al., 2006).

No.	Reaction	k (s⁻¹ or mol⁻¹ L s⁻¹)
Speciation Equilibria (I = 1)		
1	$Fe^{3+} + Cl^{-} \rightleftarrows FeCl^{2+}$	$k_1 = 4.79 \times 10^{10}$, $k_{-1} = 1 \times 10^{10}$
2	$Fe^{3+} + 2\,Cl^{-} \rightleftarrows FeCl_2^{+}$	$k_2 = 6.31 \times 10^{10}$, $k_{-2} = 1 \times 10^{10}$
Reactions of Iron Species		
3	$Fe^{2+} + Cl^{\bullet} \rightarrow Fe^{3+} + Cl^{-}$	$k_3 = 5.9 \times 10^9$
4	$Fe^{2+} + Cl_2^{\bullet-} \rightarrow Fe^{3+} + 2Cl^{-}$	$k_4 = 5 \times 10^6$
Reactions of Reactive Chlorine Species		
5	$Cl^{\bullet} + Cl^{-} \rightleftarrows Cl_2^{\bullet-}$	$k_5 = 7.8 \times 10^9$, $k_{-5} = 5.7 \times 10^4$
6	$Cl^{\bullet} + H_2O_2 \rightarrow HO_2^{\bullet} + Cl^{-} + H^{+}$	$k_6 = 1 \times 10^9$
7	$Cl_2^{\bullet-} + Cl_2^{\bullet-} \rightarrow 2Cl^{-} + Cl_2$	$k_7 = 2.8 \times 10^9$
8	$Cl_2^{\bullet-} + H_2O_2 \rightarrow HO_2^{\bullet} + 2Cl^{-} + H^{+}$	$k_8 = 1.4 \times 10^5$
9	$Cl_2^{\bullet-} + HO_2^{\bullet} \rightarrow 2Cl^{-} + H^{+} + O_2$	$k_9 = 3.1 \times 10^9$
10	$Cl^{-} + HO^{\bullet} \rightleftarrows HOCl^{\bullet-}$	$k_{10} = 4.2 \times 10^9$, $k_{-10} = 6.1 \times 10^9$
11	$HOCl^{\bullet-} + H^{+} \rightleftarrows H_2O + Cl^{\bullet}$	$k_{11} = 2.4 \times 10^{10}$, $k_{-11} = 1.8 \times 10^5$
12	$Cl_2^{\bullet-} + Cl^{\bullet} \rightarrow Cl^{-} + Cl_2$	$k_{12} = 1.4 \times 10^9$

Table 1. Set of reactions required to simulate the kinetics for production and decay of $Cl_2^{\bullet-}$ upon laser flash photolysis of Fe^{3+} in the presence of chloride ion over a wide range of pH and Fe^{3+} and NaCl concentrations.

The initial concentrations of HO$^\bullet$ and Cl$^\bullet$ produced at t = 0 by the laser pulse were calculated from the relationship (equation 9):

$$\left[X^\bullet\right]_{t=0} = \left(I_{Laser}\right) \left(\frac{A_{Fe(X)}}{A_T}\right) \left(1-10^{A_T}\right) \left(\Phi_X\right) \tag{9}$$

where X = OH or Cl and Φ_{HO} = 0.21 and Φ_{Cl} = 0.47 are, respectively, the quantum yields for production of Fe(I1) from Fe(OH)$^{2+}$ or Fe(Cl)$^{2+}$ at 355 nm. $A_{Fe(X)}$ is the absorbance of either Fe(OH)$^{2+}$ or Fe(Cl)$^{2+}$ and A_T the total absorbance of the solution, both at the laser excitation wavelength of 355 nm (in an effective optical path length of 0.5 cm). The ratio $A_{Fe(X)}/A_T$ was calculated from the initial concentrations of the species Fe(Cl)$^{2+}$, Fe(Cl)$_2^+$, Fe^{3+}, Fe(OH)$^{2+}$, Fe(OH)$_2^+$ and [Fe$_2$(H$_2$O)$_4$(OH)$_2$]$^{4+}$, together with molar absorption coefficients at 355 nm taken from the literature absorption spectra (Byrne & Kester, 1978, 1981). For each experimental condition (pH, [Cl-], [Fe(III)]), the relative concentrations of the iron(III) species were calculated with the public domain speciation program Hydra (Puigdomenech, 2010), employing the set of standard equilibrium constants for complexation supplied with the program. The incident laser intensity (I$_{laser}$) was estimated by fitting the transient absorbance of Cl$_2^{\bullet-}$ at time zero at pH 1.0 in the presence of 0.5 M NaCl, where competitive photolysis of Fe(OH)$^{2+}$ is negligible. The concentrations of HO$^\bullet$ and Cl$^\bullet$ at t = 0 calculated from equation 9 served as the initial conditions for numerical solution of the set of reactions and rate constants listed in Table 1. This kinetic model provided a quantitative fit of the observed transient decay curves (like that shown in the inset of Figure 4) over the entire range of pH and concentrations of chloride ion and Fe(III) investigated.

Having established the basic mechanistic scheme for the photoinduced steps of the photo-Fenton reaction, one can then use it to infer the course of a typical photo-Fenton degradation in the presence of chloride ion. At the beginning of the photo-Fenton reaction, when the pH is still ca. 3.0, the concentration of Fe(OH)$^{2+}$ exceeds that of FeCl^{2+} or FeCl$_2^+$, even in the presence of relatively high concentrations of chloride ion (Figure 5). The pH-dependent scavenging of HO$^\bullet$ by chloride ion (equations 10-11):

$$HO^\bullet + Cl^- \rightleftarrows HOCl^{\bullet-} \tag{10}$$

$$H^+ + HOCl^{\bullet-} \rightleftarrows H_2OCl^\bullet \rightarrow H_2O + Cl^\bullet \tag{11}$$

is also still a relatively inefficient process, so that the photochemical formation of HO$^\bullet$ should predominate. However, as the photo-Fenton process proceeds, partial degradation of the organic material causes the pH of the medium to fall to ca. pH 2.0, where photolysis of FeCl^{2+} dominates over photolysis of Fe(OH)$^{2+}$ and where the chloride ion efficiently converts any HO$^\bullet$ formed in the system into the intrinsically much less reactive Cl$_2^{\bullet-}$ radical anion (Bacardit et al., 2007; Buxton et al., 1999; De Laat et al., 2004; George & Chovelon, 2002; Kiwi et al., 2000; McElroy, 1990; Moraes, et al., 2004a; Nadtochenko & Kiwi, 1998; Pignatello, 1992; Soler et al., 2009; Truong et al., 2004; Yu, 2004; Yu et al., 2004; Yu & Barker, 2003a, 2003b; Zapata et al., 2009). As a result, virtually complete inhibition of the photo-Fenton degradation of typical organic substrates (Kiwi & Nadtochenko, 2000; Machulek et al., 2007; Moraes et al., 2004a, 2004b; Pignatello, 1992) will occur at moderate chloride ion concentrations [>0.03 M NaCl for aliphatic hydrocarbons (Moraes et al., 2004a); >0.2 M NaCl for phenols] when the pH of the medium reaches pH 2.0 or below. On the other hand, this sequence of events clearly indicates that it should be possible to circumvent the inhibition by chloride ion by simply maintaining the pH at or slightly above pH 3 throughout the degradation process (see Figure 5). This is nicely illustrated by curve d in Figure 3, which shows that pH control does indeed permit nearly complete mineralization of phenol in the presence of chloride ion (Machulek et al., 2006, 2007). Although pH control also enhances the rate of the photo-Fenton reaction in the absence of chloride ion by optimizing the concentration of $Fe(OH)^{2+}$ (see Figure 1), the effect is much less dramatic (compare curves e and f in Figure 3).

Although the mechanism outlined in Table 1 nicely rationalizes the kinetics of formation and decay of $Cl_2^{\bullet-}$ on a fast time scale, it does not permit prediction of the accumulation of Fe(II) in the same system in the absence of H_2O_2 under steady-state irradiation. For this purpose, one must include additional kinetic steps that are too slow to be relevant on the time scale of the laser flash photolysis experiments. Experimentally (Machulek et al., 2009b) it is found that the concentration of Fe^{2+} eventually reaches a plateau value at long times, more quickly in the presence of chloride ion than in its absence (Figure 6). Simulation of this behavior required the inclusion of additional kinetic steps, in particular for the back reactions that result in reoxidation of Fe^{2+}. These additional kinetic steps are listed in Table 2.

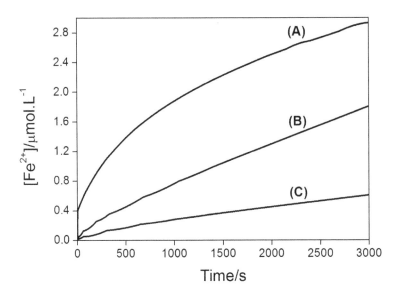

Fig. 6. The accumulation of ferrous ion during irradiation (350 nm) of 1.0 mM Fe(III) at: (A) pH 1.0 in the presence of 0.75 mol L^{-1} chloride ion; (B) at pH 3 in the absence of chloride ion; and (C) at pH 3 in the presence of 1.8 mmol L^{-1} sulfate ions.

No.	Reaction	k (s^{-1} or mol^{-1} L s^{-1})
Speciation Equilibria (I = 1)		
1	$Fe^{3+} + H_2O \rightleftarrows FeOH^{2+} + H^+$	$k_1 = 1.91 \times 10^7$, $k_{-1} = 1 \times 10^{10}$
2	$Fe^{3+} + 2H_2O \rightleftarrows Fe(OH)_2^+ + 2H^+$	$k_2 = 3.39 \times 10^3$, $k_{-2} = 1 \times 10^{10}$
3	$2Fe^{3+} + 2H_2O \rightleftarrows [Fe_2(OH)_2]^{4+} + 2H^+$	$k_3 = 1.12 \times 10^7$, $k_{-3} = 1 \times 10^{10}$
Reactions of Reactive Oxygen Radicals		
4	$HO^{\bullet} + HO^{\bullet} \rightarrow H_2O_2$	$k_4 = 6.0 \times 10^9$
5	$HO^{\bullet} + H_2O_2 \rightarrow H_2O + HO_2^{\bullet}$	$k_5 = 2.7 \times 10^7$
6	$HO_2^{\bullet} + HO_2^{\bullet} \rightarrow H_2O_2 + O_2$	$k_6 = 8.3 \times 10^5$
7	$HO_2^{\bullet} + H_2O_2 \rightarrow HO^{\bullet} + O_2 + H_2O$	$k_7 = 0.5$
Reactions of Iron Species		
8	$Fe^{2+} + HO^{\bullet} \rightarrow Fe^{3+} + OH^-$	$k_8 = 4.3 \times 10^8$
9	$Fe^{2+} + H_2O_2 \rightarrow Fe^{3+} + HO^{\bullet} + OH^-$	$k_9 = 63$
10	$Fe^{3+} + HO_2^{\bullet} \rightarrow Fe^{2+} + O_2 + H^+$	$k_{10} = 1 \times 10^6$
11	$Fe^{3+} + H_2O_2 \rightarrow Fe^{2+} + HO_2^{\bullet} + H^+$	$k_{11} = 0.01$

Table 2. Set of additional ground state reactions required to fit the accumulation of Fe^{2+} during the steady-state irradiation of Fe^{3+} in the presence and absence of chloride ion (Machulek et al., 2009b)

4.3 Inhibition of the Photo-Fenton reaction by sulfate ion

Also indicated in Figure 6 is the inhibition of the formation of Fe^{2+} in the presence of a relatively low concentration of added sulfate ions, based on the rate constants for complexation of Fe(III) by sulfate in Table 3. Sulfate ion is often present in photo-Fenton reactions as a result of the addition of iron in the form of readily available sulfate salts (De Laat & Le, 2005). Sulfate ion complexes strongly with Fe(III) over a wide pH range (Figure 7) and the quantum yield of production of Fe^{2+} plus a sulfate anion radical from photolysis of $Fe(SO_4)^+$ is only about 0.05 (Benkelberg & Warneck, 1995). Iron phosphate is even more photoinert than the iron sulfate complex (Benkelberg & Warneck, 1995; Lee et al., 2003) and should be an even more powerful inhibitor of the photo-Fenton reaction.

4.4 Catalysis of the Photo-Fenton reaction by complexation of Fe(III)

The efficiency of the photo-Fenton process can be further enhanced by using organic carboxylic acids to complex Fe(III) (Pignatello et al., 2006). A particularly important example is provided by oxalic acid. Thus, unlike the thermal Fenton reaction, in which oxalic acid is a recalcitrant intermediate, in the photo-Fenton reaction it can act as a catalyst. Thus, ferrioxalate complexes can absorb light as far out as 570 nm, i.e., well into the visible region of the spectrum. Moreover, upon irradiation, they decompose efficiently (quantum yields of the order of unity) to Fe(II) and CO_2. The net result is that, in the presence of oxalate, the photo-Fenton reaction is intrinsically more efficient, can be induced by a wider range of wavelengths of light, and results in the mineralization of the oxalate ion. Thus, for example, in a municipal water treatment system, Kim & Vogelpohl (1998) found that, with UV irradiation, the photo-Fenton process was at least 30% more energy efficient in the presence of oxalate than in its absence. Clearly, the sensitivity of the ferrioxalate-catalyzed photo-Fenton process to both UV and visible light makes it particularly attractive for applications in which the sun is employed as the radiation source (Silva et al., 2010; Trovó & Nogueira, 2011).

No.	Reaction	k (s^{-1} or mol^{-1} L s^{-1})
Speciation Equilibria (I = 1)		
1	$Fe^{3+} + SO_4^{2-} \rightleftarrows FeSO_4^+$	$k_1 = 2.09 \times 10^{12}, k_{-1} = 1 \times 10^{10}$ (I=1)
2	$Fe^{3+} + 2SO_4^{2-} \rightleftarrows Fe(SO_4)_2^-$	$k_2 = 1.95 \times 10^{13}, k_{-2} = 1 \times 10^{10}$ (I=1)
3	$Fe^{2+} + SO_4^{2-} \rightleftarrows FeSO_4$	$k_3 = 1.55 \times 10^{11}, k_{-3} = 1 \times 10^{10}$ (I=1)
Reactions of Sulfate Ions		
4	$H^+ + SO_4^{2-} \rightleftarrows HSO_4^-$	$k_4 = 2.8 \times 10^{11}, k_{-4} = 1 \times 10^{10}$ (I=1)
5	$HSO_4^- + HO^\bullet \rightarrow SO_4^{\bullet-} + H_2O$	$k_5 = 3.5 \times 10^5$
6	$SO_4^{\bullet-} + H_2O \rightarrow H^+ + SO_4^{2-} + HO^\bullet$	$k_6 = 6.6 \times 10^2$
7	$SO_4^{\bullet-} + OH^- \rightarrow SO_4^{2-} + HO^\bullet$	$k_7 = 1.4 \times 10^7$
8	$SO_4^{\bullet-} + H_2O_2 \rightarrow SO_4^{2-} + H^+ + HO_2^\bullet$	$k_8 = 1.2 \times 10^7$
9	$SO_4^{\bullet-} + HO_2^\bullet \rightarrow SO_4^{2-} + H^+ + O_2$	$k_9 = 3.5 \times 10^9$
10	$SO_4^{\bullet-} + SO_4^{\bullet-} \rightarrow S_2O_8^{2-}$	$k_{10} = 2.7 \times 10^8$
Reactions of Iron Species		
11	$Fe^{2+} + SO_4^{\bullet-} \rightarrow Fe^{3+} + SO_4^{2-}$	$k_{11} = 3.0 \times 10^8$

Table 3. Set of additional ground state reactions required to fit the accumulation of Fe^{2+} during the steady-state irradiation of Fe^{3+} in the presence of sulfate ion (Machulek et al., 2009b).

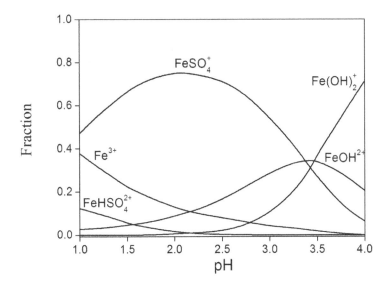

Fig. 7. Speciation of 1.0 mM Fe(III) between pH 1 and 4 at 25°C and an ionic strength of 0.2 in the presence of 1.8 mM sulfate. Mole fractions of each species were calculated with the public domain program Hydra (Puigdomenech, 2010), employing the equilibrium constants for complexation supplied with the program (excluding the insoluble iron species).

4.5 Concurrent Photo-Fenton and Fenton reactions

A final point that should be emphasized is that, under typical photo-Fenton reaction conditions, both the thermal Fenton reaction and the photo-Fenton reactions occur concurrently in the reaction mixture. Light is almost always the limiting reagent in photochemical reactions and photo-Fenton reactions are often conducted in reactors in which the solution is recirculated through the photoreactor from an external reservoir. Thus, at any given moment, the photo-Fenton reaction will be operative only in the portion of the total reaction mixture that is contained in the photoreactor *per se*, while the thermal Fenton reaction can proceed throughout the entire reaction volume. Indeed, comparison of the degradation of phenol under various conditions (Figure 3) strongly suggests that the initial phase of the reaction is dominated by the thermal reaction, with the inflection at about 80 min where the curves diverge indicating the point at which the photo-Fenton reaction becomes dominant.

Particularly interesting is the apparent insensitivity of the thermal Fenton degradation of phenol to the presence of added chloride ion, as indicated by the similarity of curves a and b of Figure 3. At the initial pH of 3 of the reaction mixture, the hydroxyl radical is inefficiently scavenged by chloride and produces, as the initial products of phenol degradation, the DHBs catechol and hydroquinone. Thus, unlike the photo-Fenton reaction, which depends on the intermediacy of the hydroxyl radical, the thermal degradation of phenol shifts to the DHB-catalyzed reaction mentioned above for the Fenton reaction until all of the DHBs have themselves been degraded (Chen and Pignatello, 1997). At that point, most of the iron(III) is rendered redox inert by complexation with oxalic acid and/or other recalcitrant aliphatic acids and the Fenton reaction ceases (Zanta et al., 2010).

5. Areas for future development

Despite significant advances in the last decade, the widespread successful application of photochemical technology for the treatment and decontamination of industrial residues and wastewaters is still not a reality. Research chemists and process engineers must work together to solve several important practical problems that limit the technological applications of the photo-Fenton reaction on a large scale (Braun et al., 1991). Since the overall effective quantum efficiency can be relatively low, long irradiation times must be employed in order to achieve total mineralization of the organic components of effluents. Electrical power consumption can then easily represent 60% of the total operating expense of a lamp-based photochemical reactor. Additional costs associated with non-solar photochemical reactors include cooling, maintenance and depreciation of the lamps and UV radiation shielding and protection for workers. One of the obvious economic benefits of the use of solar photochemical reactors is the elimination of costs associated with installation and maintenance of lamps and with electrical power consumption (Moraes et al., 2004b; Nascimento et al., 2007). In parallel with the development of solar reactors or hybrid lamp/solar reactors, an important area is the development of photocatalysts that operate effectively under visible or solar irradiation (Anpo, 2000).

Particularly in larger scale reactors, operational problems such as the formation of deposits on the reactor walls that block the incident radiation must be addressed. Optimization strategies (Braun et al., 1991, 1993; Cassano et al., 1995; Nascimento et al., 1994; Oliveros et al., 1998) must be applied to the design of reactors that homogenize the incidence of light on the reactor walls and maximize the total amount of light absorbed by the system. Reactor design must also take into account the strong interdependence between light absorption, mass transport and reaction kinetics in solutions with high optical densities and optimize the balance between the concurrent photochemical and thermal contributions to the Fenton reaction. Although the relatively long irradiation times required for the complete mineralization of organic contaminants tend to compromise the economic viability of stand-alone photo-Fenton processes, they can be viable as a pre-treatment to reduce the toxicity of the effluent to levels compatible with other remediation technologies such as conventional biological treatment. Hence, the design of photo-Fenton reactors should also take into account the necessity of facile integration with other effluent decontamination technologies. A final point that has only recently been considered in a more systematic context is the intensification of Fenton processes by increasing the temperature (Zazo et al., 2011), particularly attractive in the case of photo-Fenton reactions where a substantial part of the absorbed light energy is dissipated as heat in the solution.

6. Conclusions

Of the currently known advanced oxidation processes, the Fenton reaction and the photo-Fenton reaction appear to be the most promising for practical industrial applications on a moderate scale. The fundamental photomechanistic aspects of the photo-Fenton reaction have been clarified by studies of the photochemistry of Fe(IIII) in the absence of hydrogen peroxide in order to avoid the complications due to the competing Fenton reaction. The Fe(III) species primarily responsible for the initial photochemical generation of Fe(II) plus a hydroxyl radical is $Fe(OH)^{2+}$. Inhibition of the photo-Fenton by ions such as chloride or sulfate can be understood by considering the competitive pH-dependent complexation of Fe(III) by these ions, the competitive absorption and photochemistry of these complexes and/or the subsequent chemical reactions that can convert the hydroxyl radical into less reactive species. Adequate mechanisms, together with the relevant rate constants, are now available for the inhibition of the net photoconversion of Fe(III) to Fe(II) by both chloride and sulfate, allowing quantitative or semi-quantitative simulation of the experimentally observed inhibitory effects. As predicted by the mechanistic models, the inhibitory effect of chloride ion on real photo-Fenton degradations can be readily circumvented by simply maintaining the medium pH at ca. pH 3 throughout the reaction. Areas where additional progress would be welcome include an understanding of the role of temperature effects on the Fenton and photo-Fenton reactions and new strategies for further enhancing the efficiency of the photo-Fenton reaction. Improvements in photoreactor design are necessary in order to optimize the contribution of the photoinduced processes relative to the concurrent thermal Fenton reaction.

7. Acknowledgment

The authors acknowledge the Brazilian funding agencies CAPES and CNPq for financial and fellowship support. F.H.Q. is associated with NAP-PhotoTech, the USP Research Consortium for Photochemical Technology, and INCT-Catalysis. A.M. Jr. and J.E.F. Moraes are affiliated with INCT-EMA. A.M.Jr. thanks FUNDECT for financial support of the work at UFMS.

8. References

Aguiar, A.; Ferraz, A.; Contreras, D. & Rodríguez, J. (2007). Mecanismo e aplicação da reação de Fenton assistida por compostos fenólicos redutores de ferro. *Química Nova*, Vol. 30, No. 3, pp. 623-628, ISSN 0100-4042

Anpo, M. (2000). Utilisation of TiO_2 photocatalysts in green chemistry. *Pure and Applied Chemistry*, Vol. 72, No. 7, pp. 1265-1270, ISSN 0033-4545

Bacardit, J.; Stötzner, J.; Chamarro, E. & Esplugas, S. (2007). Effect of salinity on the photo-Fenton process. *Industrial & Engineering Chemistry Research*, Vol. 46, No. 23, pp. 7615-7619, ISSN 0888-5885

Bauer, R. & Fallmann, H. (1997). The photo-Fenton oxidation - a cheap and efficient wastewater treatment method. *Research on Chemical Intermediates*, Vol. 23, No. 4, pp. 341-354, ISSN 0922-6168

Benitez, F.J.; Beltran-Heredia, J.; Acero, J.L. & Rubio, F.J. (1999). Chemical decomposition of 2,4,6-triclorophenol by ozone, Fenton's reagent, and UV radiation. *Industrial & Engineering Chemistry Research*, Vol. 38, No. 4, pp. 1341-1349, ISSN 0888-5885

Benitez, F.J.; Acero, J.L.; Real, F.J.; Roldan, G. & Casas, F. (2011). Comparison of different chemical oxidation treatments for the removal of selected pharmaceuticals in water matrices. *Chemical Engineering Journal*, Vol. 168, No. 3, pp. 1149-1156, ISSN 1385-8947

Benkelberg, H. J. & Warneck, P. (1995). Photodecomposition of iron(III) hydroxo and sulfate complexes in aqueous solution: Wavelength dependence of OH and SO_4^- quantum yields. *Journal of Physical Chemistry*, Vol. 99, No. 14, pp. 5214-5221, ISSN 0022-3654

Bigda, R. (1995). Consider Fenton's chemistry for wastewater treatment. *Chemical Engineering Progress*, (Dec.), Vol. 91, No. 12, pp. 62-66, ISSN 03607275

Bolton, J. R. (1999). Light Compendium – Ultraviolet Principles and Applications. *Inter-American Photochemical Society Newsletter*, Vol. 22, No. 2, pp. 20-61

Bossmann, S.H.; Oliveros, E.; Göb, S.; Siegwart, S.; Dahlen, E.P.; Payawan, L.; Straub, M.; Wörner, M. & Braun, A.M. (1998). New evidence against hydroxyl radicals as reactive intermediates in the thermal and photochemically enhanced Fenton Reactions. *Journal of Physical Chemistry A*, Vol. 102, No. 28, pp. 5542-5550, ISSN 1089-5639

Braslavsky, S.E.; Braun, A.M.; Cassano, A.E.; Emeline, A.V.; Litter, M.I.; Palmisano, L.; Parmon, V.N. & Serpone, N. (2011). Glossary of terms used in photocatalysis and radiation catalysis (IUPAC recommendations 2011). *Pure and Applied Chemistry*, Vol. 83, No. 4, pp. 931-1014, ISSN 0033-4545

Braun, A.M.; Maurette, M.T. & Oliveros, E. (1991). *Photochemical Technology*. John Wiley & Sons, ISBN-13 978-0471926528, New York, USA

Braun, A.M.; Jakob, L.; Oliveros, E. & Nascimento, C.A.O. (1993). Up-scaling photochemical reactions, In: *Advances in Photochemistry*, Volman, D.H.; Hammond, G.S. & Neckers, D.C., Vol. 18, pp. 235-313, ISBN 9780471591337, John Wiley & Sons, USA

Buxton, G.V., Bydder, M. & Salmon, G.A. (1999). The reactivity of chlorine atoms in aqueous solution — Part II. The equilibrium $SO_4^{\bullet-} + Cl^- \rightarrow Cl^\bullet + SO_4^{2-}$. *Physical Chemistry Chemical Physics*, Vol. 1, No. 2, pp. 269–273, ISSN 1463-9076

Byrne, R.H. & Kester, D.R. (1978). Ultraviolet spectroscopic study of ferric hydroxide complexation. *Journal of Solution Chemistry*, Vol. 7, No. 5, pp. 373–383, ISSN 0095-9782

Byrne, R.H. & Kester, D.R. (1981). Ultraviolet spectroscopic study of ferric equilibria at high chloride concentration. *Journal of Solution Chemistry*, Vol. 10, No. 1, pp. 51–67, ISSN 0095-9782

Cassano, A.; Martín, C.A.; Brandi, R.J. & Alfano, O.M. (1995). Photoreactor analysis and design: fundamentals and applications. *Industrial & Engineering Chemistry Research*, Vol. 34, No. 7, pp. 2155-2201, ISSN 0888-5885

Chen, R. & Pignatello, J.J. (1997). Role of quinone intermediates as electron shuttles in Fenton and photoassisted Fenton oxidations of aromatic compouds. *Environmental Science & Technology*, Vol. 31, No. 8, pp. 2399-2406, ISSN 0013-936X

Dao, Y.H. & Laat, J. (2011). Hydroxyl radical involvement in the decomposition of hydrogen peroxide by ferrous and ferric-nitrilotriacetate complexes at neutral pH. *Water Research*, Vol. 45, No. 11, pp. 3309-3317, ISSN 0043-1354

De Laat, J.; Le, G.T. & Legube, B. (2004). A comparative study of the effects of chloride, sulfate and nitrate ions on the rates of decomposition of H_2O_2 and organic compounds by Fe(II)/H_2O_2 and Fe(III)/H_2O_2. *Chemosphere*, Vol. 55, No. 5, pp. 715-723, ISSN 0045-6535

De Laat, J. & Le, T. G. (2005). Kinetics and Modeling of the Fe(III)/H_2O_2 System in the Presence of Sulfate in Acidic Aqueous Solutions. *Environmental Science & Technology*, Vol. 39, No. 6, pp. 1811-1818, ISSN 0013-936X

De Laat, J. & Le, T. G. (2006). Effects of chloride ions on the iron(III)-catalyzed decomposition of hydrogen peroxide and on the efficiency of the Fenton-like oxidation process. *Applied Catalysis, B: Environmetal*, Vol. 66, No. 1-2, pp. 137-146, ISSN 0926-3373

Domínguez, C.; García, J.; Pedraz, M.A.; Torres, A. & Galán, M.A. (1998). Photocatalytic oxidation of organic pollutants in water. *Catalysis Today*, Vol. 40, No. 1, pp. 85-101, ISSN 0920-5861

Esplugas, S.; Yue, P.L. & Pervez, M.I. (1994). Degradation of 4-chlorophenol by photolytic oxidation. *Water Research*, Vol. 28, No. 6, pp. 1323-1328, ISSN 0043-1354

Fenton, H.J.H. (1894). Oxidation of tartaric acid in presence of iron. *Journal of the Chemical Society*, Vol. 65, pp. 899-901, ISSN 0368-1769

Gaya, U.I. & Abdullah, A.H. (2008). Heterogeneous photocatalytic degradation of organic contaminants over titanium dioxide: A review of fundamentals, progress and problems. *Journal of Photochemistry and Photobiology C: Photochemistry Reviews*, Vol. 9, No. 1, pp. 1-12, ISSN 1389-5567

George, C. & Chovelon, J.M. (2002). A laser flash photolysis study of the decay of SO_4^- and Cl_2^{-} radical anions in the presence of Cl- in aqueous solution. *Chemosphere*, Vol. 47, No. 4, pp. 385–393, ISSN 0045-6535

Gonzalez, M.C.; Oliveros, E.; Wörner, M. & Braun, A.M. (2004). Vacuum-ultraviolet photlolysis of aqueous reaction systems. *Journal of Photochemistry and Photobiology C: Photochemistry Reviews*, Vol. 5, No. 3, pp. 225-246, ISSN 1389-5567

Gryglik, D.; Olak, M. & Miller, J.S. (2010). Photodegradation kinetics of androgenic steroids boldenone and trenbolone in aqueous solutions. *Journal of Photochemistry and Photobiology A: Chemistry*, Vol. 212, No. 1, pp. 14-19, ISSN 1010-6030

Haddou, M.; Benoit-Marquié, F.; Maurette, M.T. & Oliveros, E. (2010). Oxidative degradation of 2,4-dihydroxybenzoic acid by Fenton and photo-Fenton process: Kinetics, mechanisms, and evidence for the substitution of H_2O_2 by O_2. *Helvetica Chimica Acta*, Vol. 93, No. 6, pp. 1067-1080, ISSN 0018-019X

Hamilton, G.A.; Friedman, J.P. & Campbell, P.M. (1966a). The hydroxylation of anisole by hydrogen peroxide in the presence of catalytic amounts of ferric ion and catechol. Scope, requirements and kinetic studies. *Journal of the American Chemical Society*, Vol. 88, No. 22, pp. 5266–5268, ISSN 0002-7863

Hamilton, G.A.; Hanifin Jr., J.W. & Friedman, J.P. (1966b). The hydroxylation of anisole by hydrogen peroxide in the presence of catalytic amounts of ferric ion and catechol. Product studies, mechanism, and relation to some enzymic reactions. *Journal of the American Chemical Society*, Vol. 88, No. 22, pp. 5269–5272, ISSN 0002-7863

Henderson, M.A (2011). A surface science perspective on TiO_2 photocatalysis. *Surface Science Reports*, Vol. 66, No. 6-7, pp. 185-297, ISSN 0167-5729

Ho, T.L. & Bolton, J.R. (1998). Toxicity changes during the UV treatment of pentachlorophenol in dilute aqueous solution. *Water Research*, Vol. 32, No. 2, pp. 489-497, ISSN 0043-1354

Huston, P.L. & Pignatello, J.J. (1999). Degradation of selected pesticide active ingredientes and commercial formulations in water by the photoassisted Fenton reaction. *Water Research*, Vol. 33, No. 5, pp. 1238-1246, ISSN 0043-1354

Jenks, W.S. (2005). The Organic Chemistry of TiO_2 Photocatalysis of Aromatic Hydrocarbons, In: *Environmental Catalysis*, Vicki H. Grassian, 307-346, CRC, ISBN 1-57444-462-X, New York, USA

Kim, S.M. & Vogelpohl, A. (1998). Degradation of organic pollutants by the photo-Fenton process. *Chemical Engineering Technology*, Vol. 21, No. 2, pp. 187-191, ISSN 0930-7516

Kiwi, J.; Pulgarin, C. & Peringer, P. (1994). Effect of Fenton and photo-Fenton reactions on the degradation and biodegradability of 2-nitrophenols and 4-nitrophenols in water treatment. *Applied Catalysis, B: Environmetal*, Vol. 3, No. 4, pp. 335-350, ISSN 0926-3373

Kiwi, J., Lopez, A. & Nadtochenko, V. (2000). Mechanism and kinetics of the OH-radical intervention during Fenton oxidation in the presence of a significant amount of radical scavenger (Cl-). *Environmental Science & Technology*, Vol. 34, No. 11, pp. 2162-2168, ISSN 0013-936X

Kwon, B.G.; Lee, D.S.; Kang, N. & Yoon, J. (1999). Characteristics of *p*-chlorophenol oxidation by Fenton's reagent. *Water Research*, Vol. 33, No. 9, pp. 2110-2118, ISSN 0043-1354

Lee, Y., J. Jeong, C. Lee, S. Kim and J. Yoon (2003) Influence of various reaction parameters on 2,4-D removal in photo/ferrioxalate/H_2O_2 process. Chemosphere, Vol. 51, No. 9, pp. 901-912, ISSN 0045-6535

Legrini, O.; Oliveros, E. & Braun, A.M. (1993). Photochemical processes for water treatment. *Chemical Reviews*, Vol. 93, No. 2, pp. 671-698, ISSN 0009-2665

Li, X.; Cubbage, J. W. & Jenks, W. S. (1999a). Photocatalytic degradation of 4-chlorophenol. 1. The 4-chlorocatechol pathway. *Journal of Organic Chemistry*, Vol. 64, No. 23, pp. 8525-8536, ISSN 0022-3263

Li, X.; Cubbage, J. W.; Tetzlaff, T. A. & Jenks, W. S., (1999b). Photocatalytic degradation of 4-chlorophenol. 1. The hydroquinone pathway. *Journal of Organic Chemistry*, Vol. 64, No. 23, pp. 8509-8524, ISSN 0022-3263

Machulek Jr., A.; Vautier-Giongo, C.; Moraes, J. E. F.; Nascimento, C. A. O. & Quina, F.H. (2006). Laser flash photolysis study of the photocatalytic step of the photo-Fenton reaction in saline solution. *Photochemestry and Photobiology*, Vol. 82, No. 1, pp. 208-212, ISSN 0031-8655

Machulek Jr., A.; Moraes, J.E.; Vautier-Giongo, C.; Silverio, C.A.; Friedrich, L.C.; Nascimento, C.A.O.; Gonzalez, M.C. & Quina, F.H. (2007). Abatement of the inhibitory effect of chloride anions in the photo-Fenton process. *Environmental Science & Technology*, Vol. 41, No. 24, pp. 8459-8463, ISSN 0013-936X

Machulek Jr., A.; Gogritcchiani, E.; Moraes, J.E.F.; Quina, F.H.; Braun, A.M. & Oliveros, E. (2009a). Kinetic and mechanistic investigation of the ozonolysis of 2,4-xylidine (2,4-dimethyl-aniline) in acid aqueous solution. *Separation and Purification Technology*, Vol. 67, No. 2, pp. 141-148, ISSN 1383-5866

Machulek Jr., A.; Moraes, J.E.F.; Okano, L.T.; Silvério, C.A. & Quina, F.H. (2009b). Photolysis of ferric íon in the presence of sulfate or chloride íons: implications for the photo-Fenton process. *Photochemical & Photobiological Sciences*, Vol. 8, No. 7, pp. 985-991, ISSN 1474-905X

Maciel, R.; Sant'Anna Jr., G. L. & Dezotti, M. (2004). Phenol removal from high salinity effluents using Fenton's reagent and photo-Fenton reactions. *Chemosphere*, Vol. 57, No. 7, pp. 711-719, ISSN 0045-6535

Marin, M.L.; Lhiaubet-Vallet, V., Santos-Juanes, L.; Soler, J.; Gomis, J.; Arques, A.; Amat, A. M. & Miranda, M.A. (2011). A photophysical approach to investigate the photooxidation mechanism of pesticides: Hydroxyl radical versus electron transfer. *Applied Catalysis, B: Environmetal*, Vol. 103, No. 1-2, pp. 48-53, ISSN 0926-3373

Martyanov, I.N.; Savinov, E.N. & Parmon, V.N. (1997). A comparative study of efficiency of photooxidation of organic contaminants in water solutions in various photochemical and photocatalytic systems. 1. Phenol photoxidation promoted by hydrogen peroxide in a flow reactor. *Journal of Photochemistry and Photobiology A: Chemistry*, Vol. 107, No. 1-3, pp. 227-231, ISSN 1010-6030

Matilainen, A. & Sillanpää, M. (2010). Removal of natural organic matter from drinking water by advanced oxidation processes. *Chemosphere*, Vol. 80, No. 4, pp. 351-365, ISSN 0045-6535

Matthews, R.W. (1992). Photocatalytic oxidation of organic contaminants in water: An aid to environmental preservation. *Pure and Applied Chemistry*, Vol. 64, No. 9, pp. 1285-1290, ISSN 0033-4545

McElroy, W.J. (1990). A laser photolysis study of the reaction of $SO_4^{\cdot-}$ with Cl^- and subsequent decay of Cl_2^- in aqueous solution. *Journal of Physical Chemistry*, Vol. 94, No. 6, pp. 2435–2441, ISSN 0022-3654

Moraes, J.E.F.; Quina, F.H.; Nascimento, C.A.O.; Silva, D.N. & Chiavone-Filho, O. (2004a). Treatment of saline wastewater contaminated with hydrocarbons by the photo-Fenton process. *Environmental Science & Technology*, Vol. 38, No. 4, pp. 1183-1187, ISSN 0013-936X

Moraes, J.E.F.; Silva, D.N.; Quina, F.H.; Chiavone-Filho, O. & Nascimento, C.A.O. (2004b). Utilization of solar energy in the photodegradation of gasoline in water and of oil-field-produced water. *Environmental Science & Technology*, Vol. 38, No. 13, pp. 3746-3751, ISSN 0013-936X

Nadtochenko, V.A. & Kiwi, J. (1998). Photolysis of Fe(OH)$^{2+}$ and Fe(Cl)$^{2+}$ in aqueous solution. Photodissociation kinetics and quantum yields. *Inorganic Chemistry*, Vol. 37, No. 20, pp. 5233-5238, ISSN 0020-1669

Nascimento, C.A.O.; Teixeira, A.C.S.C.; Guardani, R.; Quina, F.H.; Chiavone-Filho, O. & Braun, A.M. (2007). Industrial wastewater treatment by photochemical processes based on solar energy. *Journal of Solar Energy Engineering*, Vol. 129, No. 1, pp. 45-52, ISSN 0199-6231

Nascimento, C.A.O.; Oliveros, E. & Braun, A.M. (1994). Neural-network modeling of photochemical processes. *Chemical Engineering Process*, Vol. 33, No. 5, pp. 319-324, ISSN 0255-2701

Nichela, D.; Haddou, M.; Benoit-Marquié, F.; Maurette, M.T.; Oliveros, E. & Einschlag, F.S.G. (2010). Degradation kinetics of hydroxy and hydroxynitro derivates of benzoic acid by Fenton-like and photo-Fenton techniques: A comparative study. *Applied Catalysis, B: Environmetal*, Vol. 98, No. 3-4, pp. 171-179, ISSN 0926-3373

Nogueira, R.F.P.; Silva, M.R.A. & Trovó, A.G. (2005). Influence of iron source on the solar photo-Fenton degradation of different classes of organic compounds. *Solar Energy*, Vol. 79, No. 4, pp. 384-392 , ISSN 0038-092X

Oliveros, E.; Benoit-Marquie, F.; Puech-Costes, E.; Maurette, M.T. & Nascimento, C.A.O. (1998). Neural network modeling of the photocatalytic degradation of 2,4-dihydroxybenzoic acid in aqueous solution. *Analusis*, Vol. 26, No. 8, pp. 326-332, ISSN 0365-4877

Oliveros, E.; Legrini, O.; Hohl, M.; Müller, T. & Braun, A.M. (1997). Industrial waste water treatment: Large scale development of a light-enhanced Fenton reaction. *Chemical Engineering and Processing*, Vol. 36, No. 5, pp. 397-405, ISSN 0255-2701

Pignatello, J.J. (1992). Dark and photoassisted Fe^{3+}- catalysed degradation of chlorophenoxy herbicides by hydrogen-peroxide. *Environmental Science & Technology*, Vol. 26, No. 5, pp. 944-951, ISSN 0013-936X

Pignatello, J.J.; Liu, D. & Huston, P. (1999). Evidence for an additional oxidant in photoassisted Fenton reaction. *Environmental Science & Technology*, Vol. 33, No. 11, pp. 1832-1839, ISSN 0013-936X

Pignatello, J.J; Oliveros, E. & Mackay, A. (2006). Advanced oxidation processes for organic contaminant destruction based on the Fenton reaction and related chemistry. *Critical Reviews in Environmental Science & Technology*, Vol. 36, No. 1, pp. 1-84. Errata. (2007). *Critical Reviews in Environmental Science & Technology*, Vol. 37, No. 3, pp. 273-275, ISSN 1064-3389

Pontes, R.F.F.; Moraes, J.E.F; Machulek, A. & Pinto, J.M. (2010). A mechanistic kinetic model for phenol degradation by the Fenton process. *Journal of Hazardous Materials*, Vol. 176, No. 1-3, pp. 402-413, ISSN 0304-3894

Pozdnyakov. I. P.; Glebov, E. M.; Plyusnin, V. F.; Grivin, V. P.; Ivanov, Y. V.; Vorobyev, D. Y. & Bazhin, N. M. (2000). Mechanism of $Fe(OH)^{2+}_{(aq)}$ photolysis in aqueous solution. *Pure and Applied Chemistry*, Vol. 72, No. 11, pp. 2187-2197, ISSN 0033-4545

Puigdomenech, I. (April 2010). Chemical Equilibrium Diagrams, 14.07.2011, Available from http://www.kemi.kth.se/medusa

Ruppert, G.; Bauer, R. & Heisler, G. (1993). The photo-Fenton reaction - an effective photochemical wastewater treatment process. *Journal of Photochemistry and Photobiology A: Chemistry*, Vol. 73, No. 1, pp. 75-78, ISSN 1010-6030

Silva, M.R.A.; Vilegas, W.; Zanoni, M.V.B. & Nogueira, R.F.P. (2010). Photo-Fenton degradation of the herbicide tebuthiuron under solar irradiation: Iron complexation and initial intermediates. *Water Research*, Vol. 44, No. 12, pp. 3745-3753, ISSN 0043-1354

Soler, J.; García-Ripoll, A.; Hayek, N.; Miró, P.; Vicente, R.; Arques, A. & Amat, A.M. (2009). Effect of inorganic íons on the solar detoxification of water polluted with pesticides. *Water Research*, Vol. 43, No. 18, pp. 4441-4450, ISSN 0043-1354

Sonntag, C. von. (2008). Advanced oxidation processes: mechanistic aspects. *Water Science & Technology*, Vol. 58, No. 5, pp. 1015-1021, ISSN 0273-1223

Trovó, A.G. & Nogueira, R.F.P. (2011). Diclofenac abatement using modified solar photo-Fenton process with ammonium iron(III) citrate. *Journal of the Brazilian Chemical Society*, Vol. 22, No. 6, pp. 1033-1039, ISSN 0103-5053

Truong, G.L.; De Laat, J. & Legube, B. (2004). Effects of chloride and sulfate on the rate of oxidation of ferrous ion by H_2O_2. *Water Research*, Vol. 38, No. 9, pp. 2384-2394, ISSN 0043-1354

Yu, X.-Y. & Barker, J. R. (2003a). Hydrogen Peroxide Photolysis in Acidic Aqueous Solutions Containing Chloride Ions. I.Chemical Mechanism. *Journal of Physical Chemistry A*, Vol. 107, No. 9, pp. 1313-1324, ISSN 1089-5639

Yu, X.-Y. & Barker, J. R. (2003b). Hydrogen Peroxide Photolysis in Acidic Aqueous Solutions Containing Chloride Ions. II.Quantum Yield of $HO^{\bullet}$ (Aq) Radicals. *Journal of Physical Chemistry A*, Vol. 107, No. 9, pp. 1325-1332, ISSN 1089-5639

Yu, X.-Y. (2004). Critical Evolution of Rate Constants and Equilibrium Constants of Hydrogen Peroxide Photolysis in Acidic Aqueous Solutions containing Chloride Ions. *Journal of Physical and Chemical Reference Data*, Vol. 33, No. 3, pp. 747-763, ISSN 0047-2689

Yu, X.Y., Bao, Z-C. & Barker, J.R. (2004). Free radical reactions involving $Cl^{\bullet}$, $Cl2^{\bullet}-$ and $SO4^{\bullet}-$ in the 248 nm photolysis of aqueous solutions containing $S_2O_8^{2-}$ and Cl-. *Journal of Physical Chemistry A*, Vol. 108, No. 2, pp. 295–308, ISSN 1089-5639

Yue, P.L. (1993). Modelling of kinetics and reactor for water purification by photooxidation. *Chemical Engineering Science*, Vol. 48, No. 1, pp. 1-11, ISSN 0009-2509

Zanta, C.L.P.S., Friedrich, L.C., Machulek Jr., A, Higa, K.M. & Quina, F.H. (2010). Surfactant degradation by a catechol-driven Fenton reaction. *Journal of Hazardous Materials*, Vol. 178, No. 1-3, pp. 258-263, ISSN 0304-3894

Zapata, A.; Oller, I.; Bizani, E.; Sánches-Pérez, J.A.; Maldonado, M.I. & Malato, S. (2009). Evaluation of operational parameters involved in solar photo-Fenton degradation of a commercial pesticide mixture. *Catalysis Today*, Vol. 144, No. 1-2, pp. 94-99, ISSN 0920-5861

Zazo, J.A.; Pliego, G.; Blasco, S.; Casas, J.A. & Rodriguez, J.J (2011). Intensification of the Fenton process by increasing the temperature. *Industrial & Engineering Chemistry Research*, Vol. 50, No. 2, pp. 866-870, ISSN 0888-5885

Fenton´s Process for the Treatment of Mixed Waste Chemicals

Cláudia Telles Benatti[1] and Célia Regina Granhen Tavares[2]
[1]Faculdade Ingá – UNINGÁ,
[2]Universidade Estadual de Maringá – UEM,
Brazil

1. Introduction

In recent years, with an increase in the stringent water quality regulations due to environmental concerns, extensive research has focused on upgrading current water treatment technologies and developing more economical processes that can effectively deal with toxic and biologically refractory organic contaminants in wastewater. In this context, in order to avoid or mitigate the possible adverse health, safety, and environmental impacts, to grantee compliance with federal, state, and local environmental laws, or only to set an example to students, many institutions of higher education have supported researches that aim to establish a treatment process for practical and economic disposal of waste chemicals. Waste chemicals in academic laboratories are by-products of research, teaching and testing activities. Waste chemicals from academic research laboratories can be considered one of the most polluting wastewaters and they pose more problems for the treatment and subsequent adequate disposal, due to their unique characteristics. These wastes are generated by laboratory operations, such as chemical analysis and research activities, including chemical and biological treatment experiments on a wide range of synthetic and natural wastewaters, and may include an abundance of unused laboratory reagents. Thus, they may present a great diversity of composition and volume, including refractory organics, toxic compounds and heavy metals, and may offer potential hazards to both health and environment.

The ultimate destination of waste is usually a treatment, storage, and disposal facility (National Research Council, 1995). The treatment of waste chemicals is typically via chemical action, such as neutralization, precipitation and reduction to yield a less toxic waste. However, in most cases, the treatment product still cannot be safely disposed of in the sanitary sewer. Most generators also adopt the practice of land filling or direct incineration of hazardous wastes. In this scenario, the development of economical methods to achieve a high degree of wastewater treatment is highly desirable.

The development and application of several Advanced Oxidation Processes (AOPs) to destroy toxic and biologically refractory organic contaminants in aqueous solutions concentrated significant research in the field of environmental engineering during the last decades. Among AOPs, the Fenton's reagent is an interesting solution since it allows high depuration levels at room temperature and pressure conditions using innocuous and easy to handle reactants. The inorganic reactions involved in Fenton process are well established and the process has been used for the treatment of a variety of wastewaters. The high efficiency of this technique can be explained by the formation of strong hydroxyl radical (•OH) and oxidation of Fe^{2+} to Fe^{3+}. Both Fe^{2+} and Fe^{3+} ions are coagulants, so the Fenton process can, therefore, have dual function, namely oxidation and coagulation, in the treatment processes (Badawy & Ali, 2006). It is essential, though, to investigate and set the operating conditions that best suits the wastewater that are being treated in order to achieve high degradation efficiencies.

In previous research work, chemical oxidation using Fenton's reagent was tested as a treatment method for mixed waste chemicals and optimized using a response surface methodology (Benatti et al., 2006). As predicted, the process optimization leaded to high COD removal (92.3%), with a high efficiency in the removal of heavy metals as a side effect. However, its major disadvantage is the production of $Fe(OH)_3$ sludge that requires further separation and proper disposal, with the consequent increase in operation costs. To overcome the drawback, the knowledge of the treatment process residue characteristics is imperative to design a successful waste management plan that may guarantee the viability of the applied technology to the wastewater treatment.

Most studies focused on applying the Fenton process to wastewater treatment and do not take into account the generated residues. Using the previously optimized set of process variables, the effects of Fenton´s reagent treatment of mixed waste chemicals are thoroughly discussed in the current chapter. The whole process was analysed from an environmental perspective, where the focus was not only the treatment of the wastewater, but also the characterization for correct final destination of the solids originating from the treatment process. Furthermore, the process was applied to wastewaters generated in different periods in order to study the effect of wastewater composition on the process efficiency.

This chapter also discusses about the unique nature of waste chemicals, presents different applications of Fenton´s reagent on the treatment of a variety of wastewaters, and provides an insight into the Fenton´s reaction mechanisms and some background on their operations conditions.

2. Theoretical approach

Advanced oxidation processes (AOPs) are based on the generation of very reactive species such as hydroxyl radical (•OH), a nonspecific, strong oxidant which reacts with most organic and biological molecules at near diffusion-controlled rates ($>10^9$ M^{-1} s^{-1}) (Büyüksönmez et al., 1999). Common AOPs involve Fenton and Fenton "like" processes, ozonation, photochemical and electrochemical oxidation, photolysis with H_2O_2 and O_3, high voltage electrical discharge (corona) process, TiO_2 photocatalysis, radiolysis, wet oxidation, water solutions treatment by electronic beams or γ-beams and various combinations of these methods (Kušić et al., 2007). Among AOPs, Fenton's reagent has been used (either alone or in combination with other treatments) as a chemical process for the treatment of a wide range of wastewaters. Recent applications of Fenton´s reagent include the pre-treatment of olive mill wastewater (Lucas & Peres, 2009), the treatment of landfill leachate (Deng & Englehardt, 2006; Zhang et al., 2005), copper mine wastewater (Mahiroglu et al., 2009), water-based printing ink wastewater (Ma & Xia, 2009) and cosmetic wastewaters (Bautista et al., 2007), the degradation of pesticide (Li et al., 2009; Chen et al., 2007), antibiotic (Ay & Kargi, 2010; Elmolla & Chaudhuri, 2009), high-strength livestock wastewater (Lee & Shoda, 2008) and organic compounds of nuclear laundry water (Vilve et al., 2009), the oxidation of combined industrial and domestic wastewater (Badawy & Ali, 2006), the pre-oxidation of pharmaceutical wastewaters (Martínez et al., 2003), the treatment of water-based paint wastewater (Kurt et al., 2006) and cellulose bleaching effluents (Torrades et al., 2003), the degradation of the explosives 2,4,6-trinitrotoluene (TNT) and hexahydro-1,3,5-trinitro-1,3,5-triazine (RDX) after iron pre-treatment (Oh et al., 2003), and the treatment of different streams of textile wastewaters, such as the treatment of hot desizing wastewaters (Lin & Lo, 1997), and the treatment of dye wastewaters (Wang et al., 2008; Gulkaya et al., 2006).

Fenton´s reagent is also combined with biological process, as a pre-treatment to enhance the biodegradability of the recalcitrant compounds and lower the toxicity (Padoley et al., 2011, Mandal et al., 2010, Badawy et al., 2009) or as a post-treatment to improve the efficiency of the wastewater treatment (Ben et al., 2009, Yetilmezsoy & Sakar, 2008).

Fenton's reagent, which involves homogenous reaction and is environmentally acceptable (Bham & Chambers, 1997), is a system based on the generation of very reactive oxidizing free radicals, especially hydroxyl radicals, which have a stronger oxidation potential than ozone; 2.8 V for •OH and 2.07 V for ozone (Heredia et al., 2001). The Fenton's reactions at acidic pH lead to the production of ferric ion and of the hydroxyl radical (Garrido-Ramírez et al., 2010; Gallard & De Laat, 2001; Walling, 1975):

$$Fe^{2+} + H_2O_2 \rightarrow Fe^{3+} + OH^- + \bullet OH \quad E_a = 39.5 \text{ kJ mol}^{-1} \text{ } k_1 = 76 \text{ M}^{-1}\text{s}^{-1} \tag{1}$$

$$Fe^{3+} + H_2O_2 \rightarrow Fe^{2+} + HO_2 \bullet + H^+ \quad E_a = 126 \text{ kJ mol}^{-1} \text{ } k_2 = 0.001\text{-}0.01 \text{ M}^{-1}\text{s}^{-1} \tag{2}$$

Hydroxyl radicals may be scavenged by reaction with another Fe^{2+} or with H_2O_2 (Torrades et al., 2003; Chamarro et al., 2001; Lu et al., 1999):

$$\bullet OH + Fe^{+2} \rightarrow HO^- + Fe^{+3} \tag{3}$$

$$\bullet OH + H_2O_2 \rightarrow HO_2 \bullet + H_2O \tag{4}$$

Hydroxyl radicals may react with organics starting a chain reaction (Bianco et al., 2011; Dercová et al., 1999):

$$\bullet OH + RH \rightarrow H_2O + R \bullet \text{ , RH = organic substrate} \tag{5}$$

$$R \bullet + O_2 \rightarrow ROO \bullet \rightarrow \text{products of degradation} \tag{6}$$

Ferrous ions and radicals are produced during the reactions as shown below (Lu et al., 1999):

$$H_2O_2 + Fe^{3+} \rightleftarrows H^+ + FeOOH^{2+} \tag{7}$$

$$FeOOH^{2+} \rightarrow HO_2 \bullet + Fe^{2+} \tag{8}$$

$$HO_2 \bullet + Fe^{2+} \rightarrow HO_2^- + Fe^{3+} \tag{9}$$

$$HO_2 \bullet + Fe^{3+} \rightarrow O_2 + Fe^{2+} + H^+ \tag{10}$$

The efficiency of this process depends on several variables, namely temperature, pH, hydrogen peroxide, ferrous ion concentration and treatment time. The oxidizing potential of the hydroxyl radical is pH dependent, and varies from E^0 =1.8 V at neutral pH to +2.7 V in acidic solutions (Buxton et al., 1988, as cited in El-Morsi et al., 2002). Operating pH of the system has been observed to significantly affect the degradation of pollutants (Benitez et al., 2001; Kang & Hwang, 2000; Nesheiwat & Swanson, 2000; Lin & Lo, 1997). The degree of oxidation of organics with Fenton's reagent is maximum when the pH lies in the interval pH 3-5 (Lunar et al., 2000; Lin & Lo, 1997). Hydrogen peroxide is most stable in the range of pH 3-4, but the decomposition rate is rapidly increased with increasing pH above 5. Thus, the acidic pH level around 3 is usually optimum for Fenton oxidation (Gogate & Pandit, 2004; Neyens & Baeyens, 2003). However, in order to achieve high performances, these experimental conditions must be optimized (Homem et al., 2010). Some environmental applications of Fenton's reagent involve reaction modifications, including the use of high concentrations of hydrogen peroxide, the substitution of different catalysts such as ferric iron and naturally occurring iron oxides, and the use of phosphate-buffered media and metal-chelating agents. These conditions, although not as stoichiometrically efficient as the standard Fenton's reactions, are often necessary to treat industrial waste streams and contaminants in soils and groundwater (Büyüksönmez et al., 1999).

3. Materials and methods

3.1 Materials
All chemicals employed in this study were analytical grade. All solutions were prepared in distilled-deionized water. Glassware used in metals determination was washed with detergent, rinsed with tap water, soaked with HNO_3 (~50% v/v) for 24 h, and rinsed with distilled-deionized water prior to drying.
Reagent grade H_2O_2 was standardized using iodometric titration (U.S. Peroxide, 2003) and used as purchased. A 1 M $FeSO_4.7H_2O$ stock solution was prepared and standardized (Pavan et al., 1992) just before Fenton's experiments. Solutions of NaOH and H_2SO_4 were used for pH adjustments.

3.2 Sample preparation
Chemical effluents generated during the laboratory operations over a period of 17 months were monitored, collected and stored in clearly marked containers. Later on these effluents were divided into two groups and then mixed to obtain two combined samples (sample 1 and 2), produced in different periods of time that were used in the experiments of Fenton's oxidation.

3.3 Fenton´s experimental procedure
The Fenton's oxidation experiments were carried out under optimal conditions established in previous work (Benatti et al., 2006): ratio [COD]:[H_2O_2] = 1:9; ratio [H_2O_2]:[Fe^{2+}] = 4.5:1 and pH 4.
Experiments were carried out in 250 mL beakers with a solution volume of 150 mL that consisted of laboratory effluent without solids separation. The effluent was continuously mixed (100 rpm) in a jar test apparatus at room temperature. Firstly the pH of the solution was adjusted to 4 with NaOH (~30% w/v), and a sample was withdrawn and centrifuged at 2500 rpm for 5 min for the separation of suspended solids and COD_0 was determined in

supernatant. The required amount of reagents was determined according to COD_0 value. The amount of $FeSO_4$ was first added to the reaction mixture. The Fenton reaction was then initiated by sequential addition of the required amount of H_2O_2, in three steps of equal volume added at intervals of 20 min, to moderate the rise in temperature that occurs as the reaction proceeds and to minimize quenching of •OH. The pH adjustments were performed using H_2SO_4 or NaOH solutions before each reagent addition and then at each hour. The Fenton reaction time was initiated by the addition of the first required amount of H_2O_2 to the reaction mixture. After reactions were completed (4 h), precipitation of the oxidized iron as $Fe(OH)_3$ was performed by adjusting the pH to 8 and then about 15 h of clarification under quiescent conditions. Final samples of supernatant were taken for COD measurements and for determination of residual hydrogen peroxide. All experiments were performed in duplicate.

The wastewater was also characterized in terms of pH, apparent color, turbidity, total phenols, sulfate, sulfide, chloride, total phousphorous, nitrogen and metals content (Ag, Al, Ca, Cd, Co, Cr, Cu, Fe, Hg, K, Mg, Mn, Na, Ni, Pb and Zn) before and after the Fenton's oxidation experiments.

3.4 Fenton's residues characterization

The Fenton's reagent residues were characterized through leaching tests, chemical fractionation and analysis of X-ray diffraction.

Sludges obtained at the end of the Fenton's process were named for ease of notation as residues 1 and 2 according to their originating effluent samples, and each one was divided into two portions. The first portion was transferred to an ambar glass flask and preserved at 4 °C for leaching tests. The second one was filtered under vacuum, rinsed with distilled-deionized water to eliminate the excess of sodium hydroxide, and then dried at 105 °C. The dried solids were removed from the filter, grinded and homogenized in a porcelain mortar, and stored for chemical fractionation and analysis of X-ray diffraction.

The leaching tests were carried out according to the Brazilian methodology (ABNT, 1987a). In these tests, 100 g of sludge was filtered through 0.45 µm cellulose ester membranes, the liquid phase was stored at 4 °C and the membranes containing the solid phases were transferred to 500 mL glass beakers. Distilled-deionized water (1:16) was added along with a sufficient quantity of acetic acid (0.5 N) to adjust the pH to 5.0±0.2. The suspension was stirred for 24 h. A pre-calculated amount of distilled-deionized water was then added to the suspension, the separation of phases was performed by filtering in 0.45 µm cellulose ester membrane, and the obtained solution was mixed with the stored liquid phase initially obtained. The mixed liquor was then preserved at 4 °C for metals determination. After analysis, the residues were classified according to ABNT-10004 (ABNT, 1987b) with regard to the maximum limits obtained in the extracts.

The chemical fractionation of solid residues was carried out by use of the following sequential dissolution procedure:

Step 1. Exchangeable ions were removed at room temperature for 1 h with magnesium chloride solution (1M $MgCl_2$, pH 7.0) and continuous agitation in Erlenmeyer flasks, using 1 g of dried precipitate and 8 mL of extractant, according to the methodology of Tessier et al. (1979).

Step 2. Amorphous material was removed by acid ammonium oxalate (pH 3.0) in a 4 h extraction in Kjeldahl flasks kept in the dark and at room temperature, using 1 g of dried precipitate and 100 mL of extractant, according to the methodology proposed by Camargo et al. (1986).

Step 3. Free Fe-oxides were removed from 0.5 g of dried samples by Na–dithionite–citrate–bicarbonate system in Kjeldahl flasks according to the methodology of Mehra & Jackson (1960).

Step 4. The residues remaining from Step 3 as well as the original samples were digested with a 2:1 mixture of nitric and perchloric acids.

Between each successive extraction, separation was performed by filtering the suspension through a 0.45 μm cellulose ester membrane. During filtration the residues were washed with distilled-deionized water. The liquid phase was transferred to volumetric flasks, diluted to mark, transferred to amber glass flasks, acidified to pH< 2 with concentrated nitric acid and preserved at 4 °C for metal determinations. The solids were transferred with a porcelain spoon and rinses of distilled-deionized water to a tare porcelain crucible and dried at 105 °C, cooled under vacuum in a desiccator, and weighed for weigh loss determinations. The solids were then crushed, manually homogenized using a porcelain spoon, dried again at 105 °C for moisture removal, and stored in a desiccator until ready for use in the next extraction step. These procedures were performed in three or more replicates.

The original dried solids as well as the solids from each extraction phase were analyzed by X-ray diffraction (XRD) using a Shimadzu D6000 diffractometer (Cu Kα radiation and λ = 1.54178 Å; scanning speed 2° 2θ min^{-1} for the ranges 5–70° 2θ). The mineralogical phases identification in generated X-ray diffractograms was performed by the position and intensity of diffraction planes. The amorphous phase diffractograms were determined by DXRD according to Schulze (1994), by subtraction of diffractograms intensity of original samples and of residues obtained after amorphous phase extraction.

3.5 Analytical methods

pH was measured with a Digimed-DM-20 pH meter calibrated with pH 4.01 and 6.86 Digimed standard buffers. The analyses of COD, settable solids, total phosphorous and soluble sulfate (turbidimetric method) were performed in accordance with standard methods (APHA-AWWA-WEF, 1998). Residual hydrogen peroxide was determined by the iodometric titration method (U.S. Peroxide, 2003). Total phenols were measured according to the colorimetric method of Folin-Ciocalteu reagent (Scalbert et al., 1989). Real and apparent color, turbidity, and sulfide were measured spectrophotometrically (spectrophotometer DR/2010, HACH, Loveland, CO) using the APHA platinum–cobalt standard method, the attenuated radiation method (direct reading), and the HACH sulfide test, respectively. Chloride was determined by the silver nitrate titration method (APHA-AWWA-WEF, 1998). Nitrogen (Kjeldahl) was measured according to Adolfo Lutz Institute Analytical Norms (Normas Analíticas do Instituto Adolfo Lutz, 1985). Metallic elements (Ag, Al, Ca, Cd, Co, Cr, Cu, Fe, Hg, K, Mg, Mn, Na, Ni, Pb and Zn) were determined in extracts by atomic absorption spectroscopy (Varian SpectrAA – 10 Plus).

The COD measured in the samples taken from Fenton's reactor was converted according to Eq. (11) to prevent the interference of H_2O_2 on COD analysis (Talini & Anderson, 1992).

$$COD = COD_M - R_p \times 0.25 \qquad (11)$$

where COD, COD value in the sample (mg O_2/L); COD_M, measured COD (mg O_2/L); R_p, residual hydrogen peroxide in the sample (mg/L).

The percent of COD removal was then determined through the following equation:

$$\eta(\%) = (\frac{COD_0 - COD_E}{COD_0}) * 100 \tag{12}$$

where η, percentage of COD removal; COD_0, measured COD in supernatant before oxidation (mg O_2/L); COD_E, COD value in clarified supernatant after precipitation (mg O_2/L).

4. Results and discussion

Throughout the monitoring period waste chemicals were generated as a consequence of chemical analyses and other research activities. Part of these residues is originated from chemical oxygen demand (COD), total phenols, nitrogen, protein, phosphate and sulfide determinations. Another significant part is constituted by diluted metal solutions containing Ag, Al, Ca, Cd, Co, Cr, Cu, Fe, Hg, K, Mg, Mn, Na, Ni, Pb and Zn, solutions standardization, and an abundance of unused laboratory reagents. As a consequence, the samples could be described as a quite complex wastewater which comprises different chemical species in dissolved, colloidal and particulate form. The main characteristics of the raw laboratory wastewater are presented in Table 1. Results in Table 1 show that the wastewater has high organic load (COD up to 2.7 g/L) and extremely low pH (<1.0). All raw samples contained an amount of brown solids presented as settable solids.

Parameter	Unity	Sample 1	Sample 2
pH		< 1	< 1
Apparent color	Pt/Co	12700	10250
Real color	Pt/Co	7150	1020
Turbidity	NTU	4010	2960
Total COD	mg O_2/L	2345	2676
Total phenols	mg/L	58.6	37.9
Settable solids	mL/L	4.5	4.0

Table 1. Characteristics of raw waste chemicals

Table 2 presents the results of chemical oxidation by Fenton´s reagent for both samples under optimum operation conditions. The characterization of the wastewater before and after Fenton's treatment is presented in Table 3.

Sample	Initial	End of the oxidation stage		End of the precipitation stage			
	COD mg O_2/L	COD mg O_2/L	% COD removal	residual H_2O_2 mg/L	Catalytic sludge mL	COD mg O_2/L	% COD removal
1	898	232	74,1	13,4	40	93	89,7
2	769	176	77,1	30,0	40	166	78,5

Table 2. Results of the waste chemical treatment by Fenton´s reagent.

Parameter	Unity	Sample 1			Sample 2		
		Raw supernatant	Initial effluent[a]	Final effluent at optimized conditions	Raw supernatant	Initial effluent[a]	Final effluent at optimized conditions
pH		<1	4	8	<1	4	8
Apparent color	Pt/Co	308	2205	137	503	1615	352
Turbidity	NTU	N.D.	42	10	N.D.	5	6
Soluble COD	mg O_2/L	1145	898	93	2576	769	166
Total phenols	mg/L	27.3	27.0	N.D.	39.8	35.9	N.D.
Sulfate	g/L	263	142	152	296	151	164
Sulfide	mg/L	0.04	0.07	0.01	0.05	0.12	0.01
Chloride[b]	mg/L	638	638	567	6523	6523	5247
Phosphorous	mg/L	203.8	90.5	0.1	394.2	233.2	0.4
Nitrogen	mg/L	1434	1038	566	38	38	N.D.
Metals:							
Ag	mg/L	1.1	1.1	0.8	1.1	1.1	1.0
Al	mg/L	18.8	5.2	1.4	6.8	6.4	4.5
Ca	mg/L	59.1	39.0	30.4	37.7	26.3	21.4
Cd	mg/L	0.3	0.3	0.3	6.9	2.3	0.5
Co	mg/L	1.2	0.8	0.8	1.3	0.8	1.0
Cr	mg/L	301.4	155.2	1.0	541.6	266.9	12.1
Cu	mg/L	1.7	0.3	0.6	4.5	2.0	0.3
Fe	mg/L	131.9	29.6	4.4	114.0	38.7	1.7
Hg	mg/L	1815.6	148.4	75.1	2769.3	70.7	99.8
K	mg/L	405.2	278.7	187.4	1084.8	646.4	645.3
Mg	mg/L	3.5	2.6	1.7	10.6	8.6	3.0
Mn	mg/L	0.7	0.6	0.5	0.8	0.4	0.1
Na	g/L	10.2	73.9	77.9	9.1	98.5	101.7
Ni	mg/L	0.9	1.2	1.1	1.1	1.3	1.1
Pb	mg/L	1.9	1.3	1.1	1.6	1.3	1.1
Zn	mg/L	0.9	0.5	N.D.	0.1	N.D.	N.D.

[a] Sample after pH adjustment to 4 with NaOH (30% w/v).
[b] Chloride content was determined at pH 8.

Table 3. Characteristics of the laboratory wastewater supernatant before (raw and after pH adjustment) and after oxidation at optimized conditions (N.D.: not detected).

One set of experiments was conducted as control experiments (without any addition of hydrogen peroxide or ferrous sulfate). The simple pH adjustment of the effluent to 8 did not contribute for significant COD or phenol removal. Though it was capable of minimizing the toxic metal content in solution as metal hydroxide precipitation (data not shown), it presented an undesirable effect. The raise of pH of the mixed waste chemicals to high values (pH=8) resulted, under temperatures below 20 °C, in a high instable solution, with the formation of a crystal solid phase simply by its manipulation. This is probably due to the formation of several inorganic substances, and for such substances solubility decreases with decreasing temperature.

Thus, Fenton's oxidation was conducted on wastewater without a previous precipitation of the metals. This process configuration favours the elimination of a pH adjustment step since the laboratory wastewater is highly acidic (see Tables 1 and 3) and the Fenton's reaction is also conducted under acidic conditions, with the raise of pH to 8 at the end of the process. Fenton's oxidation conducted under different conditions suggests that the presence of others metals do not interfere in the efficiency of oxidation of organic compounds by the Fenton's reagent.

Bidga (1995) describes the Fenton method as a process divided into four stages. First, pH is adjusted to low acidity, at pH value of 3-5. In this study, it was adjusted to optimum pH (pH=4). Then, main oxidation reaction takes place. The wastewater is then neutralized at pH of 7-8 and finally precipitation occurs. In this study pH=8 was adopted in order to favour the precipitation of metal species as hydroxides. Wastewater characteristics at different process stages are presented in Table 2.

During the Fenton oxidation process and, mainly, during the pH adjustment to 8, a large amount of flocks of various sizes in the wastewater were observed. According to Walling & Kato (1971), the small flocs were ferric hydroxo complexes formed by complex chain reactions of ferrous and hydroxide ions. After a period of natural sedimentation (about 15 hours), all flocs settled out in wastewater forming a catalytic sludge, whose volume is also presented in Table 2.

The supernatant separated by decantation revealed that the Fenton's reagent oxidation was efficient in degrading organic matter in both samples, reaching 89.7 and 78.5% COD removal in samples 1 and 2, respectively. Moreover, the total phenols presence was not detected after the application of Fenton's reagent. Moreover, the treated liquor under the optimized conditions showed Fe concentrations up to 4.4 mg/L. Thus, the Fe concentration could be kept below 15 mg/L, which is the maximum value for Fe concentration in disposed effluents imposed by CONAMA (Brazilian National Environmental Council) standards (CONAMA, 2008).

It is worth mentioning that great amount of NaOH were necessary to raise the pH from below 1 to 2. From this point on, the pH showed to be more sensitive to the sequential addition of NaOH solution. As a consequence, the effluent presented a high sodium concentration that increases with the raise of pH (see Table 3).

Regarding the sulfate content, the raw wastewater at pH 4 presented an initial sulfate concentration of 142 and 151 g/L for samples 1 and 2, respectively (see Table 3). Because sulfuric acid and sodium hydroxide solutions were used for pH adjustments, and ferrous sulfate was used as a catalyst in the Fenton's process, further amounts of sulfate resulted from the wastewater oxidation. Thus, the sulfate concentration still remained extremely high at the end of the process, reaching 152 and 164 g/L for samples 1 and 2, respectively. Although the Brazilian legislation does not directly limit its concentration for effluent discharge, the CONAMA Resolution No. 357 (CONAMA/2005) states that effluents must not give to the receiving waters characteristics different from those used in their classification. This resolution does not include the sulfate concentration as a parameter to be monitored, probably due to the fact that the damage caused by sulfate emissions is not direct, since sulfate is a chemically inert, non-volatile, and non-toxic compound. However, high sulfate concentrations can unbalance the natural sulfur cycle (Silva et al., 2002; Lens et al., 1998). The accumulation of sulfate rich sediments in lakes, rivers and seas may cause the release of toxic sulfides that can provoke damages to the environment (Ghigliazza et al.,

2000). In addition, the release of sulfate-rich wastewaters in sewage systems may cause the inhibition or even the collapse of the biological treatment system. Thus, a post treatment system is required to bring the sulfate levels of the treated wastewater to values that are less than the maximum allowable limit set by the local regulatory authority (1000 mg/L) in order to allow its discharged directly to the municipal biological treatment facilities.

Regarding the metal content, the Fenton's reagent process presented as a side effect the removal of certain elements, such as chromium (up to 99.4%) and aluminum (up to 73.1%).

Fu et al. (2009) studied the removal of heavy metal ions in metal–EDTA complexes by Fenton and Fenton-like reaction followed by hydroxide precipitation, and achieved high removal efficiencies while conventional technologies, such as hydroxide, sulfide and dithiocarbamate-type precipitants could hardly work for it. Although Fenton-like process presented higher efficiency than Fenton process, at optimal operation conditions ($[H_2O_2]_0$ = 141 mM, $[Fe^{2+}]_0$ = 1.0 mM, $[Fe^{3+}]_0$ = 1.0 mM, initial pH 3.0 and precipitation pH 11.0), the removal efficiency of Ni(II) were above 92% for the two systems.

Mahiroglu et al. (2009) investigated the treatability of combined acid mine drainage (AMD) – Flotation circuit effluents from copper mine via Fenton's process and pointed out that heavy metals in the AMD could also be reduced to very low levels via Fenton reactions by taking part in oxidation steps.

Despite the results, the heavy metals content in the treated liquor still exceeded the maximum allowable limits for effluent discharge according to the CONAMA Resolution No. 397 (CONAMA/2008), especially for mercury, lead, chromium and silver, whose limit for disposal are 0.01, 0.5, 0.5 and 0.1 mg/L, respectively This stresses the need for an additional treatment for the removal of the remaining heavy metals prior to discharge to the environment.

Finally, the application of the Fenton's reagent in the destruction of organic compounds in mixed waste chemicals generated a dark brown slurry phase mainly constituted by heavy metals-Fe(III)-iron sludge, with a formation of 267 mL of sludge per liter of oxidized wastewater. This slurry phase presented a percentage of total suspended solids between 2.4 and 2.5% (w/w).

Thus, the overall result of the waste chemicals treatment by Fenton´s Reagent is the production of an aqueous solution with a substantially lower total carbonaceous load. However, the treated liquor still presented levels of heavy metals and sulfate that were too high to meet discharge standards. Thus, the Fenton's reagent cannot be applied as a stand-alone treatment option, but it can be used in combination with other treatment techniques. Finally, it is noteworthy that a potentially hazardous solid residue is obtained as a by-product of the Fenton's treatment. Hence, it is important to characterize this material for proper waste management, satisfying environmental and health related criteria.

As a first approach to characterize the solids originating from the Fenton´s process, this study focused on identifying the metals present in the residues, on their crystal structure, on the specific chemical bond in which electrons are delocalized and mobile, and on their magnetic properties. As a second approach, and in order to complement the physical characterization of the residues, leaching tests were also performed in order to obtain correct predictions of elements possible mobilization process in the environment.

The results of metals determination in these residues (Table 4) showed that the predominant metals are silver, chromium, mercury and iron. The main source of the first three ones is COD analysis. Iron is mainly a result of the Fenton's process, once it is used as a catalyst of the reaction and undergoes precipitation after oxidation. Regarding cadmium, the results of effluent characterization (Table 3) showed that sample 1 presented an initial concentration of 0.3 mg/L, which was kept constant after the equalization and oxidation stages of effluent treatment. In consequence, this element was not detected in the originating residue from the Fenton's process (Residue 1).

Metal	Residue 1	Residue 2
Ag	30157.1	38678.3
Al	2184.5	362.0
Ca	1827.7	1839.4
Cd	ND	480.0
Co	61.2	50.2
Cr	20747.2	28395.0
Cu	210.0	538.8
Fe	400065.2	324171.4
Hg	70430.1	96926.7
Mn	221.7	282.6
Ni	49.8	36.4
Pb	57.2	11.3
Zn	96.3	50.1

Table 4. Total element determinations (mg/kg) in samples (N.D.: not detected).

Fig. 1 shows the X-ray diffractograms obtained for the studied residues. This figure shows that the amorphous phase is predominant in the material. Besides the amorphous material identified as 2-line ferrihydrite (Cornell & Schuwertmann, 1996), the presence of the crystalline phases HgCl (calomel) and AgCl (chlorargyrite) was identified in both residues. The three solid phases detected by X-ray diffraction are also the ones that present the highest concentration considering the total metal content presented in Table 3. Although the studied residues were produced by the oxidation of mixed laboratory wastewaters generated in different periods, they present similar characteristics.

The knowledge of the total concentration of metals in residues of Fenton´s process is, however, not enough to evaluate their environmental impact. Addressing the chemical form of the element instead of the total trace element concentration renders the information gained through careful analysis much more valuable (Cornelis & Nordberg, 2007). The mobility of trace metals, as well as their bioavailability and related ecotoxicity to plants, depends strongly on their specific chemical forms or ways of binding (Quevauviller et al., 1997). Consequently, it is necessary to determine the solubility of these metals, which depends on the way of their association in residue.

Generally speaking, the sequential extraction of trace elements adopted in this study uses a series of chemical reagents, each time stronger, under specific conditions, to dissolve one or more specific phases from the solid sample, while preserving others. In the extract it is possible to speciate elements solubilised during the dissolution. By studying the distribution of the metals between the different phases, their contamination risks can be ascertained, because it is possible to divide a specific metal into fractions of increasing stability.

Thus, in order to study the distribution of the metals in residues and the phases to which they are bound, a sequential extraction procedure was designed based on the characteristics of the Fenton's process as well as on the analysis of the X-ray diffractograms presented in Fig. 1. In this procedure, the total metal content is divided into four fractions of increasing stability: exchangeable (Fraction 1), amorphous iron oxide (Fraction 2), crystalline iron oxide (Fraction 3), and inert or residual (Fraction 4). The solubility of the metals in the residues can be associated with their extraction, decreasing in the order of extraction sequence. In this sense, samples with a higher metal content in Fraction 1 will be potentially more dangerous than those that present a lower content in this fraction.

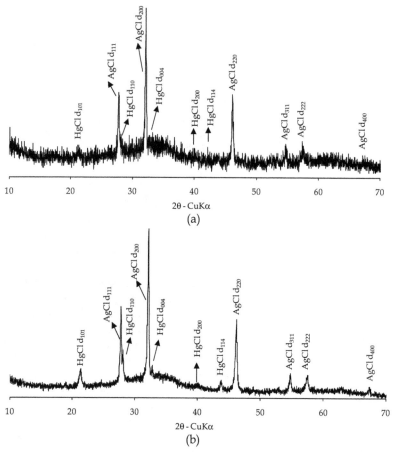

Fig. 1. X-ray diffractograms of the residue of Fenton's process: (a) Residue 1 and (b) Residue 2.

Metals extracted in Fraction 1 (exchangeable) correspond to those weakly absorbed, in particular to those retained in the residue surface with a weak electrostatic force. Changes in water ionic composition greatly affect the sorption–desorption processes of this fraction. Fraction 2 contains metals bound to amorphous iron oxide and corresponds to the reactive part of the iron compounds of the residues. Fraction 3 contains metals bound to crystalline iron oxide. This extraction favours the preferential orientation of remaining metals when submitted to X-ray diffractometry. Fraction 4 (inert or residual) contains primary and secondary minerals, mostly silicates, titanium and aluminum oxides, which can retain metals in their crystalline structure, removed from the laboratory wastewaters. These minerals are not expected to be released in a reasonable space of time in nature's normal conditions. Metals of this fraction are chemically stable and biologically inactive.

Fig. 2 shows the weight loss of the sample after each stage of the sequential extraction method adopted. As can be seen, the last two phases of the sequential extraction (crystalline iron oxide and residual), the less reactive fraction of the residues, represent only a small percentage in mass, about 11.2 and 16.6% for Residues 1 and 2, respectively. In fact, the residues originating from the Fenton's process is mainly constituted of amorphous material (over 80%) and most metals are co-precipitated to this fraction.

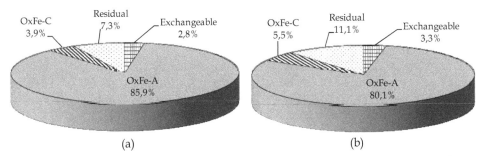

Fig. 2. Mass fractionation (% in mass) of Residues 1 (a) and 2 (b). Exchangeable, amorphous iron oxide (OxFe-A), crystalline iron oxide (OxFe-C), and final residual.

Table 5 presents the results of metals concentration in the different fractions of the studied residues. Fig. 3 shows the metal partitioning, in percentage of the total amount, found in the residues fractions. As can be seen, the sum of the four fractions (Table 5) is reasonably similar to the total contents obtained after digestion of the original samples (Table 4), indicating that no significant loss occurred during the sequential dissolution procedures, with recoveries of 96–100% in all cases. Thus, it is reasonable to assume that the dissolution procedure adopted was reliable for metals partitioning, increasing the confidence of the obtained data.

Several elements were detected in Fraction 1 (exchangeable), such as silver (0.2%), chromium (0.1%), copper (3.8%), iron (0.004%) and mercury (20.5%) of Residue 1; silver (0.1%), cadmium (5.2%), chromium (1.1%), copper (1.1%), iron (0.01%), mercury (31.6%) and zinc (0.2%) of Residue 2. These metals were associated to soluble salts as chlorites or sulfates, or simply absorbed to the residues surface.

Generally, both studied residues presented a similar distribution of metals at the different fractions obtained by the sequential dissolution procedure employed. Fig. 3 shows the chemical fractionation of each residue. With regard to silver, this element was mainly associated to the residual fraction of both residues (over 97% extraction). Elements like aluminum, calcium, cobalt, chromium, iron, manganese, nickel and zinc were predominantly found in the amorphous phase of both residues. In the case of cadmium, which was detected only in Residue 2, 86.9% of the total was found in the amorphous phase and only 6.8% was found in the most stable phases, which means in the crystalline iron oxide and residual phases. With regard to mercury, besides the exchangeable phase previously mentioned, 21.7 and 14.8% were found in the amorphous phase, 0.2 and 0.02% in the crystalline fraction and 53.5 and 49.8% in final residual, in Residues 1 and 2, respectively. In Residue 1, lead was found predominantly in the amorphous phase, representing 73.1% of the total content, and the remaining content was distributed between the crystalline (7.4%) and residual (19.3%) fractions. In Residue 2, though, lead was mainly associated with the residual fraction (99.7%); its availability was therefore low since most of it was in the insoluble form.

Sample	Metal	Exchangeable	OxFe-A	OxFe-C	Residual	Sum	% Recovery
	Ag	75.0	35.1	107.0	29344.0	29561.1	98.0
	Al	N.D.	2051.4	68.6	56.4	2176.4	99.6
	Ca	N.D.	1739.3	16.6	64.6	1820.5	99.6
	Cd	N.D.	N.D.	N.D.	N.D.	0.0	-
	Co	N.D.	57	2.6	1.5	61.1	99.9
	Cr	28.5	19195.5	1122.5	318.0	20664.5	99.6
Residue 1	Cu	7.9	190.9	0.3	6.2	205.3	97.7
	Fe	15.7	370959.6	19230.1	8634.0	398839.4	99.7
	Hg	14450.1	15263.0	140.0	37688.3	67541.4	95.9
	Mn	N.D.	210.2	6.5	4.0	220.7	99.5
	Ni	N.D.	45.6	1.3	1.2	48.1	96.7
	Pb	N.D.	41.9	4.3	11.1	57.3	99.9
	Zn	N.D.	75.4	N.D.	20.2	95.6	99.3
	Ag	9.7	4.9	12.4	38571.0	38598.0	99.8
	Al	N.D.	323.8	5.8	29.1	358.7	99.1
	Ca	N.D.	1467.2	287.9	82.9	1838.0	99.9
	Cd	25.1	417.0	12.0	20.3	474.4	98.8
	Co	N.D.	42.5	5.3	0.7	48.5	96.7
	Cr	310.1	26745.1	725.6	145.6	27926.4	98.3
Residue 2	Cu	5.6	529.6	0.1	3.2	538.5	100.0
	Fe	19.9	315243.4	4616.7	2934.5	322814.5	99.6
	Hg	30648.7	14296.5	23.6	48263.1	93231.9	96.2
	Mn	N.D.	273.1	2.4	0.4	275.9	97.7
	Ni	N.D.	33.7	1.3	0.8	35.8	98.5
	Pb	N.D.	N.D.	N.D.	11.2	11.2	99.7
	Zn	0.1	40.8	N.D.	7.9	48.8	97.3

Table 5. Metals concentration (mg/kg) in the different fractions (exchangeable, amorphous iron oxide (OxFe-A), crystalline iron oxide (OxFe-C), and final residual) of the studied residues (N.D.: not detected).

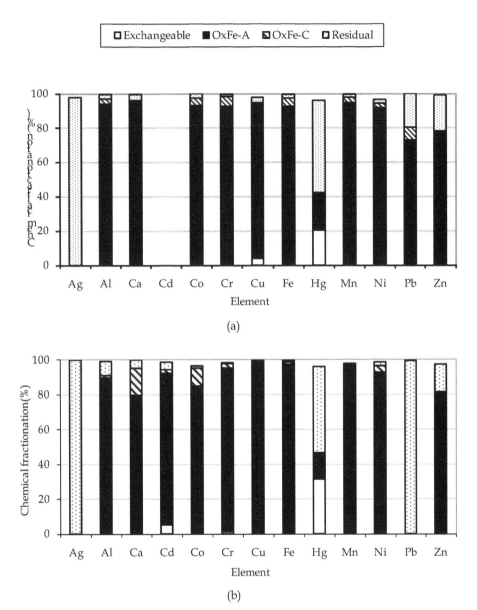

Fig. 3. Chemical fractionation of solids originating from the Fenton's process: (a) Residue 1 and (b) Residue 2. Exchangeable, amorphous iron oxide (OxFe-A), crystalline iron oxide (OxFe-C), and final residual.

Figs. 4 and 5 show the X-ray diffractograms of the studied residues. Fig. 6 shows the differential X-ray diffractograms for the amorphous phase of Residues 1 and 2.

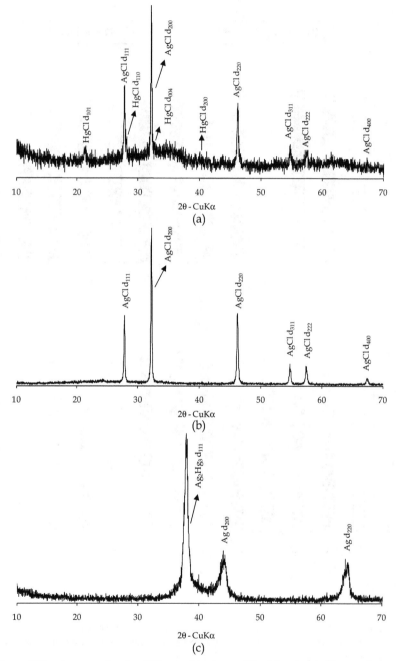

Fig. 4. X-ray diffractograms for the different fractions of Residue 1. (a) After exchangeable extraction; (b) after amorphous iron oxide extraction and (c) after crystalline iron oxide extraction.

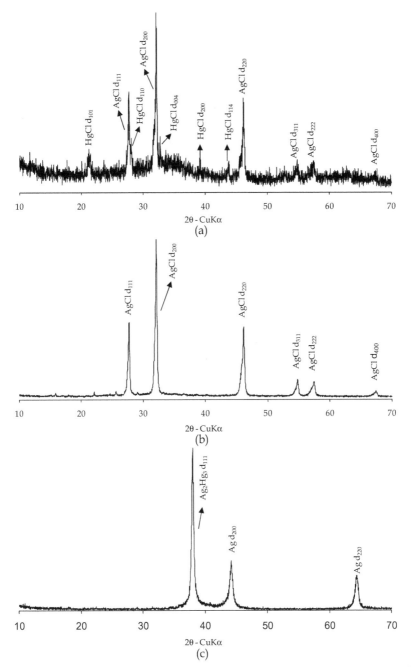

Fig. 5. X-ray diffractograms for the different fractions of Residue 2. (a) After exchangeable extraction; (b) after amorphous iron oxide extraction and (c) after crystalline iron oxide extraction.

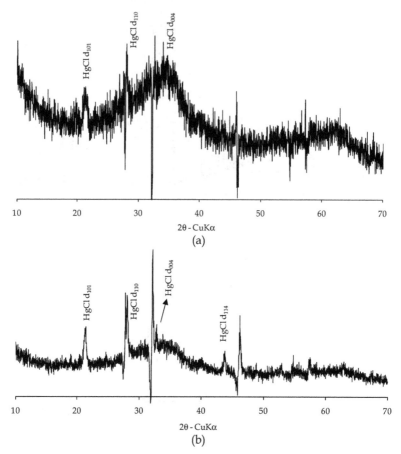

Fig. 6. Differential X-ray diffractograms for the amorphous phase and in the presence of HgCl: (a) Residue 1 and (b) Residue 2.

The position and intensity of diffraction planes indicate that there are no differences between the two solid samples. Therefore, although the wastewater samples were generated in distinct periods of time, their differences in chemical composition did not affect the mineralogical phases present in the solids resulting from their treatment by Fenton's reactions. From the interpretation of the main diffraction planes observed in the diffractograms, it was possible to identify the most likely compounds related to them. The solid residues originated from the Fenton's process presented a typical profile of material constituted predominantly by amorphous phase (2-line ferrihydrite (Schwertmann et al., 1982)), and silver and mercury compounds with well-defined crystalline structure were identified as chlorargyrite and calomel, respectively. The residual or inert fraction of the residues presented characteristics of a metallic league, formed mostly by elementary silver and mercury with minor amounts of other metals (Table 5). The Fenton's residues were originally brown, and did not exhibit detectable magnetic properties by the presence of a magnetic field, and a black coloured residual was obtained at the end of the sequential extraction procedure. The diffractograms in Fig. 6 present the amorphous phase of the studied residues with the presence of calomel, obtained by the subtraction of difractograms intensities of original samples (Fig. 1a and b) and of residues obtained after amorphous phase extraction (Figs. 4b and 5b). Therefore, differently to that observed for soils, acid ammonium oxalate extracted a portion of poorly crystalline crystals of calomel, besides most iron associated to 2-line ferrihydrite. The use of dithionite at the crystalline iron extraction was capable of reducing remaining mercury and silver of the selective dissolution procedure and formed a metallic league that presents diffraction planes of Ag and Ag_2Hg_3.

From the results of the selective dissolution, the solids originated from the Fenton's process are not stable, being composed fundamentally by high toxicity metals like chromium and mercury. Hence, the results suggest that leaching of metals into the environment can occur even under mild environmental conditions.

Finally, Table 6 presents the results of metal leachability for both studied residues determined by ABNT-NBR 10005 method (ABNT, 1987a). For comparative purposes, the same table shows the maximum levels of heavy metals allowed in Brazilian non-hazardous wastes (ABNT, 1987b). Based on the obtained chemical analytical results and on the classificatory procedure proposed by ABNT-NBR 10004 (ABNT, 1987b), the solid residues from the Fenton's process can be classified as Class I—hazardous. Both residues are characterized as toxic TL (leaking test). Residue 1 received the identification codes D009 and D011, due to its mean metal concentration above the maxima permitted for total chromium and mercury, and Residue 2, D007, D009 and D011 due to its cadmium, total chromium and mercury concentration, respectively.

Metals	Element code	Residue 1	Residue 2	Limits*
Cd	D007	N.D.	0.8	0.5
Pb	D008	N.D.	N.D.	5.0
Cr	D009	13.6	18.4	5.0
Hg	D011	49.8	243.1	0.1
Ag	D012	0.2	0.2	5.0

* Identification codes and maximum established limits according to ABNT-NBR 10004 (ABNT, 1987b).

Table 6. Results of metals leachability (mg/L) for both studied residues (N.D.: not detected).

Thus, the obtained results indicate a great potential for soil, surface and underground waters contamination by heavy metals (chromium, cadmium and mercury), if the residues originated from the Fenton's process are disposed of improperly. A stabilization process of the residues is necessary prior to their disposal in the environment. Furthermore, the results indicated a high potential of silver and mercury recovery, which represent a large part of the studied residues.

5. Conclusion

Chemical oxidation using Fenton's reagent under optimum conditions has proven to be a viable alternative to the oxidative destruction of organic pollutants in mixed waste chemicals, with a COD removal of 89.7 e 78.5% in samples 1 and 2, respectively. Moreover, total phenols were not detected in the treated liquor. However, the reported results also indicated that no single technology could be applied to mixed waste chemicals as a stand-alone treatment option once the concentration of certain inorganic constituents, such as heavy metals and sulfate, still remained high for effluent discharge.

Regarding to the Fenton's residues, they were classified as hazardous according to Brazilian waste regulations. The application of the sequential dissolution procedure indicated that the metals in the Fenton's residues are mainly constituted of amorphous material (over 80%). Furthermore, the reactive fractions of the residues (exchangeable and amorphous iron oxide fractions) retain most of remaining metals. Therefore, the Fenton's residues present great potential for environmental contamination, and require an administration system and control of their final disposal. However, Fenton's residues present great potential for silver and mercury recovery, since these elements represent a great portion of the studied residues.

6. Acknowledgment

The support provided by CAPES and by the State University of Maringá is gratefully acknowledged.

7. References

ABNT – Associação Brasileira de Normas Técnicas. (1987a). Lixiviação de Resíduos – Procedimento, NBR 10005, Rio de Janeiro, R.J.

ABNT – Associação Brasileira de Normas Técnicas (1987b). Resíduos Sólidos–Classificação, NBR 10004, Rio de Janeiro, R.J.

APHA-AWWA-WEF—American Public Health Association, American Water Works Association, Water Environment Federation. (1998). *Standard Methods for the Examination of Water and Wastewater* (20th ed), ISBN 0875530478, Washington, DC.

Ay, F., Kargi, F. (2010). Advanced oxidation of amoxicillin by Fenton's reagent treatment. *Journal of Hazardous Materials*, Vol. 179, pp. 622–627, ISSN 0304-3894.

Badawy, M., Wahaab, R.A., El-Kalliny, A.S. (2009). Fenton-biological treatment processes for the removal of some pharmaceuticals from industrial wastewater. *Journal of Hazardous Materials*, Vol. 167, pp. 567–574, ISSN 0304-3894.

Badawy, M.I.; Ali, M.E.M. (2006). Fenton's peroxidation and coagulation processes for the treatment of combined industrial and domestic wastewater. *Journal of Hazardous Materials*, Vol. B136, pp. 961–966, ISSN 0304-3894.

Bautista, P., Mohedano, A.F., Gilarranz, M.A., Casas, J.A., Rodriguez, J.J. (2007). Application of Fenton oxidation to cosmetic wastewaters treatment. *Journal of Hazardous Materials*, Vol. 143, pp. 128–134, ISSN 0304-3894.

Ben, W., Qiang, Z., Pan, X., Chen, M. (2009). Removal of veterinary antibiotics from sequencing batch reactor (SBR) pretreated swine wastewater by Fenton's reagent. *Water Research*, Vol. 43, pp. 4392–4402, ISSN 0043-1354.

Benatti, C.T.; Tavares, C.R.G.; Guedes, T.A. (2006). Optimization of Fenton's oxidation of chemical laboratory wastewaters using the response surface methodology. *Journal of Environmental Management*, Vol. 80, pp. 66–74, ISSN 0301-4797.

Benitez, F.J., Acero, J.L., Real, F.J., Rubio, F.J., Leal A.I. (2001). The Role of Hydroxyl Radicals for the Decomposition of p-hydroxy Phenylacetic Acid in Aqueous Solutions. *Water Research*, Vol. 35, No. 5, pp. 1338-1343, ISSN 0043-1354.

Bham, A.A., Chambers, R.P. (1997). Degradation of high molecular weight chlorinated aromatics and aliphatics in bleach plant effluent by Fenton's Reagent. *Advances in Environmental Research*, Vol. 1, pp. 135–143, ISSN 1093-7927.

Bianco, B., Michelis, I., Vegliò, F. (2011). Fenton treatment of complex industrial wastewater: Optimization of process conditions by surface response method. *Journal of Hazardous Materials*, Vol. 186, pp. 1733–1738, ISSN 0304-3894.

Bigda, R.J. (1995). Consider Fenton's Chemistry for Wastewater Treatment. *Chemical Engineering Progress*, December, pp. 62–66, ISSN 0360-7275.

Büyüksönmez, F., Hess, T.F., Crawford, R.L., Paszczynski, A., Watts, R.J. (1999). Optimization of Simultaneous Chemical and Biological Mineralization of Perchloroethylene. *Applied and Environmental Microbiology*, Vol. 65, No. 6 (Jun), pp. 2784-2788, ISSN 0099-2240.

Camargo, O.A. ; Moniz, A.C. ; Jorge, J.A. ; Valadares, J.M.A.S. (1986). *Métodos de análise química, Mineralógica e física de solos do Instituto Agronômico de Campinas*, Boletim Técnico no. 106, Instituto Agronômico, ISBN 8585564, Campinas, S.P.

Chamarro, E., Marco, A., Esplugas, S., 2001. Use of Fenton reagent to improve organic chemical biodegradability. *Water Research*, Vol. 35, No. 4, pp. 1047–1051, ISSN 0043-1354.

Chen, S., Sun, D., Chung, J.S. (2007). Treatment of pesticide wastewater by moving-bed biofilm reactor combined with Fenton-coagulation pre-treatment. *Journal of Hazardous Materials*, Vol. 144, pp. 577–584, ISSN 0304-3894.

CONAMA – Conselho Nacional do Meio Ambiente. (2005). Resolution No. 357, March 17, 2005, SãoPaulo, Brazil.

CONAMA – Conselho Nacional do Meio Ambiente. (2008). Resolution No. 397, April 03, 2008, SãoPaulo, Brazil.

Cornelis, R., Nordberg, M. (2007). General Chemistry, Sampling, Analytical Methods and Speciation, In: *Handbook on the toxicology of metals*, Nordberg, G.F. (Eds.), pp. 11-38, Academic Press, ISBN 9780123694133, USA.

Cornell, R.M., Schuwertmann, U. (1996). *The Iron Oxides: Structure, Properties, Reactions, Occurrence and Uses*. VCH Publishers, ISBN 9783527302741, Weinheim, Germany.

Deng, Y., Englehardt, J.D. (2006). Treatment of landfill leachate by the Fenton process. *Water Research*, Vol. 40, pp. 3683–3694, ISSN 0043-1354.

Dercová, K., Vrana, B., Tandlich, R., Šubová, L. (1999). Fenton's type reaction and chemical pretreatment of PCBs. *Chemosphere*, Vol. 39, No. 15, pp. 2621–2628, ISSN 0045-6535.

Elmolla, E., Chaudhuri, M. (2009). Optimization of Fenton process for treatment of amoxicillin, ampicillin and cloxacillin antibiotics in aqueous solution. *Journal of Hazardous Materials*, Vol. 170, pp. 666–672, ISSN 0304-3894.

El-Morsi, T.M., Emara, M.M., Abd El Bary, H.M.H., Abd-El-Aziz, A.S., Friesen, K.J. (2002). Homogeneous degradation of 1,2,9,10-tetrachlorodecane in aqueous solutions using hydrogen peroxide, iron and UV light. *Chemosphere*, Vol. 47, pp. 343–348, ISSN 0045-6535.

Fu, F., Wang, Q., Tang, B. (2009). Fenton and Fenton-like reaction followed by hydroxide precipitation in the removal of Ni(II) from NiEDTA wastewater: A comparative study. *Chemical Engineering Journal*, Vol. 155, pp. 769–774, ISSN 1385-8947.

Gallard, H., De Laat, J. (2001). Kinetics of oxidation of chlorobenzenes and phenyl-ureas by Fe(II)/H₂O₂ and Fe(III)/H₂O₂. Evidence of reduction and oxidation reactions of intermediates by Fe(II) or Fe(III). *Chemosphere*, Vol. 42, pp. 405-413, ISSN 0045-6535.

Garrido-Ramírez, E.G., Theng, B.K.G, Mora, M.L. (2010). Clays and oxide minerals as catalysts and nanocatalysts in Fenton-like reactions — A review. *Applied Clay Science*, Vol. 47, pp. 182-192, ISSN 0169-1317.

Ghigliazza, R., Lodi, A., Rovatti, M. (2000). Kinetic and process considerations on biological reduction of soluble and scarcely soluble sulfates. *Resources, Conservation and Recycling*, Vol. 29, pp. 181-194, ISSN 0921-3449.

Gogate, P.R., Pandit, A.B. (2004). A Review of Imperative Technologies for Wastewater Treatment I: Oxidation Technologies at Ambient Conditions. *Advances in Environmental Research*, v.8, pp. 501-551, ISSN 1093-7927.

Gulkaya, I., Surucu, G.A., Dilek, F.B. (2006). Importance of H₂O₂/Fe²⁺ ratio in Fenton's treatment of a carpet dyeing wastewater. *Journal of Hazardous Materials*, Vol. B136, pp. 763-769, ISSN 0304-3894.

Heredia, J.B., Terregrosa, J., Dominguez, J.R., Peres, J.A. (2001). Kinetic model for phenolic compound oxidation by Fenton's reagent. *Chemosphere*, Vol. 45, pp. 85-90, ISSN 0045-6535.

Homem, V., Alves, A., Santos, L. (2010). Amoxicillin degradation at ppb levels by Fenton's oxidation using design of experiments. *Science of the Total Environment*, Vol. 408, pp. 6272-6280, ISSN 0048-9697.

Kang, Y.W., Hwang, K.Y. (2000). Effects of Reaction Conditions on the Efficiency in the Fenton Process. *Water Research*, Vol. 34, No. 10, pp. 2786-2790, ISSN 0043-1354.

Kurt, U., Avsar, Y., Gonullu, M.T. (2006). Treatability of water-based paint wastewater with Fenton process in different reactor types. *Chemosphere*, Vol. 64, pp. 1536-1540, ISSN 0045-6535.

Kušić, H., Božić, A.L., Koprivanac, N. (2007). Fenton type processes for minimization of organic content in coloured wastewaters: Part I: Processes optimization. *Dyes and Pigments*, Vol. 74, pp. 380-387, ISSN 0143-7208.

Lee, H., Shoda, M. (2008). Removal of COD and colour from livestock wastewater by the Fenton method. *Journal of Hazardous Materials*, Vol. 153, pp. 1314-1319, ISSN 0304-3894.

Lens, P.N.L., Visser, A., Janssen, A.J.H., Hulshoff Pol, L.W., Lettinga, G. (1998). Biotechnological treatment of sulfate-rich wastewaters. *Critical Reviews in Environmental Science and Technology*, Vol. 28, No. 1, pp. 41-88, ISSN 1064-3389.

Li, R., Yang, C., Chen, H., Zeng, G., Yu, G., Guo, J. (2009). Removal of triazophos pesticide from wastewater with Fenton reagent. *Journal of Hazardous Materials*, Vol. 167, pp. 1028-1032, ISSN 0304-3894.

Lin, S.H., Lo, C.C. (1997). Fenton Process for Treatment of Desizing Wastewater. *Water Research*, Vol. 31, No. 8, pp. 2050-2056, ISSN 0043-1354.

Lu, M.C., Chen, J.N., Chang, C.P. (1999). Oxidation of dichlorvos with hydrogen peroxide using ferrous ion as catalyst. *Journal of Hazardous Materials*, Vol. B65, pp. 277-288, ISSN 0304-3894.

Lucas, M.S., Peres, J.A. (2009). Removal of COD from olive mill wastewater by Fenton's reagent: Kinetic study. *Journal of Hazardous Materials*, Vol. 168, pp. 1253-1259, ISSN 0304-3894.

Lunar, L., Sicilia, D., Rubio, S., Rez-Bendito, D., Nickel, U. (2000). Degradation of Photographic Developers by Fenton's Reagent: Condition Optimization and Kinetics for Metol Oxidation. *Water Research*, Vol. 34, No. 6, pp. 1791-1802, ISSN 0043-1354.

Ma X.J., Xia, H.L. (2009). Treatment of water-based printing ink wastewater by Fenton process combined with coagulation. *Journal of Hazardous Materials*, Vol. 162, pp. 386–390, ISSN 0304-3894.

Mahiroglu, A., Tarlan-Yel, E., Sevimli, M.F. (2009). Treatment of combined acid mine drainage (AMD)—Flotation circuit effluents from copper mine via Fenton's process. *Journal of Hazardous Materials*, Vol. 166, pp. 782–787, ISSN 0304-3894.

Mandal, T., Dasgupta, D., Mandal, S., Datta, S. (2010) Treatment of leather industry wastewater by aerobic biological and Fenton oxidation process. *Journal of Hazardous Materials*, Vol. 180, pp. 204–211, ISSN 0304-3894.

Martínez, N.S.S., Fernández, J.F., Segura, X.F., Ferrer, A.S. (2003). Preoxidation of an extremely polluted industrial wastewater by the Fenton's reagent. *Journal of Hazardous Materials*, Vol. B101, pp. 315–322, ISSN 0304-3894.

Mehra, O.P.; Jackson, M. (1960). Iron oxide removal from soils and clays by a dithionite-citrate system buffered with sodium bicarbonate, *Proceedings of the 7th National Clay Conference*, Pergamon Press, New York, pp. 317–327.

National Research Council, 1995. Prudent Practices in the Laboratory Handling and Disposal of Chemicals. National Academy Press, Washington, DC. ISBN 0309052297.

Nesheiwat, F.K., Swanson, A.G. (2000). Clean Contaminated Sites using Fenton's Reagent. *Chemical Engineering Progress*, Vol. 96, No. 4, pp. 61-66, ISSN 0360-7275.

Neyens, E., Baeyens, J. (2003). A review of classic Fenton's peroxidation as an advanced oxidation technique. *Journal of Hazardous Materials, Vol.* 98(B), pp. 33–50, ISSN 0304-3894.

Normas Analíticas do Instituto Adolfo Lutz. (1985). Métodos químicos e físicos para análises de alimentos (3rd ed). Editoração Débora D. Estrella Rebocho,São Paulo, BR.

Oh, S.Y., Chiu, P.C., Kim, B.J., Cha, D.K. (2003). Enhancing Fenton oxidation of TNT and RDX through pretreatment with zero-valent iron. *Water Research*, Vol. 37, pp. 4275–4283, ISSN 0043-1354.

Padoley, K.V., Mudliar, S.N. , Banerjee, S.K., Deshmukh, S.C., Pandey, R.A. (2011). Fenton oxidation: A pretreatment option for improved biological treatment of pyridine and 3-cyanopyridine plant wastewater. *Chemical Engineering Journal*, Vol. 166, pp. 1–9, ISSN 1385-8947.

Pavan, M.A., Bloch, M.F., Zempulski, H.C., Miyakawa, M., Zocoler, D.C. (1992). *Manual de análise química de solo e controle de qualidade*. Instituto Agronômico do Paraná e IAPAR, Londrina, BR.

Quevauviller, Ph. Rauret, G. López-Sánchez, J.-F. Rubio, R. Ure, A. Muntau, H. (1997). Certification of trace metal extractible contents in a sediment reference material (CRM 601) following a three-step sequential extraction procedure. *Science of the Total Environment*, Vol. 205, pp. 223–234, ISSN 0048-9697.

Scalbert, A.; Monties, B.; Janin, G. (1989). Tannins in wood: comparison of different estimation methods. *Journal of Agricultural and Food Chemistry*, Vol. 37, pp. 1324–1329, ISSN 0021-8561.

Schulze, D.G. (1994). Differential X-ray diffraction analysis of soil minerals, In: *Quantitative Methods in Soil Mineralogy*, J.E. Amonette, L.W. Zelazny (Eds.), pp. 412–429, Soil Sci. Soc. Amer. Misc. Publ, ISBN 0891188061, Madison.

Schwertmann, U. Schulze, D.G. Murad, E. (1982). Identification of ferrihydrite in soils by dissolution kinetics, differential X-ray diffraction and Mössbauer spectroscopy. *Soil Science Society of America Journal*, Vol. 46, pp. 869–875, ISSN 0361-5995.

Silva, A.J., Varesche, M.B., Foresti, E., Zaiat, M. (2002). Sulphate removal from industrial wastewater using a packed-bed anaerobic reactor. *Process Biochemistry*, Vol. 37, pp. 927-935, ISSN 0032-9592.

Talini, I., Anderson, G.K. (1992). Interference of Hydrogen Peroxide on the Standard COD Test. *Water Research*, Vol. 26, No. 1, pp. 107-110, ISSN 0043-1354.

Tessier, A.; Campbell, P.G.C.; Bisson, M. (1979). Sequential extraction procedure for the speciation of particulate trace metals. *Analytical Chemistry*, Vol. 51, pp. 844–851, ISSN 0003-2700.

Torrades, F., Pérez, M., Mansilla, H.D., Peral, J. (2003). Experimental design of Fenton and photo-Fenton reactions for the treatment of cellulose bleaching effluents. *Chemosphere*, Vol. 53, pp. 1211–1220, ISSN 0045-6535.

U.S. Peroxide (2003). Methods for residual peroxide determination: iodometric titration. Available from: <http://www.h2o2.com/intro/analytical.html>.

Vilve, M., Hirvonen, A., Sillanpää, M. (2009). Effects of reaction conditions on nuclear laundry water treatment in Fenton process. *Journal of Hazardous Materials*, Vol. 164, pp. 1468–1473, ISSN 0304-3894.

Walling, C., 1975. Fenton's reagent revisited. *Accounts of Chemical Research*, Vol. 8, pp. 125–131, ISSN 0001-4842.

Walling, C., Kato, S. (1971). The oxidation of alcohols by Fenton's reagent: the effect of copper ion. *Journal of American Chemical Society*, Vol. 93, pp. 4275–4281, ISSN 0002-7863.

Wang, X., Zenga, G.,Zhua, J. (2008). Treatment of jean-wash wastewater by combined coagulation, hydrolysis/acidification and Fenton oxidation. *Journal of Hazardous Materials*, Vol. 153, pp. 810–816, ISSN 0304-3894.

Yetilmezsoy, K., Sakar, S. (2008). Improvement of COD and color removal from UASB treated poultry manure wastewater using Fenton's oxidation. *Journal of Hazardous Materials*, Vol. 151, pp. 547–558, ISSN 0304-3894.

Zhang, H., Choi, H.J., Huang, C.P. (2005). Optimization of Fenton process for the treatment of landfill leachate. *Journal of Hazardous Materials*, Vol. B125, pp. 166–174, ISSN 0304-3894.

3

Photocatalytic Degradation of Organic Pollutants: Mechanisms and Kinetics

Malik Mohibbul Haque[1], Detlef Bahnemann[2] and Mohammad Muneer[1]
[1]Department of Chemistry, Aligarh Muslim University,
[2]Institut fuer Technische Chemie, Leibniz Universität Hannover,
[1]India
[2]Germany

1. Introduction

A wide variety of organic pollutants are introduced into the water system from various sources such as industrial effluents, agricultural runoff and chemical spills (Muszkat et al., 1994; Cohen et al., 1986). Their toxicity, stability to natural decomposition and persistence in the environment has been the cause of much concern to the societies and regulation authorities around the world (Dowd et al., 1998).

Development of appropriate methods for the degradation of contaminated drinking, ground, surface waters, wastewaters containing toxic or nonbiodegradable compounds is necessary. Among many processes proposed and/or being developed for the destruction of the organic contaminants, biodegradation has received the greatest attention. However, many organic chemicals, especially which are toxic or refractory, are not amendable to microbial degradation. Researcher showed their interest and started the intensive studies on heterogeneous photocatalysis, after the discovery of the photo-induced splitting of water on TiO_2 electrodes (Fujishima and Honda, 1972).

Semiconductor particles have been found to act as heterogeneous photocatalysts in a number of environmentally important reactions (Blake, 2001; Pirkanniemi, & Sillanpää, 2002; Gaya & Abdullah, 2008). Materials such as colloidal TiO_2 and CdS have been found to be efficient in laboratory-scale pollution abatement systems (Barni et al., 1995; Bellobono et al., 1994; Legrini et al., 1993; Mills & Hunte, 1997; Halmann, 1996), reducing both organic [e.g. halogenocarbons (Gupta & Tanaka, 1995; Martin et al., 1994; Read et al., 1996;), benzene derivatives (Blanco et al., 1996; Mao et al., 1996) detergents (Rao & Dube, 1996), PCB's (Huang et al., 1996), pesticides (Gianturco et al., 1997; Minero et al., 1996; Lobedank et al., 1997; Haque & Muneer 2003; Muneer & Bahnemann, 2002), explosives (Schmelling et al., 1996), dyes (Vinodgopal et al., 1996), cyanobacterial toxins (Liu et al., 2002)] and inorganic [e.g. N_2 (Ranjit et al., 1996), NO_3^- and NO_2^- (Mills et al., 1994; Ranjit et al., 1995; Kosanic & Topalov, 1990; Pollema et al., 1992), cyanides (Mihaylov et al., 1993; Frank & Bard 1977), thiocyanates (Draper & Fox, 1990), cyanates (Bravo et al., 1994), bromates (Mills et al., 1996) etc.] pollutants/impurities to harmless species. Semiconductor photocatalysts have been shown to be useful as carbon dioxide (Irvine et al., 1990) and nitrogen (Khan & Rao, 1991) fixatives and for the decomposition of O_3 (Ohtani et al., 1992), destruction of micro-organisms such as bacteria (Matsunaga & Okochi, 1995; Zhang et al., 1994; Dunlop et al.,

2002), viruses (Lee et al., 1997), for the killing of malignant cancer-cells (Kubota et al., 1994), for the photo splitting of water (Pleskov & Krotova, 1993; Grätzel, 1981), for the cleanup of oil spills (Gerisher & Heller, 1992; Berry & Mueller, 1994), and for the control of the quality of meat freshness in food industry (Funazaki et al., 1995). Photocatalytic semiconductor films have been studied for the purposes of laboratory scale up of some of the above applications (Dunlop et al., 2002; Byrne et al., 2002) and, in the case of TiO_2 films, for their photo-induced wettability properties (Wang et al., 1998, 1999; Sakai et al., 1998; Yu et al., 2002). This latter phenomenon, termed superhydrophilicity, is being explored for applications in the development of self-cleaning and antifogging surfaces.

Our research group at the Department of Chemistry, Aligarh Muslim University, Aligarh, India in collaboration with the Institut fuer Technische Chemie, Leibniz Universität Hannover, Hannover, Germany, have been actively involved in studying the photocatalytic degradation of variety of priority organic pollutants in aqueous suspensions.

1.1 Mechanism of photooxidation process

The acceleration of a chemical transformation by the presence of a catalyst with light is called photocatalysis. The catalyst may accelerate the photoreaction by interaction with the substrate in its ground or excited state and/or with a primary photoproduct, depending upon the mechanism of the photoreaction and itself remaining unaltered at the end of each catalytic cycle. Heterogeneous photocatalysis is a process in which two active phases solid and liquid are present. The solid phase is a catalyst, usually a semiconductor. The molecular orbital of semiconductors has a band structure. The bands of interest in photocatalysis are the populated valence band (VB) and it's largely vacant conduction band (CB), which is commonly characterized by band gap energy (E_{bg}). The semiconductors may be photoexcited to form electron-donor sites (reducing sites) and electron-acceptor sites (oxidising sites), providing great scope for redox reaction. When the semiconductor is illuminated with light (hv) of greater energy than that of the band gap, an electron is promoted from the VB to the CB leaving a positive hole in the valence band and an electron in the conduction band as illustrated in Figure 1.

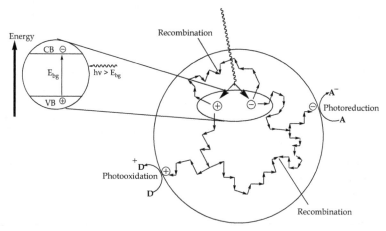

Fig. 1. Photoexcitation of semiconductor leading to charge separation / oxidation / reduction sites.

If charge separation is maintained, the electron and hole may migrate to the catalyst surface where they participate in redox reactions with sorbed species. Specially, h^+_{vb} may react with surface-bound H_2O or OH^- to produce the hydroxyl radical and e^-_{cb} is picked up by oxygen to generate superoxide radical anion ($O_2^{\cdot-}$), as indicated in the following equations 1-3;

- absorption of efficient photons by titania ($h\, v \geq E_{bg} = 3.2$ ev)

$$TiO_2 + h\, v \rightarrow e^-_{cb} + h^+_{vb} \tag{1}$$

- formation of superoxide radical anion

$$O_2 + e^-_{cb} \rightarrow O_2^{\cdot-} \tag{2}$$

- neutralization of OH. group into $\cdot OH$ by the hole

$$(H_2O \Leftrightarrow H_+ + OH.)_{ads} + h_{vb+} \rightarrow \cdot OH + H_+ \tag{3}$$

It has been suggested that the hydroxyl radical ($\cdot OH$) and superoxide radical anions ($O_2^{\cdot-}$) are the primary oxidizing species in the photocatalytic oxidation processes. These oxidative reactions would results in the degradation of the pollutants as shown in the following equations 4-5;

- oxidation of the organic pollutants via successive attack by $\cdot OH$ radicals

$$R + \cdot OH \rightarrow R'\cdot + H_2O \tag{4}$$

- or by direct reaction with holes

$$R + h^+ \rightarrow R^{\cdot+} \rightarrow degradation\ products \tag{5}$$

For oxidation reactions to occur, the VB must have a higher oxidation potential than the material under consideration. The redox potential of the VB and the CB for different semiconductors varies between +4.0 and -1.5 volts versus Normal Hydrogen Electrode (NHE) respectively. The VB and CB energies of the TiO_2 are estimated to be +3.1 and -0.1 volts, respectively, which means that its band gap energy is 3.2 eV and therefore absorbs in the near UV light ($\lambda < 387$ nm). Many organic compounds have a potential above that of the TiO_2 valence band and therefore can be oxidized. In contrast, fewer organic compounds can be reduced since a smaller number of them have a potential below that of the TiO_2 conduction band.

1.2 Use of semiconductor (TiO₂) in various fields

Due to non-toxic, easily available, inexpensive, biologically and chemically inert and stable to photo and chemical corrosion, TiO_2 is used in various fields as shown in Figure 2.

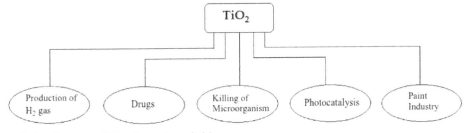

Fig. 2. Application of TiO_2 in various fields.

The process of photocatalysis is also widely being contributed to various sub-discipline of Chemistry as shown in Figure 3.

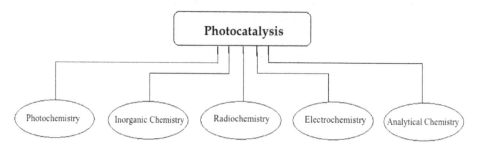

Fig. 3. Application of photocatalysis in various sub-divisions of Chemistry.

1.3 Our research focus

We have studied the photocatalysed degradation of a large variety of organic pollutants under different reaction conditions to determine the detailed degradation kinetics and product identification in few selected systems for better mechanistic understanding. The different class of organic pollutants studied by our research group are shown below in chart 1.

S. No.	Compound Studied	Reference
1	Dinoterb	2011; Dar et al., Res. Chem. Intermed., DOI 10.1007/s11164-011-0299-6
2	Fenoprop and Dichloroprop-P	2010; Faisal et al., Adv. Sci. Lett. 3, 512,
3	Acetamiprid	2010; Khan et al., Desalination, 261, 169.
4	Glyphosate	2008; Muneer & Boxall, Int. J. Photoenergy, article ID 197346
5	4-chlorophenoxyacetic acid	2007; Singh et al., J. Hazard. Mat., 142, 374.
6	Phenoxyacetic acid and 2,4,5-trichlorophenoxyacetic acid	2007; Singh et al., J. Mol. Catal. A: Chem., 264, 66.
7	Uracil and 5-bromouracil	2007; Singh et al., J. Hazard. Mat., 142, 425.
8	4-bromoaniline, 3-nitroaniline, pentachlorophenol, 1,2,3-trichlorobenzene and diphenylamine	2007; Abu Tariq et al., J. Mol. Catal. A: Chem., 265, 231.
9	Chlorotoluron	2006; Haque et al., Environ. Sci. Technol., 40, 4765.
10	Trichlopyr and Daminozid	2006; Qamar et al., J. Environ. Manag., 80, 99.

11	Acephate	2005; Atiqur Rahman et al., J. Adv. Oxid. Technol., 9, 1.
12	Dichlorvos and Phosphamidon	2005; Atiqur Rahman & Muneer, Desalination, 181, 161.
13	Propham, Propachlor and Tebuthiuron	2005; Muneer et al., Chemosphere, 61, 457.
14	Dichlone, 2-amino-5-chloropyridine, benzoyl peroxide and 3-chloro perbenzoic acid	2005; Qamar et al., Res. Chem. Intermed., 31, 807.
15	2,2'-dinitro biphenyl , N,N'-dimethyl-4-nitroso aniline , 4-dimethyl amino benzaldehyde, phthalaldehyde and tetramethyl benzoquinone	2005; Muneer et al., Appl. Catal. A: General, 289, 224.
16	Indole-3-acetic acid and Indole-3-butyric acid	2005; Qamar & Muneer, J. Hazard. Mat., 120, 219.
17	Thiram	2005; Haque & Muneer, Indian J. Chem. Technol., 12, 68.
18	Picloram, Dicamba and Floumeturon	2005; Atiqur Rahman & Muneer, J. Environ. Sci. & Health, B40, 257.
19	2,4-dichlorophenoxy Acetic Acid	2004; Singh & Muneer Res. Chem. Intermed., 30, 317.
20	Maleic Hydrazide	2004; Singh et al., J. Adv. Oxid. Tech., 7, 184.
21	Methoxychlor, Chlorothalonil and Disulfoton	2004; Muneer et al., Res. Chem. Intermed., 30, 663.
22	Dimethyl Terephthalate	2003; Atiqur Rahman et al., Res. Chem. Intermed., 29, 35.
23	Isoproturon	2003; Haque & Muneer, J. Environ. Manag., 69, 169.
24	Bromacil	2003; Singh et al., Photochem. Photobiol. Sci., 2, 151.
25	Diphenamid	2003; Atiqur Rahman et al., J. Adv. Oxid. Technol., 6, 100.
26	Benzidine and 1,2-diphenyl hydrazine	2002; Muneer et al., Chemosphere, 49, 193.
27	Terbacil and 2,4,5-tribromoimidazole	2002; Muneer & Bahnemann, Appl. Catal. B: Environ., 36, 95.
28	1,2-diethyphthalate	2001; Muneer et al., J. Photochem. Photobiol., A: Chem., 143, 213.
29	Diuron	1999; Muneer et al., Res. Chem. Intermed., 25, 667.

Chart 1.

The following text describes the results of the photocatalysed degradation of different pollutants under variety of conditions for degradation kinetics and product identification.

1.4 Procedure for conducting the degradation experiments

Stock solutions of the pollutants containing the desired concentration were prepared in double distilled water. An immersion well photochemical reactor made of Pyrex glass equipped with a magnetic stirring bar, water circulating jacket and an opening for supply of molecular oxygen was used. A simplified diagram of the reactor system is shown in Figure 4.

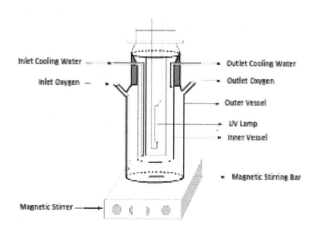

Fig. 4. Simplified diagram of Photochemical Reaction Vessel.

For irradiation experiment an aqueous solution of the pollutants with desired concentration is taken into the photoreactor and required amount of photocatalyst was added. Then the solution was stirred and bubbled with molecular / atmospheric oxygen for at least 15 minutes in the dark to allow equilibration of the system so that the loss of compound due to adsorption can be taken into account. The zero time reading was obtained from blank solution kept in the dark in the presence of TiO_2 and oxygen but otherwise treated similarly to the irradiated solution. The suspensions were continuously purged with molecular / atmospheric oxygen throughout each experiment. Irradiations were carried out using medium pressure mercury lamp. The light intensity was measured using UV-light intensity detector. IR-radiation and short-wavelength UV-radiation were eliminated by water circulating Pyrex glass jacket. Samples (10 mL) were collected before and at regular intervals during irradiation and analyzed after centrifugation.

The sunlight experiments were carried out between 9:00 A.M to 2:30 P.M. on a sunny day. The light intensity was measured using UV-light intensity detector (Lutron UV-340), which was found to be in the range of 0.370 to 0.480 mW/cm². Reactions were carried out in the same reaction vessel as described above. Aqueous solution of desired concentration of the model compound containing required amount of photocatalyst was taken and stirred for 15 min. in the dark in presence of oxygen for equilibration. The solution was then placed on flat platform under sunlight with continuous stirring and purging of molecular oxygen. Samples (10 mL) were collected before and at regular intervals during the illumination and analyzed after centrifugation.

2. Analysis

2.1 Photomineralization of organic pollutants
The photomineralization of the pesticide was measured using Total Organic Carbon analyzer (Shimadzu TOC 5000 A). The main principle of TOC analyzer involves the use of carrier gas (oxygen), which is flow-regulated (150 ml / min) and allows to flow through the total carbon (TC) combustion tube, which is packed with catalyst, and kept at 680^0 C. When the sample enters the TC combustion tube, TC in the sample is oxidized to carbon dioxide. The carrier gas containing the combustion products from the TC combustion tube flows through the inorganic carbon (IC) reaction vessel, dehumidifier, halogen scrubber and finally reaches the sample cell of the nondispersive infrared (NDIR) detector which measures the carbon dioxide content. The output signal (analog) of the NDIR detector is displayed as peaks. The peak areas are measured and processed by the data processing unit. Since the peak areas are proportional to the total carbon concentration, the total carbon in a sample may be easily determined from the calibration curve prepared using standard solution of known carbon content. Total carbon is the sum of TOC (Total Organic Carbon) and IC (Inorganic Carbon).

2.2 Photodegradation of the pesticide
The photodegradation of the pesticide was measured using UV-Vis spectrophotometry or HPLC analysis techniques.

2.3 Characterization of intermediate products
Intermediate product formed during the photooxidation process was characterized by monitoring the reaction as a function of time using GC/MS analysis technique. For GC/MS analysis a Shimadzu Gas Chromatograph and Mass Spectrometer (GCMS-QP 5050) equipped with a 25m CP SIL 19 CB (d=0.25mm) capillary column, operating temperature programmed (220^0C for 40 min at the rate of 10^0C min^{-1}) in splitless mode injection volume (1.0 µL) with helium as a carrier gas was used.

3. Photocatalysis of organic pollutants under different conditions

3.1 In the presence of TiO$_2$
Irradiation of an aqueous suspension of desired organic pollutants in the presence of TiO$_2$ lead to decrease in absorption intensity and depletion in TOC content as a function of time.

As a representative example Figure 5 shows the change in absorption intensity and depletion in TOC as a function of time on irradiation of an aqueous solution of isoproturon (0.5 mM, 250 ml) in the presence and absence of photocatalyst (Degussa P25, 1 gL^{-1}) by the "Pyrex" filtered output of a 125 W medium Pressure mercury lamp (radiant flux 4.860 mW/cm^2).

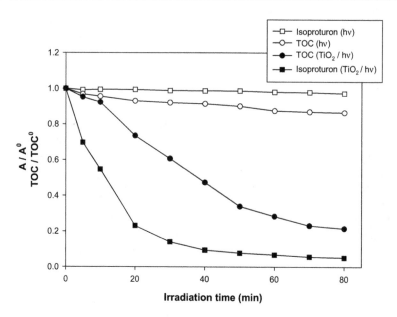

Fig. 5. Depletion in TOC and change in absorption intensity as a function of irradiation time for an aqueous solution of isoproturon in the presence and absence of the photocatalyst. Experimental conditions: 0.5 mM isoproturon, V=250 mL, photocatalyst: TiO_2 (Degussa P25, 1 gL^{-1}), immersion well photoreactor, 125 W medium pressure Hg lamp, absorbance was followed at 238 nm after 75% dilution, cont. O_2 purging and stirring, irradiation time = 80 min.

It could be seen that 94.8% degradation and 78.45 % mineralization of the isoproturon takes place after 80 min of illumination. On the other hand no observable loss of the compound was found when the irradiations were carried out in the absence of the photocatalysts (Haque & Muneer, 2003).

Both the mineralization (depletion of TOC Vs irradiation time) and decomposition (decrease in absorption intensity/concentration Vs irradiation time) curves can be fitted reasonably well by an exponential decay curve suggesting the first order kinetics. Figure 6 shows the linear regression curve fit for the natural logarithm of the degradation of isoproturon Vs irradiation time for first order reaction. For each experiment, the degradation rate constant was calculated from the plot of the natural logarithm of the TOC depletion or absorbance / concentration of the pesticide as a function of irradiation time (the concentration of the pollutant was calculated by taking the absorbance of the pollutant at its λ_{max} of standard concentration and then by plotting the graph of absorbance Vs standard concentration of the pollutant). The degradation rate for the mineralization and for the decomposition of the pollutants for first order reaction was calculated using formula given below;

$$-d[TOC]/dt = kc^n \tag{6}$$

$$-d[c]/dt = kc^n \tag{7}$$

TOC = Total Organic Carbon, k = rate constant, c = concentration of the pollutant, n = order of reaction.

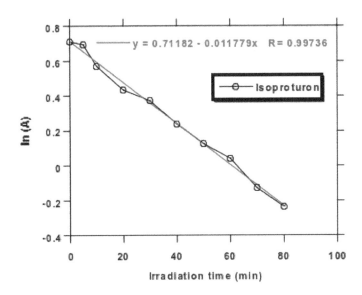

Fig. 6. Plot showing the linear regression curve fit for the natural logarithm of absorbance of the pesticide concentration of isoproturon against irradiation time for the first order reaction.

3.2 Comparison of different photocatalysts

Titanium dioxide is known to be the semiconductor with the highest photocatalytic activity and stable in aqueous solution. Several reviews have been written, regarding the mechanistic and kinetic details as well as the influence of experimental parameters. It has been demonstrated that degradation by photocatalysis can be more efficient than by other wet-oxidation technique (Weichrebe and Vogelpohl, 1995).

We have tested the photocatalytic activity of four different commercially available TiO_2 powders (namely Degussa P25, Sachtleben Hombikat UV100, Milenium Inorganic PC500 and Travancore Titanium Product, India) on the degradation kinetics of the pollutants.

In most of the cases it has been observed that the degradation of pollutants under investigation proceed much more rapidly in the presence of Degussa P25 as compared with other TiO_2 samples (Muneer and Bahnemann, 2001; Haque & Muneer, 2003; Bahnemann et al., 2007) as shown in Figure 7 (isoproturon). The better photocatalytic activity of Degussa P25 has also been reported by Pizzaro et al., in 2005. While in some cases Hombikat UV 100 is found to be better than Degussa P25 for the degradation of benzidine, eosine yellowish, and Remazol brilliant blue R (Muneer et al., 2002; Saquib and Muneer, 2002, 2003) as shown in Figure 8 (eosine yellowish). In an earlier study Lindner et al. 1995, showed that Hombikat UV100 was almost four times more effective than P25 when dichloroacetic acid was used as the model pollutant.

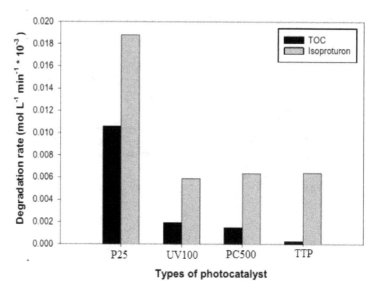

Fig. 7. Comparison of degradation rate for the mineralization and for the decomposition of isoproturon in the presence of different photocatalysts. Experimental conditions: 0.5 mM isoproturon, V=250 mL, photocatalysts: TiO_2 Degussa P25 (1 gL-1), Sachtleben Hombikat UV100 (1 gL-1), PC500 (1 gL-1), TTP (1 gL-1).

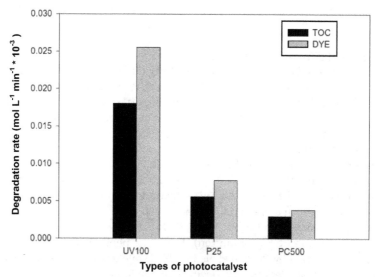

Fig. 8. Comparison of degradation rate for the mineralization and for the decomposition of eosine yellowish under different photocatalysts. Experimental conditions: 0.25 mM dye concentration, V=250 mL, photocatalysts: TiO_2 Degussa P25 (1 gL-1), Sachtleben Hombikat UV100 (1 gL-1) and PC500 (1 gL-1).

The differences in the photocatalytic activity of TiO_2 are likely to be due to differences in the BET-surface, impurities, lattice mismatches or density of hydroxyl groups present on the catalyst's surface. Since they will affect the adsorption behaviour of a pollutant or intermediate molecule and the lifetime and recombination rate of electron-hole pairs.

The reason for the better photocatalytic activity of Degussa P25, could be attributed to the fact that P25 being composed of small nano-crystallites of rutile being dispersed within an anatase matrix. The smaller band gap of rutile "catches" the photons, generating electron-hole pairs. The electron transfer, from the rutile conduction band to electron traps in anatase phase takes place. Recombination is thus inhibited allowing the hole to move to the surface of the particle and react (Hurum et al., 2003). The better efficiency of photocatalyst Degussa P25 may also due to 'quantum size effect' (Nozik et al., 1993; weller, 1993). When the particles become too small, there is a 'blue shift' with an increase of the band gap energy, detrimental to the near UV-photon absorption, and an increase of the electron-hole recombination.

3.3 Influence of pH on the degradation kinetics

An important parameter in the photocatalytic reactions taking place on the particulate surfaces is the pH of the solution, since it dictates the surface charge properties of the photocatalyst and size of aggregates it forms. Employing Degussa P25 as photocatalyst the decomposition and mineralization of pollutant in aqueous suspensions of TiO_2 was studied as a function of pH.

The degradation rate for the decomposition and mineralization of acid red 29 was found to increase with the increase in pH from 3 to 11 as shown in Figure 9 (Qamar et al., 2005) while the degradation rate for the decomposition and mineralization of the pesticide derivative propachlor was found to decrease with the increase in pH from 3 to 11 as shown in Figure 10 (Muneer et al, 2005). Similar results on pH effect have also been reported earlier by Vaz et al., 1998 (for the degradation of uracil and 5-halogenouracil) and Lu et al., 1995 (for the degradation of dichlorvos, propoxur and 2,4-D).

The interpretation of pH effects on the photocatalytic process is very difficult task because of its multiple roles such as electrostatic interactions between the semiconductor surface, solvent molecules, substrate and charged radicals formed during the reaction process. The ionization state of the surface of the photocatalyst can be protonated and deprotonated under acidic and alkaline conditions, respectively, as shown in following equations:

$$TiOH + H^+ \longrightarrow TiOH_2^+ \tag{8}$$

$$TiOH + {}^-OH \longrightarrow TiO^- + H_2O \tag{9}$$

The point of zero charge (pzc) of the TiO_2 (Degussa P25) is widely reported at pH ~ 6.25 (Augustynski, 1988). Thus, the TiO_2 surface will remain positively charged in acidic medium (pH < 6.25) and negatively charged in alkaline medium (pH > 6.25). The functional group present on the pollutants can be protonated and deprotonated depending on the pH of the reaction mixture. The better degradation rate in acidic or basic pH may also be attributed on the basis of the fact that the structural orientation of the molecule is favoured for the attack of the reactive species under that condition.

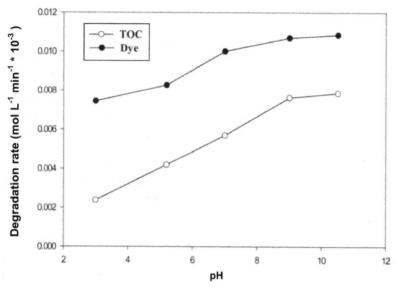

Fig. 9. Influence of pH on the degradation rate for the mineralization and for the decomposition of acid red 29. Experimental conditions: Reaction pH (3, 5.2, 7, 9 and 10.5), 0.25 mM dye concentration, V=250 mL, photocatalyst: TiO_2 (Degussa P25, 1 gL^{-1}).

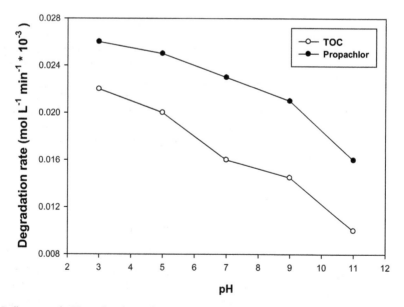

Fig. 10. Influence of pH on the degradation rate for the mineralization and for the decomposition of propachlor. Experimental conditions: Reaction pH (3, 5, 7, 9 and 11), 0.6 mM propachlor, V=250 mL, photocatalyst TiO_2 (Degussa P25, 1 gL^{-1}).

3.4 Effect of substrate concentration

It is important both from mechanistic and from application point of view to study the dependence of substrate concentration in the photocatalytic reaction rate. Effect of substrate concentration on the degradation of the pollutants was studied at different concentrations. As a representative example Figure 11 shows the degradation rate for the TOC depletion and for the decomposition of pollutant as a function of substrate concentration employing Degussa P25 as photocatalyst (Muneer et al, 2005). It was found that the degradation rate for the decomposition and for the mineralization increases gradually with the increase in substrate concentrations. Similar trend was found in most of the colourless organic pollutants as reported earlier (Sabin et al., 1992; Krosley et al., 1993; O'Shea et al., 1997). In coloured compound it has been found that the degradation rate increase upto a certain limit and after that a further increase in substrate concentration lead to decrease in the degradation rate. This may be due to the fact that as the initial concentrations of the pollutant increases, the irradiating mixture becomes more and more intense which prevents the penetration of light to the surface of the catalyst. Hence, the generation of relative amount of $\cdot OH$ and $O_2{}^{\cdot-}$ on the surface of the catalyst do not increase as the intensity of light and irradiation times are constant. Conversely, their concentrations will decrease with increase in concentration of the pollutant as the light photons are largely absorbed and prevented from reaching the catalyst surface by the substrate molecules.

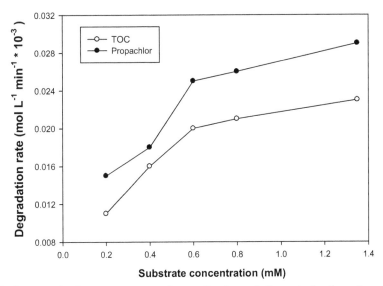

Fig. 11. Influence of substrate concentration on the degradation rate for the mineralization and for the decomposition of propachlor. Experimental conditions: Substrate concentrations (0.20, 0.40, 0.60, 0.80 and 1.35 mM), V=250 mL, photocatalyst: TiO_2 (Degussa P25, 1 gL^{-1}).

As oxidation proceeds, less and less of the surface of the TiO_2 particle is covered as the pollutant is decomposed. Evidently, at total decomposition, the rate of degradation is zero and a decreased photocatalytic rate is to be expected with increasing irradiation time. It has been agreed, with minor variation that the expression for the rate of photodegradation of organic substrates with irradiated TiO_2 follows the Langmuir Hinshelwood (L-H) law for the four possible situations, i.e., (1) the reaction takes place between two adsorbed substances, (2) the reaction occurs between a radical in solution and an adsorbed substrate molecule, (3) the reaction takes place between a radical linked to the surface and a substrate molecule in solution, and (4) the reaction occurs with both the species being in solution. In all cases, the expression for the rate equation is similar to that derived from the L-H model, which has been useful in modelling the process, although it is not possible to find out whether the process takes place on the surface, in the solution or at the interface. Our results, on the effect of the initial concentration on the degradation rate are in agreement with the assumption of the Langmuir Hinshelwood model.

3.5 Effect of catalyst concentration

The effect of photocatalyst concentration on the degradation kinetics of pollutant under investigation was studied employing different concentrations of Degussa P25 varying from 0.5 to 7.5 gL^{-1}. As a representative example the degradation rate for the TOC depletion and for the decomposition of the tebuthiuron as function of catalyst loading is shown in Figure 12 (Muneer et al., 2005). It was observed that the degradation rate was found to increase with the increase in catalyst concentration upto 5 gL^{-1} and on subsequent addition of catalyst lead to the levelling off the degradation rate.

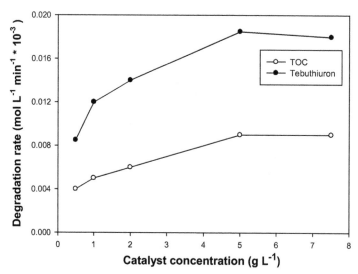

Fig. 12. Influence of catalyst concentration on the degradation rate for the mineralization and for the decomposition of tebuthiuron. Experimental conditions: Photocatalyst concentrations: TiO_2 Degussa P25 (0. 5, 1.0, 2.0, 5.0 and 7.5 gL^{-1}), V=250 mL.

Whether in static, slurry, or dynamic flow reactors, the initial reaction rates were found to be directly proportional to catalyst concentration, indicating a heterogeneous regime. However, it was observed that above a certain concentration, the reaction rate decreases and becomes independent of the catalyst concentration. This limit depends on the geometry and working conditions of the photoreactor and for a definite amount of TiO_2 in which all the particles, i.e., surface exposed, are totally illuminated. When the catalyst concentration is very high, after travelling a certain distance on an optical path, turbidity impedes further penetration of light in the reactor. In any given application, this optimum catalyst concentration [$(TiO_2)_{OPT}$] has to be found, in order to avoid excess catalyst and insure total absorption of efficient photons. Our results on the effect of catalyst concentration on the degradation rate for the TOC depletion and decomposition of pollutants under investigation are in agreement with numerous studies reported in the literature (Shifu and Yunzhang 2007, Daneshvar et al., 2004).

It is believed that both the number of photons absorbed as well as the solute molecules adsorbed increases with increase in number of TiO_2 particles upto the optimum value. Any further increase in TiO_2 concentration beyond optimum value may cause scattering and screening effects which reduces the specific activity of the catalyst (Evgenidou et al., 2007). The highly turbid suspension may prevent the catalyst farthest from being illuminated (Rahman and Muneer, 2005). Higher amount of catalyst may lead to aggregation of TiO_2 particles which may decrease the catalytic activity (Garcia and Takashima, 2003). The optimum value of catalyst has been found to vary with different initial solute concentrations (Sakthivel et al., 2003).

3.6 Effect of different electron acceptors

One practical problem in using TiO_2 as a photocatalyst is the undesired electron / hole recombination, which, in the absence of proper electron acceptor or donor, is extremely efficient and represent the major energy - wasting step thus limiting the achievable quantum yield. One strategy to inhibit electron - hole pair recombination is to add other (irreversible) electron acceptors to the reaction. They could have several different effects such as, i.e., (1) to increase the number of trapped electrons and, consequently, avoid recombination, (2) to generate more radicals and other oxidizing species, (3) to increase the oxidation rate of intermediate compounds and (4) to avoid problems caused by low oxygen concentration. In highly toxic wastewater where the degradation of organic pollutants is the major concern, the addition of electron acceptors to enhance the degradation rate may often be justified. With this view, the electron acceptor such as potassium persulphate, potassium bromate and hydrogen peroxide were added in the solution.

As a representative example Figure 13 shows the degradation rate for the mineralization and decomposition of the pollutants in the presence of different electron acceptors. The electron acceptors such as hydrogen peroxide, bromate and persulphate ions are known to generate hydroxyl radicals by the mechanisms shown in equations 10-14;

$$H_2O_2 + e^- \longrightarrow {}^\bullet OH + {}^-OH \tag{10}$$
$$BrO_3^- + 2H + e_{cb}^- \longrightarrow BrO_2^\bullet + H_2O \tag{11}$$
$$BrO_3^- + 6H + 6e_{cb}^- \longrightarrow [BrO_2^-, HOBr] \longrightarrow Br^- + 3H_2O \tag{12}$$
$$S_2O_8^{2-} + e_{cb}^- \longrightarrow SO_4^{2-} + SO_4^\bullet{}^- \tag{13}$$
$$SO_4^\bullet{}^- + H_2O \longrightarrow SO_4^{2-} + {}^\bullet OH + H^+ \tag{14}$$

The respective one-electron reduction potentials of different species are: E ($O_2/O_2^{\bullet-}$) = -155mV, E ($H_2O_2/HO^\bullet$) = 800mV, E ($BrO_3^-/BrO_2^\bullet$) = 1150 mV, and E ($S_2O_8^{2-}/SO_4^-$) = 1100 mV (Wardman, 1989). From the thermodynamic point of view all employed additives should therefore be more efficient electron acceptors than molecular oxygen.

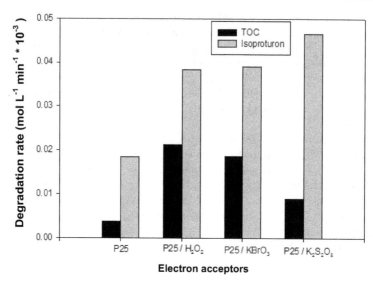

Fig. 13. Comparison of degradation rate for the mineralization and for the decomposition of isoproturon in the presence of different electron acceptors. Experimental conditions: Electron acceptors: $KBrO_3$ (5mM), $K_2S_2O_8$ (5mM), H_2O_2 (10mM), 1.5 mM isoproturon, V=250mL, photocatalyst: TiO_2 (Degussa P25, 1 g L^{-1}).

The effective electron acceptor ability of $KBrO_3$ has been observed in number of studies before (Nevim et al., 2001; Muneer and Bahnemann, 2001). The reason can be attributed to the number of electrons it reacts as shown in eq. 12. Another possible explanation might be a change in the reaction mechanism of the photocatalytic degradation. Since the reduction of bromate ions by electrons does not lead directly to the formation of hydroxyl radicals, but rather to the formation of other reactive radicals or oxidizing agents eg. BrO_2^- and HOBr. Furthermore, bromate ions by themselves can act as oxidizing agents. Linder has proposed a mechanism for the photocatalytic degradation of 4-chlorophenol in the presence of bromate ions considering direct oxidation of the substrate by bromate ions (Linder, 1997).

The enhanced effect of persulphate ion on the degradation of pollutants may be accounted on the basis that persulphate is a beneficial oxidizing agent in photocatalytic detoxification because SO_4^- is formed from the oxidant by reaction with the electron generated at conduction band (e^-_{cb}) of the semiconductor as shown in eq. 13. This sulphate radical anion (SO_4^-) is a strong oxidant (E_0 = 2.6 eV) and it can react with the organic pollutants in three possible modes (1) by abstracting a hydrogen atom from saturated carbon, (2) by adding to unsaturated or aromatic carbon and (3) by removing one electron from the carboxylate anions and from certain neutral molecules. In spite of this sulphate radical anion can trap the photogenerated electrons and/ or generated hydroxyl radical as shown in eq. 14. The formation of sulphate radical anion and hydroxyl radical are powerful oxidant, which can degrade the organic pollutants at a faster rate.

The effect of H_2O_2 has been investigated in numerous studies and it was observed that it increases the photodegradation rates of organic pollutants (Hallamn, 1992). The enhancement of the degradation rate on addition of H_2O_2 can be rationalized in terms of several reason. Firstly, it increase the rate by removing the surface-trapped electrons, thereby lowering the electron-hole recombination rate and increasing the efficiency of hole utilization for reactions such as ($^-$OH + h$^+$ ➡ $^\bullet$OH). Secondly, H_2O_2 may split photolytically to produce hydroxyl radicals ($^\bullet$OH) directly, as cited in studies of homogeneous photooxidation using UV / (H_2O_2 + O_2) (Peyton and Glaze 1988). Thirdly, the solution phase may at times be oxygen starved, because of either oxygen consumption or slow oxygen mass transfer, and peroxide addition thereby increases the reaction rate.

During the photocatalytic degradation the free radical formed serve a dual function. They are not only the strong oxidants but also at the same time their formation and subsequent rapid oxidation reactions inhibit the electron hole pair recombination.

3.7 Comparison of degradation rate under UV and sunlight

For practical applications of wastewater treatment based on these processes, the utilization of sunlight is preferred. Hence the aqueous suspension of TiO$_2$ containing organic pollutants was exposed to solar radiation. As a representative example Figure 14 shows the change in concentration as a function of irradiation time on illumination of an aqueous suspension of acetamiprid (0.1 mM) in the presence of TiO$_2$ (Degussa P25, 1gL^{-1}) under sunlight and UV light source (Khan et al., 2010). It was found that the degradation of the model compound proceeds reasonably fast under sunlight as well.

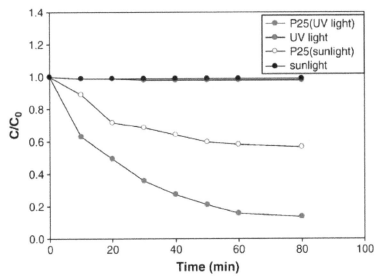

Fig. 14. Change in concentration on irradiation of an aqueous suspension in the presence and absence of TiO$_2$ (Degussa P25, 1 gL^{-1}) containing acetamiprid (0.1 mM), Light source: Pyrex filtered output of a medium pressure mercury lamp and sunlight.

3.8 Characterization of intermediate products

The identification and characterization of by-products formed during the photodegradation of organic pollutants has been of great interest among the people working in this area around the world. A brief summary showing the starting material and by-products under photolytic condition is shown below in Chart 2.

S. No.	Compound	By-products	Reference
1.	Pyridaben	Fragmented products	Zhu et al., 2004
2	Urea derivative	Hydroxylated, Phenyl Hydroxy Ureas, Aniline Derivatives	Lhomme et al., 2005; Vulliet et al., 2002; Maurino et al., 1999; Pramauro et al., 1993; Kinkennon et al., 1995; Richard & Bengana, 1996; Parra et al., 2000, 2002
3	Chlorophenols derivatives	Hydroxylated, Dechlorinated, Chloro Derivatives	Minero et al., 1996; Jardim et al., 1997; Tseng & Huang, 1991
4	Phenol	Catechol, Hydroquinone, 3-phenyl-2-propenal	Azevedo et al., 2009
5	Triazines derivative	Amido, Dealkylated, Hydroxylated, Ammeline, Cyanouric Acid	Goutailler et al., 2001; Konstantinou et al., 2001a; Pelizzetti et al., 1990, 1992a, 1992b; Minero et al., 1996; Muszkat et al., 1995; Sanlaville et al., 1996, Sleiman et al., 2006
6	Aniline and Amide derivative	Amines, Dechlorinated, Dealkylated, Cyclized, Aliphatics, Cyclized	Konstantinou et al., 2001b, 2002; Sakkas et al., 2004; Peñuela and Barceló, 1996; Pathirana & Maithreepala, 1997
7	Thicarbamate derivative	Amine, Carboxy, Sulfoxide, Dealkylated	Vidal et al., 1999; Sturini et al., 1996; Vidal & Martin, 2001
8	Phenoxy-acids derivatives	Hydroxylated, Carboxylated, Chlorophenols, Quinonidal	Herrmann et al., 1998; Topalov et al., 2000; Barbeni et al., 1987; Poulios et al., 1998;
9	Organophosphorus derivatives	Hydroxy, Oxon, Phenol, Dialkylated, Trialkyl esters, Fragmented products	Herrmann, 1999; Konstantinou et al., 2001a, Oncescu et al., 2010; Herrmann et al., 1999; Hua et al., 1995; Sakkas et al., 2002; Dominguez et al., 1998
10	Carbamate derivative	Hydroxylated, Decarboxylated, Phenolic, Dealkylated, Cyclized	Tamimi et al., 2006; Percerancier et al., 1995; Pramauro et al., 1997; Tanaka et al., 1999; Marinas et al., 2001; Bianco Prevot et al., 1999;
11	Organochlorine derivative	Hydroxy, Dechlorinated	Guillard et al., 1996; Vidal, 1998; Zalenska et al., 2000; Peñuela & Barceló, 1998a, 1998b; Pichat 1997
12	Pyridines	Fragmented, Hydroxylated	Stapleton et al., 2010

Chart 2.

We have also made an attempt to identify the intermediate products formed in the photocatalytic degradation of variety of pesticide derivatives in aqueous suspensions of titanium dioxide through GC-MS analysis technique. Results on the photocatalytic degradation of few selected pesticide derivatives for product analysis are shown below;

3.8.1 Photocatalysis of propham (1)

Irradiation of an aqueous solution of propham (1) in the presence of Degussa P25 TiO_2 and analysis of the irradiated mixture at different time intervals through GC-MS analysis showed the formation of several intermediate products out of which, two products appearing at retention times (t_R) 12.95 and 14.81 min., respectively were characterized as products 3 and 8, on the basis of molecular ion and fragmentation pattern (Muneer et al., 2005). A probable mechanism for the formation of these products from propham (1) involving electron transfer reactions and reactions with hydroxyl radical and superoxide radical anions formed in photocatalytic system, is shown in Scheme 1. The model compound 1, upon the transfer of an electron can form the radical anion 2, which may undergo addition of a hydroxyl radical either at ortho or para position followed by loss of a proton to give the observed product 3. This compound on further transfer of an electron followed by addition of a hydroxyl group can undergo cleavage reaction to give 4. This intermediate on subsequent transfer of an electron can undergo addition of a hydroxyl group to give the observed product 8.

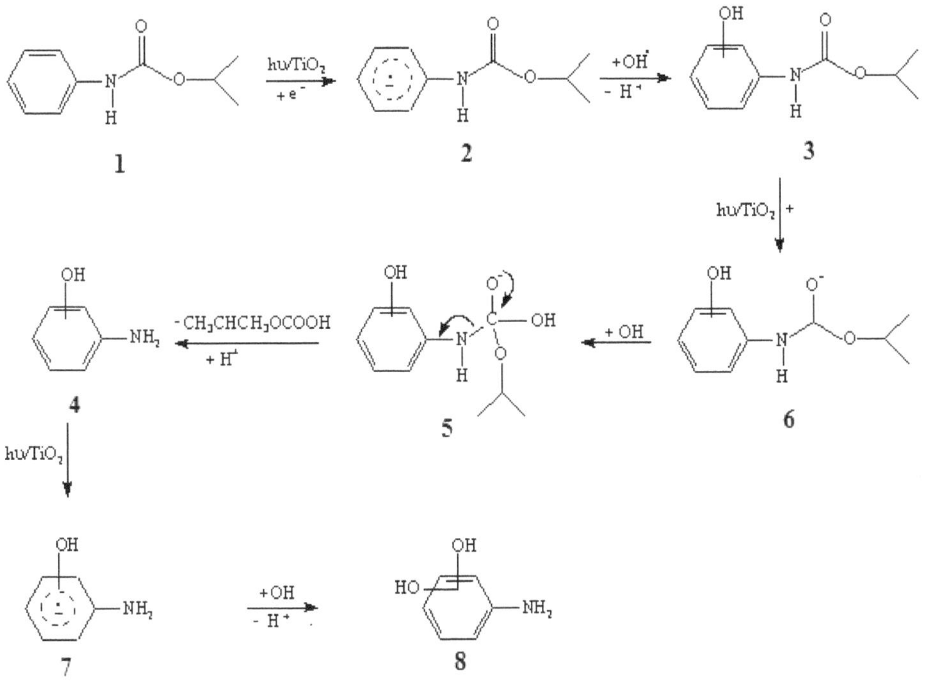

Scheme 1.

3.8.2 Photocatalysis of propachlor (9)

The GC-MS analysis of an irradiated mixture of propachlor (9) in the presence of TiO_2 showed the formation of several by-products. A plausible mechanism for the formation of the different products such as N-Isopropyl-N-phenyl-acetamide (11), 2-Hydroxy-N-isopropyl-N-phenyl acetamide (13), hydroxyl inserted product (15) and N-phenyl-acetamide (16) from propachlor (9) involving similar reactive species, is proposed in Scheme 2 (Muneer et al., 2005). The model compound, 9 on transfer of an electron can form the radical species 10 on removal of chloride ion. This radical species can lead to the observed products 11 and 13, on abstraction of a hydrogen atom or hydroxyl radical. The product 11, on further transfer of an electron can form the radical species 14, which may lose isopropyl to give the observed product 16. The formation of 15 could be understood in terms of pathways shown in Scheme 2 through electron transfer to give radical anion species 12 followed by addition of hydroxyl radical.

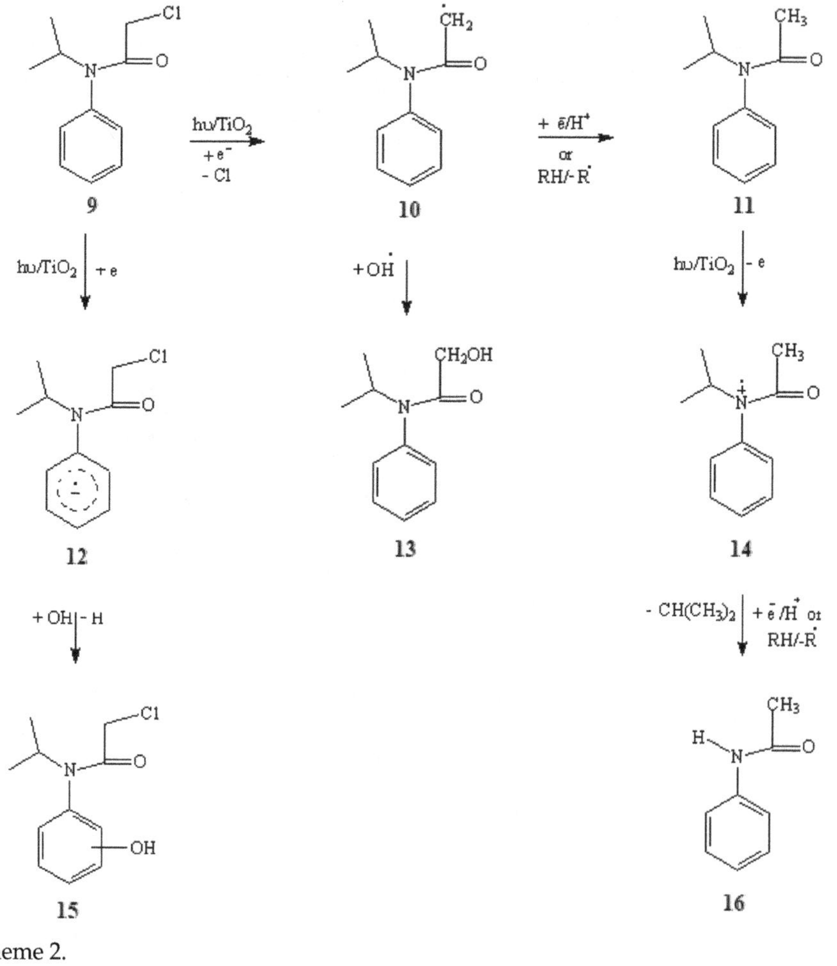

Scheme 2.

3.8.3 Photocatalysis of chlorotoluron (17)

The steady state photolysis of an aqueous suspension of chlorotoluron (17) in the presence of TiO$_2$ under constant bubbling of atmospheric oxygen for 2 h showed the formation of several products, of which, three products such as 3-(3-Hydroxy-4-methylphenyl)-1,1-dimethylurea (23) (3-chloro-4-methylphenyl) urea (24) and 3-chloro-4-methylphenylamine (25) appearing at retention times (t$_R$) 10.310 min, 10.255 min and 5.520 min., respectively were identified on the basis of molecular ion and fragmentation pattern with those reported in the GC/MS NIST library, along with some unchanged starting material (17) appearing at (t$_R$) 5.097 min. (Haque et al., 2006). A possible mechanism for the formation of products 23, 24 and 25 from chlorotoluron (17) involving electron transfer reaction and reaction with hydroxyl radicals could be understood in terms of pathways as shown in Scheme 3. The model compound 17, upon the transfer of an electron followed by the loss of methyl radical and abstraction of a proton to give species 21. This species on subsequent similar reaction can give the observed product 24. The observed product 25 from 17 may be arising through the transfer of an electron to give radical anionic species 19, which may undergo addition of a hydroxyl radical to give 22. This species may lose N(CH$_3$)$_2$COOH followed by abstraction of a proton to give the observed product 25 as shown in the Scheme 3. The product 23 could be formed upon the transfer of an electron followed by the loss of chloride ion with subsequent addition of hydroxyl radical as shown in scheme below.

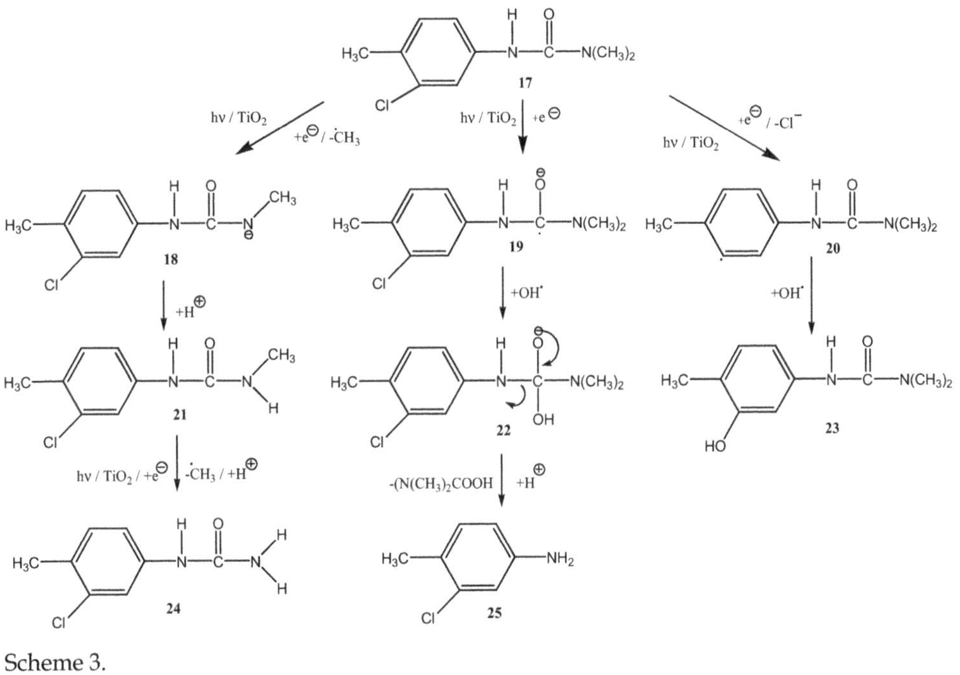

Scheme 3.

3.8.4 Photocatalysis of phenoxyacetic acid (26)

The desired concentration of an aqueous solution of phenoxyacetic acid (26) was irradiated in the presence of Degussa P25 and the analysis of the reaction mixture under analogous conditions, showed the formation of several intermediate products of which two intermediate products phenol (28) and 1,2-diphenoxyethane (30) appearing at retention time (t_R) 2.92 and 9.84 min., respectively (Singh et al., 2007a). These products were identified by comparing its molecular ion and mass fragmentation peaks with those reported in the NIST library. A plausible mechanism for the formation of these products involving reactions with hydroxyl radical and superoxide radical anion formed in the photocatalytic process is shown in Scheme 4. The model compound undergoes addition of a hydroxyl radical leading to the formation of a radical species 27, which may undergo loss of ($^\bullet OCH_2COOH$) to give the observed product phenol (28). Alternatively, the model compound 26 on addition of hydroxyl radical followed by loss of CO_2 may lead to the formation of the radical species 29. This species can either lose formaldehyde molecule followed by addition of hydroxyl radical led to the formation of phenol or can combine with another radical species leading to the formation of the 1,2-diphenoxyethane (30).

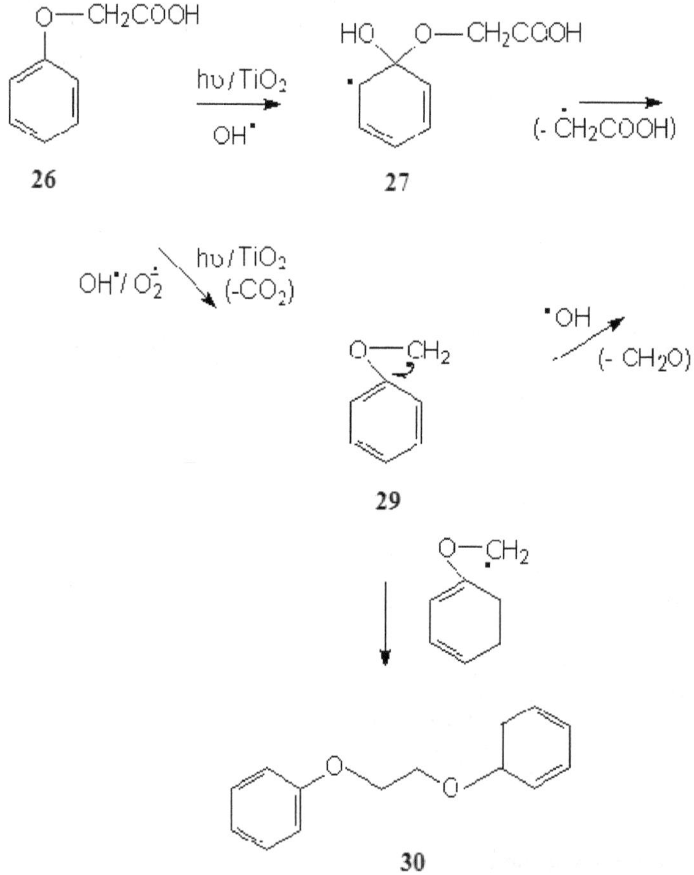

Scheme 4.

3.8.5 Photocatalysis of 2,4,5-trichlorophenoxy acetic acid (31)

Analysis of the irradiated aqueous mixture of 2,4,5-trichlorophenoxy acetic acid (31) in the presence of photocatalyst showed the formation of three intermediate products namely, 2,4,5-trichlorophenol (33), 2,4-dichlorophenol (37) and 1,2,4-trichloro-5-methoxy benzene (35) appearing at retention times 4.11, 6.3 and 12.86 min., respectively and were identified by comparing their molecular ion and mass fragmentation peaks with those reported in the NIST library (Singh et al., 2007a). The model compound 31 on addition of a hydroxyl radical can lead to the formation of a radical species 32, which may undergo loss of ($\cdot$OCH$_2$COOH) group to give the observed product 2,4,5-trichlorophenol (33). Alternatively, the model compound 31 upon the transfer of an electron followed by loss of CO$_2$ may lead to the formation of the radical species 34. This species abstracts a hydrogen atom to form the observed product 1,2,4-trichloro-5-methoxy benzene (35). The product, 33 on further transfer of an electron may undergo loss of chlorine atom followed hydrogen atom abstraction giving 2,4-dichlorophenol (37) as shown in Scheme 5.

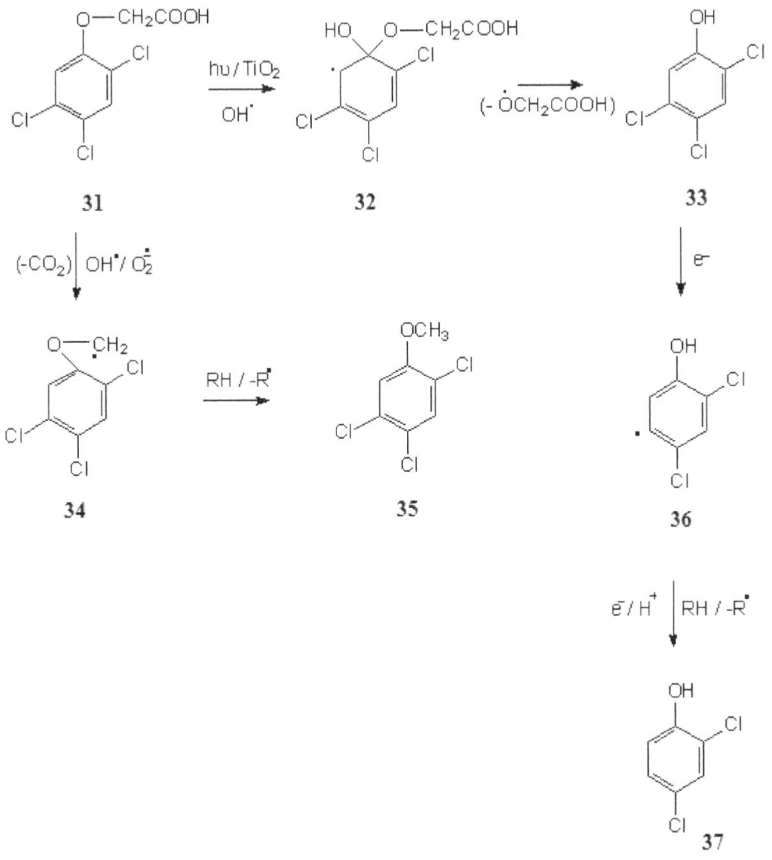

Scheme 5.

3.8.6 Photocatalysis of 4-chlorophenoxy acetic acid (38)

Irradiation of 4-chlorophenoxyacetic acid (4-CPA, 38) in the presence of titanium dioxide showed the formation of 4-chlorophenol (40), which was identified by comparing its molecular ion and fragment ion peak with those reported in the NIST library (Singh et al., 2007b). A plausible mechanism for the formation of product, 40 involving reactions with hydroxyl radicals formed in the photocatalytic process is shown in Scheme 6. The model compound, 4-CPA undergoes addition of a hydroxyl radical formed in the photocatalytic process leading to the formation of a radical species 39 as an intermediate, which may undergo loss of ($\cdot$OCH$_2$COOH) group to give the observed product 4-chlorophenol (40). Alternatively, the model compound 4-CPA (38) on addition of hydroxyl radical followed by loss of CO_2 may lead to the formation of the intermediate radical species 41, which upon loss of formaldehyde molecule followed by addition of hydroxyl radical led to the formation of the observed product 4-chlorophenol (40).

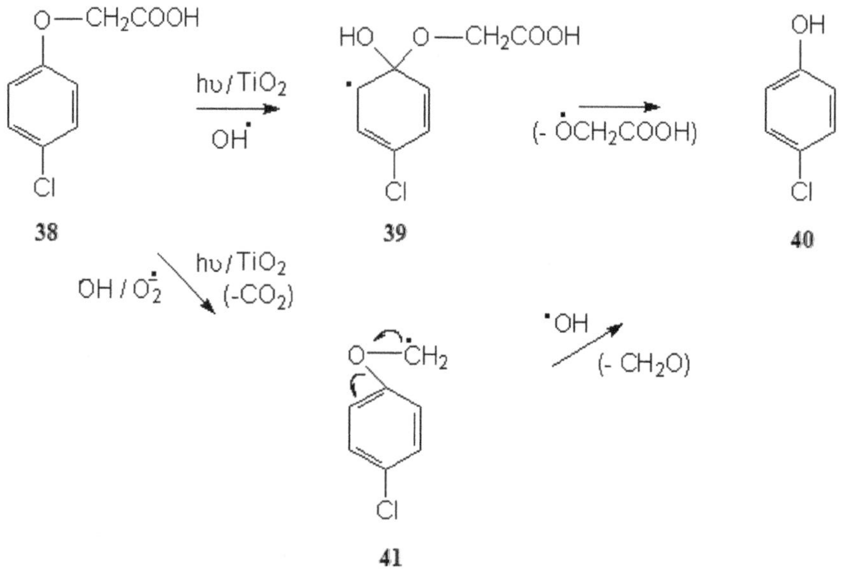

Scheme 6.

4. Conclusions

The results of these studies clearly indicate that TiO_2 can efficiently catalyse the photodegradation and photomineralization of the pollutants in the presence of light and oxygen. In most of the cases the photocatalyst Degussa P25 was found to be more efficient as compare to other photocatalyst TiO_2 powders. But in few cases Hombikat UV100 has also found to be more efficient photocatalyst. The addition of electron acceptor enhanced the degradation rate of the pollutants. The results also indicate that degradation rates could be influenced not only by the different parameters such as type of photocatalyst, catalyst concentration, substrate concentration, pH and additives and their concentration but also by the model pollutants. The observations of these investigations clearly demonstrate the importance of choosing the optimum degradation parameters to obtain high degradation rate, which is essential for any practical application of photocatalytic oxidation processes. The intermediate products formed during the process are also a useful source of information for the degradation pathways.

5. Acknowledgements

Financial support from Alexander Humboldt Foundation, Germany, Volkswagen Foundation, Germany, Marie Curie Foundation, UK, Third World Academy of Sciences, Italy, Department of Science and Technology and UGC, New Delhi, CSTUP, Lucknow, Department of Chemistry, AMU, Aligarh and the Institut füer Technische Chemie, Leibniz Universität Hannover, Hannover, Germany are gratefully acknowledge.

6. References

Augustynski, J. (1988) In: *Structural Bonding*, Springer, Berlin, New York, p. 69 (Chapter1).

Azevedo, E.B., Tôrres, A.R., Aquino Neto, F.R., & Dezotti, M., (2009). TiO$_2$-Photocatalyzed degradation of phenol in saline media in an annular reactor: hydrodynamics, lumped kinetics, intermediates, and acute toxicity. *Brazilian J. Chem. Engg.*, . 26, 75 – 87.

Bahnemann, W., Muneer, M., & Haque, M.M., (2007). Titanium dioxide-mediated photocatalysed degradation of few selected organic pollutants in aqueous suspensions. *Catal. Today.* 124, 133-148.

Barbeni, M., Morello, M., Pramauro, E., & Pelizzetti, E., (1987). Sunlight photodegradation of 2,4,5-trichlorophenoxy-acetic acid and 2,4,5,trichlorophenol on TiO$_2$. Identification of intermediates and degradation pathway. *Chemosphere*, 16, 1165-1179.

Barni, B., Cavicchioli, A., Riva, E., Zanoni, L., Bignoli, F., Bellobono, I.R., Gianturco, F., De Giorgi, A., Muntau, H., Montanarella, L., Facchetti, S., & Castellano, L., (1995). Pilot-plant-scale photodegradation of phenol in aqueous solution by photocatalytic membranes immobilizing titanium dioxide (PHOTOPERM® process). *Chemosphere*, 30, 1861-1874.

Bellobono, I.R., Carrara, A., Barni, B., & Gazzotti, A., (1994). Laboratory- and pilot-plant-scale photodegradation of chloroaliphatics in aqueous solution by photocatalytic membranes immobilizing titanium dioxide. *J. Photochem. Photobiol. A:Chem.*, 84, 83-90.

Berry, R.J., & Mueller, M.R., (1994). Photocatalytic decomposition of crude oil slicks using TiO$_2$ on a floating substrate. *Microchem. J.*, 50, 28-32.

Bianco Prevot, A., Pramauro, E., & de la Guardia, M., (1999). Photocatalytic degradation of carbaryl in aqueous TiO$_2$ suspensions containing surfactants. *Chemosphere*, 39, 493-502.

Blake, D. M. (2001). Bibliography of work on the photocatalytic Removal of Hazardous Compounds from Water and Air, National Renewal Energy Laboratory, USA.

Blanco, J., Avila, P., Bahamonde, A., Alvarez, E., Sanchez, B., & Romero, M., (1996). Photocatalytic destruction of Toluene and Xylene at gas-phase on a titania based monolithic catalyst. *Catal Today.*, 29, 437-442.

Bravo, A., Garcia, J., Domenech, X., & Peral, J., (1994). Some observations about the photocatalytic oxidation of cyanate to nitrate over TiO_2. *Electrochim. Acta*, 39, 2461-2463.

Byrne, J.A., Davidson, A., Dunlop, P.S.M., & Eggins, B.R., (2002). Water treatment using nano-crystalline TiO_2 electrodes. *J. Photochem. Photobiol. A:Chem.*, 148, 365-374.

Cohen, Z.Z., Eiden, C., Lober, M.N., (1986). In: Gerner, W.Y., (Ed.), *Evaluation of Pesticide in Ground Water*, ACS Symposium Series 315, American Chemical Society, Washington, DC, p. 170.

Daneshvar, N., Salari, D., & Khataee, A.R., (2004). Photocatalytic degradation of azo dye acid red 14 in water on ZnO as an alternative catalyst to TiO_2. *J. Photochem. Photobiol. A Chem.*, 162, 317-322.

Dominguez, C., Garcia, J., Pedraz, M.A., Torres, A., & Galan, M.A., (1998).Photocatalytic oxidation of organic pollutants in water. *Catal. Today*, 40, 85-101.

Dowd, R. M., Anderson, M. P., & Johnson, M. L. (1998). Proceedings of the Second National Outdoor Action Conference on Aquifer Restoration, Groundwater Monitoring Geophysical Methods, National Water Well Association, Dublin, OH, p. 1365.

Draper, R.B., & Fox, M.A ., (1990). Titanium dioxide photooxidation of thiocyanate: (SCN)2.cntdot.- studied by diffuse reflectance flash photolysis. *J. Phys. Chem.*, 94, 4628-4634.

Dunlop, P.S.M., Byrne, J.A., Manga, N., & Eggins, B.R., (2002). The photocatalytic removal of bacterial pollutants from drinking water. *J. Photochem. Photobiol. A:Chem:*, 148, 355-363.

Evgenidou, E., Bizani, E., Christophoridis, C., & Fytianos, K. (2007). Heterogeneous photocatalytic degradation of prometryn in aqueous solutions under UV/Vis irradiation.*Chemosphere*, 68, 1877-1882.

Frank, S.N., & Bard, A.J., (1977). Heterogeneous photocatalytic oxidation of cyanide and sulfite in aqueous solutions at semiconductor powders. *J. Phys. Chem.*, 81, 1484-1488.

Fujishima, A., & Honda, K., (1972). Electrochemical Photolysis of Water at a Semiconductor Electrode. *Nature*, 238, 37-38.

Funazaki, N., Hemmi, A., Ito, S., Asano, Y., Yano, Y., Miura, N., & Yamazoe, N., (1995). Application of semiconductor gas sensor to quality control of meat freshness in food industry. *Sensors &Actuators, B* 24-25, 797-800.

Garcia, J.C., & Takashima, K. (2003). Photocatalytic degradation of imazaquin in an aqueous suspension of titanium dioxide. *J. Photochem. Photobiol. A: Chem.*, 155, 215-222.

Gaya, U.I., & Abdullah, A.H., (2008). Heterogeneous photocatalytic degradation of organic contaminants over titanium dioxide: A review of fundamentals, progress and problems. *J. Photochem. Photobiol. C: Photochem. Rev.*, 9, 1-12.

Gerisher, H., & Heller, A., (1992). Photocatalytic oxidation of organic molecules at TiO_2 particles by sunlight in aerated water. *J. Electrochem. Soc.*, 139, 113-118.

Gianturco, F., Chiodaroli, C.M., Bellobono, I.R., Raimondi, M.L., Moroni, A., & Gawlik, B., (1997). Pilot-plant photomineralization of atrazine in aqueous solution by photocatalytic membranes immobilising titanium dioxide and promoting photocatalysts. *Fresenius Environ. Bull.*, 6, 461-468.

Goutailler, G., Valette, J.C., Guillard, C., Paissé, O., & Faure, R., (2001). Photocatalysed degradation of cyromazine in aqueous titanium dioxide suspensions: comparison with photolysis. *J. Photochem. Photobiol. A: Chem.*, 141, 79-84.

Grätzel, M., (1981). Artificial photosynthesis: water cleavage into hydrogen and oxygen by visible light. *Acc. Chem. Res.*, 14, 376-384.

Guillard, C., Pichat, P., Huber, G., & Hoang-Van, C., (1996). The GC-MS analysis of organic intermediates from the TiO$_2$ photocatalytic treatment of water contaminated by Lindane (1α,2α,3β,4α,5α,6β,-hexachlorocyclohexane). *J. Adv. Oxid. Technol.*, 1, 53-60.

Gupta, H., & Tanaka, S., (1995). Photocatalytic Mineralisation of. Perchloroethylene Using Titanium Dioxide. *Wat. Sci. Technol.*, 31, 47-54.

Hallamn, M. (1992). Photodegradation of di-n-butyl-ortho-phthalate in aqueous solution. *J. Photochem. Photobiol. A: Chem.*, 66, 215-223.

Halmann, M.M., (1996). *Photodegradation of Water Pollutants*, CRC Press Inc, Boca Raton, Florida

Haque, M.M., & Muneer, M., (2003). Heterogeneous photocatalysed degradation of a herbicide derivative, isoproturon in aqueous suspension of titanium dioxide. *J. Environ. Manag.*, 69, 169-176.

Haque, M.M., Muneer, M., & Bahnemann, D., (2006). Semiconductor-Mediated photocatalysed degradation of a herbicide derivative, chlorotoluron, in aqueous suspensions. *Environ. Sci. Technol.*, 40, 4765-4770.

Herrmann, J.M., (1999). Heterogeneous photocatalysis: fundamentals and applications to the removal of various types of aqueous pollutants. *Catal. Today*, 53,115-129.

Herrmann, J.M., Disdier, Pichat, P., Malato, S., & Blanco, J., (1998). TiO$_2$-based solar photocatalytic detoxification of water organic pollutants. Case studies of 2,4-dichlorophenooxyacetic acid (2,4-D) and of benzofuran. *J., Appl. Catal. B: Environ.*, 17, 15-23.

Herrmann, J.M., Guillard, C., Arguello, M., Agüera, A., Tejedor, A., Piedra, L., & Fernández-Alba, A., (1999). Photocatalytic degradation of pesticide pirimiphos-methyl: Determination of the reaction pathway and identification of intermediate products by various analytical methods.. *Catal. Today*, 54, 353–367.

Hua, Z., Manping, Z., Zongfeng, X., & Low, G.K.C., (1995). Titanium dioxide mediated photocatalytic degradation of monocrotophos. *Water Res.* 29, 2681-2688.

Huang, I.W., Hong, C.S., & Bush, B., (1996). Photocatalytic degradation of PCBs in TiO$_2$ aqueous suspensions. *Chemosphere*, 32, 1869-1881.

Hurum, D.C., Agrios, A.G., Gray, K.A., Rajh T., & Thurnauer, M.C., (2003). Explaining the enhanced photocatalytic activity of Degussa P25 mixed-phase TiO$_2$ using EPR. *J. Phy. Chem. B*, 107, 4545-4549.

Irvine., J.T.S., Eggins, B.R., & Grimshaw, J., (1990). Solar energy fixation of carbon dioxide via cadmium sulphide and other semiconductor photocatalysts. *J. Sol. Energy*, 45, 27-33.

Jardim, W.F. , Moraes, S.G., & Takiyama, M.M.K., (1997). Photocatalytic degradation of aromatic chlorinated compounds using TiO$_2$: toxicity of intermediates. *Water Res.*31, 1728-1732.

Khan, A., Haque, M.M., Mir, N.A., Muneer, M., & Boxall, C. (2010). Heterogeneous photocatalysed degradation of an insecticide derivative acetamiprid in aqueous suspensions of semiconductor. *Desalination*, 261, 169-174.

Khan, M.M.T., & Rao, N.N., (1991). Stepwise reduction of coordinated dinitrogen to ammonia via diazinido and hydrazido intermediates on a visible light irradiated Pt/CdS · Ag$_2$S/RuO$_2$ particulate system suspended in an aqueous solution of K[Ru(EDTA-H)Cl]2H$_2$O. *J. Photochem. Photobiol. A: Chem.*, 56, 101-111.

Kinkennon, A.E., Green, D.B., & Hutchinson, B., (1995). The use of simulated or concentrated natural solar radiation for the TiO$_2$-mediated photodecomposition of basagran, diquat, and diuron. *Chemosphere*,31, 3663-3671.

Konstantinou, I.K., Sakellarides, T.M.., Sakkas, V.A., & Albanis, T.A., (2001a). Photocatalytic Degradation of Selected s-Triazine Herbicides and Organophosphorus Insecticides over Aqueous TiO$_2$ Suspensions. *Environ. Sci. Technol.* 35, 398-405.

Konstantinou, I.K., Sakkas, V.A., & Albanis, T.A., (2001b). Photocatalytic degradation of the herbicides propanil and molinate over aqueous TiO2 suspensions: identification of intermediates and the reaction pathway. *Appl. Catal. B: Environ.*, 34, 227-239.

Konstantinou, I.K., Sakkas, V.A., & Albanis, T.A., (2002). Photocatalytic degradation of propachlor in aqueous TiO$_2$ suspensions. Determination of the reaction pathway and identification of intermediate products by various analytical methods. *Wat. Res.*, 36, 2733-2742.

Kosanic, M.M., & Topalov, A.S., (1990). Photochemical hydrogen production from CdS/RhOx/Na$_2$S dispersions. *Int. J. Hydrogen Energy*, 15, 319-323.

Krosley, K.W., Collard, D.M., Adamson, J., Fox, M.A., (1993). Degradation of organophosphonic acids catalyzed by irradiated titanium dioxide. *J. Photochem. Photobiol. A:Chem.*, 69, 357-360.

Kubota, Y., Shuin, T., Kawasaki, C., Hosaka, M., Kitamura, H., Cai, R., Sakai, H., Hashimoto, K., & Fujishima, A., (1994). Photokilling of T-24 human bladder cancer cells with titanium dioxide. *British J. Cancer*, 70, 1107-1111.

Lee, S., Nishida, K., Otaki, M., & Ohgaki, S., (1997). Photocatalytic inactivation of phage Q□ by immobilized titanium dioxide mediated photocatalyst. *Wat. Sci. Technol.*, 35, 101-106.

Legrini, O., Oliveros, E., & Braun, A.M., (1993). Photochemical processes for water treatment. *Chem Rev.*, 93, 671-698.

Lhomme, L., Brosillon, S., Wolbert, D., & Dussaud, J., (2005). Photocatalytic degradation of a phenylurea, chlortoluron, in water using an industrial titanium dioxide coated media. *Appl. Catal B: Environ.*, 61, 227- 235.

Linder, M., (1997) Ph.D. Thesis, Department of Chemistry, University of Hannover, Hannover, Germany.

Lindner, M., Bahnemann, D., Hirthe, B., & Griebler, W.D. (1995). *Novel TiO$_2$ powders as highly active photocatalysts. In Solar Water Detoxification; Solar Engineering*, Stine, W. B., Tanaka, T., Claridge, D.E., Eds.; ASME: New York, 339.

Liu, I., Lawson, L.A., Cornish, B., & Robertson, P.K.J., (2002). Mechanistic and toxicity studies of the photocatalytic oxidation of microcystin-LR. *J. Photochem. Photobiol. A:Chem.*, 148, 349-354.

Lobedank, J., Bellmann, E., & Bendig, J., (1997). Sensitized photocatalytic oxidation of herbicides using natural sunlight. *J. Photochem. Photobiol. A:Chem.*, 108, 89-93.

Lu, M.C., Roam, G.D., Chen, J.N., & Huang, C.P., (1995). Photocatalytic mineralization of toxic chemicals with illuminated TiO_2. *Chem. Eng. Commun.*, 139, 1-13.

Mao, Y., & Bakac, A. (1996). Photocatalytic Oxidation of Aromatic Hydrocarbons. *Inorg. Chem.*, 35, 3925-3930.

Marinas, A., Guillard, C., Marinas, J.M., Fernández-Alba, A., Agüera, A., & Herrmann, J.M., (2001). Photocatalytic degradation of pesticide–acaricide formetanate in aqueous suspension of TiO_2. *Appl. Catal. B: Environ.*, 34, 241-252.

Martin, C.A., Baltanas, M.A., Cassano, A.E., (1994). Photocatalytic reactors.3. kinetics of the decomposition of chloroform including effects. *Environ. Sci .Technol.*, 30, 2355-2364.

Matsunaga, T., & Okochi, M., (1995). TiO_2-Mediated Photochemical Disinfection of Escherichia coli Using Optical Fibers. *Environ. Sci. Technol.*, 29, 501-505.

Maurino, V., Minero, C., Pelizzetti, E., & Vincenti, M., (1999). Photocatalytic transformation of sulfonylurea herbicides over irradiated titanium dioxide particles. *Coll.Surf. A* 151, 329-338.

Mihaylov, B.V., Hendrix, J.L., & Nelson, J.H., (1993). Comparative catalytic activity of selected metal oxides and sulphides for the photo-oxidation of cyanide. *J. Photochem. Photobiol. A:Chem.*, 72, 173-177.

Mills, A., & Le Hunte, S. (1997). An overview of semiconductor photocatalysis. *J. Photochem. Photobiol. A:Chem.*, 108, 1-35.

Mills, A., Belghazi, A., & Rodman, D., (1996). Bromate removal from drinking water by semiconductor photocatalysis. *Wat. Res.*, 30, 1973-1978.

Mills, A., Peral, J., Domenech, X., & Navio, J.A., (1994). Heterogeneous photocatalytic oxidation of nitrite over iron-doped TiO_2 samples. *J. Mol. Catal.*, 87, 67-74.

Minero, C., Pelizzetti, E., Malato, S., & Blanco, J. (1996). Large solar plant photocatalytic water decontamination: degradation of atrazine. *Solar Energy*, 56, 411-419.

Muneer, M., & Bahnemann, D., (2001). Semiconductor mediated photocatalysed degradation of two selected pesticide derivatives, terbacil and 2,4,5-tribromoimidazole in aqueous suspension. *Water Sci. Technol.*, 144, 331-337.

Muneer, M., & Bahnemann, D., (2002). Semiconductor mediated photocatalysed degradation of two selected pesticide derivatives, terbacil and 2,4,5-tribromoimidazole in aqueous suspension. *Appl. Catal. B: Environ.*, 36, 95-111.

Muneer, M., Qamar, M., Saquib, M., & Bahnemann, D. (2005). Heterogeneous photocatalysed reaction of three selected pesticide derivatives, propham, propachlor and tebuthiuron in aqueous suspension of titanium dioxide. *Chemosphere*, 61, 457-468.

Muneer, M., Singh, H.K., & Bahnemann, D., (2002). Photocatalysed degradation of two selected priority organic pollutants, benzidine and 1,2-diphenyl hydrazine in aqueous suspensions of titanium dioxide. *Chemosphere*, 49, 193-203.

Muszkat, L., Bir, L., & Feigelson, L., (1995). Solar photocatalytic mineralization of pesticides in polluted waters. *J. Photochem. Photobiol.A: Chem.* 87, 85-88.

Muszkat, L., Raucher, D., Magaritz, M., Ronen, D., (1994). In: Zoller, U., (Ed.), *Groundwater Contamination and Control*, Marcel Dekker, p.257.

Nevim, S., Arzu, H., Gulin K., & Cinar, Z., (2001). Prediction of primary intermediates and the photodegradation kinetics of 3-aminophenol in aqueous TiO$_2$ suspensions. *J. Photochem. Photobiol. A: Chem.*, 139, 225-232.

Nozik, A. J. in : Ollis, D. F., & EL-Ekabi, H., (Eds.), (1993). *Photocatalytic Purification and Treatment of water and Air*, Elsevier, Amsterdam, p. 391.

O'Shea, K.E., Garcia, I., & Aguilar, M., (1997). TiO$_2$ Photocatalytic degradation of dimethyl- and diethyl-methylphosphonate, effects of catalyst and environmental factors. *Res. Chem. Intermed.*, 23, 325-339.

Ohtani, B., Zhang, S.W., Nishimoto, S., & Kagiya, T., (1992). Catalytic and photocatalytic decomposition of ozone at room temperature over titanium(IV) oxide. *J. Chem. Soc. Faraday Trans.*, 88, 1049-1053.

Oncescu, T., Stefan, M.I., & Oancea, P., (2010). Photocatalytic degradation of dichlorvos in aqueous TiO$_2$ Suspensions. *Environ. Sci. Pollut. Res.*, 17, 1158-1166.

Parra, S., Olivero, J., & Pulgarin, C., (2002). Relationships between physicochemical properties and photoreactivity of four biorecalcitrant phenylurea herbicides inaqueous TiO$_2$ suspension. *Appl. Catal. B: Environ.*36, 75-85.

Parra, S., Sarria, V., Malato, S., Péringer, P., & Pulgarin, C., (2000). Photochemical versus coupled photochemical–biological flow system for the treatment of two biorecalcitrant herbicides: metobromuron and isoproturon. *Appl. Catal. B: Environ.*, 27, 153-168.

Pathirana, H.M.K.K., & Maithreepala, R.A., (1997). Photodegradation of 3,4-dichloropropionamide in aqueous TiO2 suspensions. *J. Photochem.Photobiol. A: Chem.*, 102, 273-277.

Pelizzetti, E., Carlin, V., Minero, C., & Pramauro, E., (1992a). Degradation pathways of atrazine under solar light and in the presence of TiO$_2$ colloidal particles. *Sci. Total Environ.*, 123/124, 161-169.

Pelizzetti, E., Maurino, V., Minero, C., Carlin, V., Pramauro, E., Zerbinati, O., & Tosato, M.L., (1990). Degradation of atrazine and other S-triazine herbicides. *Environ. Sci. Technol.*, 24, 1559-1565.

Pelizzetti, E., Minero, C., Carlin, V., Vincenti, M., & Pramauro, E., (1992b). Identification of photocatalytic degradation pathways of 2-Cl-s-triazine herbicides and detection of their decomposition intermediates. *Chemosphere*, 24, 891-910.

Peñuela, G.A., & Barceló, D., (1996). Comparative degradation kinetics of alachlor in water by photocatalysis with FeCl$_3$ and photolysis, studied by solid-phase disk extraction followed by gas chromatographic techniques. *J. Chromatogr. A*, 754, 187-195.

Peñuela, G.A., & Barceló, D., (1998a). Photodegradation and stability of chlorothalonil in water studied by solid-phase disk extraction, followed by gas chromatographic techniques . *J. Chromatogr. A* 823, 81.

Peñuela, G.A., & Barceló, D., (1998b). Application of C$_{18}$ disks followed by gas chromatography techniques to degradation kinetics, stability and monitoring of endosulfan in water. *J. Chromatogr. A* 795, 93-104.

Percerancier, J.P., Chapelon, R., & Pouyet, B., (1995). Semiconductor-sensitized photodegradation of pesticides in water: the case of carbetamide. *J. Photochem. Photobiol. A: Chem.*, 87, 261-266.

Peyton, G.R., & Glaze, W.H. (1988). Destruction of pollutants in water with ozone in combination with ultraviolet radiation. 3. Photolysis of aqueous ozone. *Environ. Sci. Technol.*, 22, 761-767.

Pichat, P., (1997). Photocatalytic degradation of aromatic and alicyclic pollutants in water: by-products, pathways and mechanisms. *Water Sci. Tech.* 35, 73-78.

Pirkanniemi, K., & Sillanpää, M., (2002). Heterogeneous water phase catalysis as an environmental application: a review. *Chemosphere*, 48, 1047-1060.

Pizarro, P., Guillard, C., Perol, N., & Herrmann, J.-M., (2005). Photocatalytic degradation of imazapyr in water: Comparison of activities of different supported and unsupported TiO_2-based catalysts. *Catal. Today.*, 101, 211-218.

Pleskov, Y.V., & Krotova, M.D., (1993). Photosplitting of water in a photoelectrolyser with solid polymer electrolyte. *Electrochim Acta*, 38, 107-109.

Pollema, C.H., Hendrix, J.L., Milosavljevic, E.B., Solujic, L., & Nelson, J.H., (1992). Photocatalytic oxidation of cyanide to nitrate at TiO_2 particles. *J. Photochem. Photobiol. A. Chem.*, 66, 235-244.

Poulios, I., Kositzi, M., & Kouras, A., (1998). Photocatalytic decomposition of triclopyr over aqueous semiconductor suspensions . *J. Photochem. Photobiol.A: Chem.*, 115, 175-183.

Pramauro, E., Bianco Prevot, A., Vinceti, M., & Brizzolesi, G., (1997). Photocatalytic degradation of cabranyl in aqueous solutions containig TiO_2 Suspensions. *Environ. Sci. Technol.*, 31, 3126-3131.

Pramauro, E., Vincenti, M., Augugliaro, V., & Palmisano, L., (1993). Photocatalytic degradation of monuron in aqueous TiO_2 dispersions. Environ. Sci. Technol., 27, 1790-1795.

Qamar, M., Saquib, M., & Muneer, M., (2005). TiO_2-mediated photocatalysed degradation of two selected azo dye derivative chryosidine R and acid red 29 in aqueous suspension. *Desalination,* 186, 255-271.

Rahman, M.A., & Muneer, M., (2005). Photocatalysed degradation of two selected pesticide derivatives, dichlorvos and phosphamidon in aqueous suspension of titanium dioxide. *Desalination,* 181, 161-172.

Ranjit, K.T., Varadarajan, T.K., & Viswanathan, B., (1995). Photocatalytic reduction of nitrite and nitrate ions to ammonia on Ru/TiO_2 catalysts. *J. Photochem. Photobiol. A:Chem.,* 89, 67-68.

Ranjit, K.T., Varadarajan, T.K., & Viswanathan, B., (1996). Photocatalytic reduction of dinitrogen to ammonia over noble-metal-loaded TiO_2. *J. Photochem. Photobiol. A:Chem.,* 96, 181-185.

Rao, N.N., & Dube, S., (1996). Photocatalytic degradation of mixed surfactants and some commercial soap/detergent products using suspended TiO_2 catalysts. *J. Mol. Catal. A:Chem.,* 104, L197-L199.

Read, H.W., Fu, X., Clark , L.A., Anderson, M.A., & Jarosch, T. (1996). Field Trials of a TiO_2 Pellet-based Photocatalytic Reactor for Off-gas Treatment of a Soil Vapor Extraction Well. *J. Soil Contam.*, 5, 187-202.

Richard, C., & Bengana, S., (1996). pH effect in the photocatalytic transformation of a phenyl-urea herbicide. *Chemosphere,* 33, 635-641.

Sabin, F., Turk, T., & Vogler, A., (1992). Photo-oxidation of organic compound in the presence of titanium dioxide: determination of the efficiency. *J. Photochem. Photobiol. A:Chem.*, 63, 99-106.

Sakai, N., Wang, R., Fujishima, A., Watanabe, T., & Hashimoto, K., (1998). Effect of Ultrasonic Treatment on Highly Hydrophilic TiO_2 Surfaces. *Langmuir*, 14, 5918-5920.

Sakkas, V.A., Arabatzis, I., Konstantinou, I.K., Dimou, T.A., Albanis, T.A., & Falaras, P., (2004). Metolachlor photocatalytic degradation using TiO_2 photocatalysts. *Appl. Catal B: Environ.*, 49, 195-205.

Sakkas, V.A., Lambropoulou, D.A., Sakellarides, T.M., & Albanis, T.A., (2002). Application of solid phase microextraction (SPME) during the photocatalytic treatment of fenthion and ethyl parathion in aqueous TiO_2 suspensions. *Anal. Chim. Acta*, 467, 233-243.

Sakthivel, S., Neppolian, B., Shankar, M.V., Arabindoo, B., Palanichamy, M., & Murugesan, V., (2003). Solar photocatalytic degradation of azo dye: Comparison of photocatalytic efficiency of ZnO and TiO_2. *Sol. Energy Mater. Sol. Cells.*, 77, 65-82.

Sanlaville, Y., Guittonneau, S., Mansour, M., Feicht, E.A., Meallier, P., & Kettrup, A., (1996). Photosensitized degradation of terbuthylazine in water. *Chemosphere*, 33, 353-362.

Saquib, M., & Muneer, M., (2002). Semiconductor mediated photocatalysed degradation of an anthraquinone dye, remazol brilliant blue R under sunlight and artificial light source. *Dyes and Pigm.*, 53, 237-249.

Saquib, M., & Muneer, M., (2003). Photocatalytic photocatalytic degradation of two selected textile dye derivatives, eosin yellowish and p-rosaniline in aqueous suspensions of titanium dioxide. *J Environ Sci Health A.*, 38, 2581-2598.

Schmelling, D.C., Gray, K.A., & Kamat, P.V., (1996). The role of reduction in the photocatalytic degradation of TNT. *Environ. Sci. Technol.*, 30, 2547-2555.

Shifu, C., & Yunzhang, L., (2007). Study on the photocatalytic degradation of glyphosate by TiO_2 photocatalyst. *Chemosphere.*, 67, 1010-1017.

Singh, H.K., Saquib, M., Haque, M.M., & Muneer, M., (2007b). Heterogeneous photocatalysed degradation of 4-chlorophenoxyacetic acid in aqueous suspension. *J. Hazard. Mater.*, 142, 374-380.

Singh, H.K., Saquib, M., Haque, M.M., Muneer, M., & Bahnemann, D.W., (2007a). Titanium dioxide mediated photocatalysed degradation of phenoxyacetic acid and 2,4,5-trichlorophenoxyacetic acid in aqueous suspensions. *J. Mol. Catal. A: Chem.*, 264, 66-72.

Sleiman, M., Ferronato, C., Fenet, B., Baudot, R., Jaber, F., and Jean-Marc C., (2006). Development of HPLC/ESI-MS and HPLC/¹H NMR Methods for the Identification of Photocatalytic Degradation Products of Iodosulfuron. *Anal. Chem.*, 78, 2957-2966

Stapleton, D.R., Konstantinou, I.K., Mantzavinos, D., Hela, D., & Papadaki, M., (2010). On the kinetics and mechanisms of photolytic/TiO_2-photocatalytic degradation of substituted pyridines in aqueous solutions. *Appl. Catal. B: Environ.*, 95, 100-109.

Sturini, M., Fasani, E., Prandi, C., Casaschi, A., & Albini, A., (1996). Titanium-dioxide photocatalysed decomposition of some thiocarbamates in water. *J. Photochem. Photobiol. A: Chem.* 101, 251-255.

Tamimi, M., Qourzal, S., Assabbane, A., Chovelon, J.-M., Ferronato, C., & Ait-Ichou, Y., (2006). Photocatalytic degradation of pesticide methomyl: determination of the reaction pathway and identification of intermediate products. *Photochem. Photobiol. Sci.*, 5, 477-482.

Tanaka, K., Robledo, S.M., Hisanaga, T., Ali, R., Ramli, Z., & Bakar, W.A., (1999), Photocatalytic degradation of 3,4-xylyl N-methylcarbamate MPMC/ and other carbamate pesticides in aqueous TiO_2 Suspensions. *J. Mol. Catal. A: Chem.*, 144, 425-430.

Topalov, A., Molnár-Gábor, D., Kosani, M., & Abramovi, B., (2000). Photomineralization of the herbicide mecoprop dissolved in water sensitized by TiO_2. *Water. Res.*, 34, 1473-1478.

Tseng, J.M., & Huang, C.P., (1991). Removal of Chlorophenols from Water by Photocatalytic Oxidation. *Water Sci. Technol.* 23, 377-387.

Vaz, J.L., Boussaoud, B., Ichou, Y.A., & Petit-Ramel, M., (1998). Photomineralization on titanium dioxide of uracil and 5-halogenouracils. Influence of pH and some anions on the photodegradation of uracil. *Analysis*, 26, 83-87.

Vidal, (1998). Developments in solar photocatalysis for water purification. *Chemosphere*, 36, 2593-2606.

Vidal, A. Dinya, Z., Mogyorodi Jr, F., & Mogyorodi, F., (1999). Photocatalytic of thiocarbamate herbicide active ingredients in water. *Appl. Catal. B: Environ.*, 21, 259-267.

Vidal, A., & Martin Luengo, M.A., (2001). Inactivation of titanium dioxide by sulpur: Photocatalytic degradation of vapam. *Appl. Catal. B: Environ.*, 32, 1-9.

Vinodgopal, K., Wynkoop, D.E., & Kamat, P.V., (1996). Environmental Photochemistry on semiconductor surfaces: A photosensitization approach for the degradation of a textile azo dye, Acid Orange 7. *Environ. Sci. Technol.*, 30, 1660-1666.

Vulliet, E., Emmelin, C., Chovelon, J.M., Guillard, C., & Herrmann, J.M., (2002). Photocatalytic degradation of sulfonylurea herbicides in aqueous TiO_2. *Appl. Catal. B: Environ.* 38, 127-137.

Wang, R., Hashimoto, K., Fujishima, A., Chikuni, M., Kojima, E., Kitamura, A., Shimohigoshi, M., & Watanabe, T., (1998). Photogeneration of Highly Amphiphilic TiO_2 Surfaces. *Adv. Mat.*, 10, 135-138.

Wang, R., Sakai, N., Fujishima, A., Watanabe, T., & Hashimoto, K., (1999). Studies of Surface Wettability Conversion on TiO_2 Single-Crystal Surfaces. *J. Phys. Chem. B:*, 103, 2188-2194.

Wardman, P. (1989). Reduction potentials of one-electron couples involving free radicals in aqueous solution. *J. Phys. Chem.*, Ref. Data, 18, 1637-1755.

Weichgrebe, D., & Vogelpohl, A. (1995) Stratgie zur Auswahl geeigneter Oxidationsverfahren. 2. Fachtagung Naboxidative Abwasserbehadlung, Clausthal.

Weller, H., (1993). Kolloidale Halbleiter-Q-Teilchen: Chemie im Übergangsbereich zwischen Festkörper und Molekül. *Angew. Chem. Int. Ed. Eng.*, 32, 41-53.

Yu, J.C., Yu, J., Ho, W., & Zhao, J., (2002). Light-induced super-hydrophilicity and photocatalytic activity of mesoporous TiO2 thin films. *J. Photochem. Photobiol. A:Chem.*, 148, 331-339.

Zalenska, A., Hupta, J., Wiergowski, M., & Biziuk, M., (2000). Photocatalytic degradation of lindane, p,p'-DDT and methoxychlor in an aqueous environment . *J.Photochem. Photobiol. A: Chem.* 135, 213-220.

Zhang, P., Scrudato, R.J., & Germano, G., (1994). Solarcatalytic inactivation of *Escherichia coli* in aqueous solutions using TiO_2 as catalyst. *Chemosphere,* 28, 607-611.

Zhu, X., Feng, X., Yuan, C., Cao, X., & Li, J., (2004). Photocatalytic degradation of pesticide pyridaben in suspension of TiO_2: identification of intermediates and degradation pathways. *J. Mol. Catal. A: Chem.,* 214, 293-300.

4

Study on Sono-Photocatalytic Degradation of POPs: A Case Study Hydrating Polyacrylamide in Wastewater

Fanxiu Li

College of Chemical & Environmental Engineering,
Yangtze University, Jingzhou, Hubei,
China

1. Introduction

1.1 Production and possible hazards of hydrating polyacrylamide (HPAM) pollutants

In recent years, many Chinese oilfields have been in their mid- or final-stage of development. The oil recovery cannot be improved further with water flooding. Enhanced oil recovery (EOR) by means of polymer flooding is an important technology for the strategic development of oilfields in China. In order to improve the oil recovery, polymer flooding (injected water containing polymer), alkaline-surfactant-polymer flooding (injected water containing alkaline, surfactant and polymer, ASP) and surfactant-polymer flooding (injected water containing surfactant and polymer) have subsequently been used in oil production, which is often called tertiary oil extraction (Han et al., 1999). The liquid which is produced from stratum should be dehydrated using three-phase separators, and then, the crude oil of upper layers will be carried to the oil refinery, and the produced water (oily water) of under layer is generated. A large part of produced water should be injected back into the stratum for reuse (Taylor et al., 1998; Zhang et al., 2010), and the rest will be discharged into water bodies or surrounding soils.

In crude oil exploitation, water-soluble, polyacrylamide (PAM) is one of the most widely used polymers to enhance oil recovery in the east oilfields of China. The produced water from polymer flooding (PWPF) which contains a lot of residual hydrolyzed polyacrylamide (HPAM) is the wastewater of polymer flooding. And PWPF which is also characterized by its high temperature, heavy metals, high mineralization, low biodegradability and high content of oil and other oilfield chemicals (OCs), is different from that produced from water flooding. Moreover, partially hydrolyzed polyacrylamide (HPAM) present in production water causes some problems. For example, after polymer flooding, HPAM will remain in the produced water generated by oilfields, increasing the difficulty in oil-water separation. Consequently, the oil content in sewage is greatly increased, and there is a high probability that the wastewater will exceed the local discharge limit. When HPAM enters an oil reservoir with injected water, it can also hardly avoid infiltrating groundwater horizontally in connection with strata configuration. In addition, the costs and difficulties of produced water treatments will be increased because of the high concentration of the HPAM remaining in the wastewater. Furthermore, the residual HPAM in the wastewater can slowly degrade into the toxic acrylamide monomer naturally. The toxicity of acrylamide monomer has been studied by numerous researchers all over the world (Bao et al., 2010). Since HPAM can remain in surface water and groundwater for a long period of time, it may endanger human health. Therefore, it is necessary to conduct studies on transforming HPAM into innocuous substance effectively and rapidly.

1.2 Study progress on treatment of HPAM pollutants in wastewater

With the amount of HPAM residue increasing in the produced fluid, the separation of oil from water is more and more difficult and the treating difficulty becomes stern. Therefore, the effective treatment on the wastewater became urgent and important. Some methods have been used to treat it in some oil fields, such as gravity settling, floatation (Thoma et al., 1999), de-emulsification (Bilstad & Espedal, 1996) and membrane separation (Cheryan & Rajagopalan, 1998; Kong & Li, 1999; Scholz & Fuchs, 2000) and biotechnology (Li et al., 2005; Zhao et al., 2006; Su, 2007). The research progress of HPAM wastewater treatment methods is reviewed in this paper, which include the coagulation treatment method, membrane treatment method, photocatalysis degradation method and photo-Fenton treatment method.

1.2.1 Coagulation treatment process of HPAM in wastewater

At present, the most applicable and effective method is flocculation. The general operation is adding flocculant to the settling tanks in the existing treatment systems to accelerate oil-water separation. There are two kinds of flocculants, inorganic and organic. The typical inorganic flocculant is polyaluminum chloride (PAC) and the organic is cationic polyacrylamide (CPAM), which play an important role on wastewater treatment (Zhao et al, 2008).

The influences of HPAM residue on the flocculation behavior of wastewater from polymer flooding had been investigated by Zhao (Zhao et al, 2008). The main conclusions from their work were listed as the following:

1. Using PAC as inorganic flocculant, the flocculation performance improved with the increase of temperature under the same dosage. At 37°C and 40°C, flocculation results were markedly better than that of 30°C and 33°C. The floc formed quickly and the treating cost was low. However, the floc was much, tiny, loose and unstable.

2. Using CPAM as organic flocculant, the flocculation performance decreased with the increase of temperature under the same dosage. Compared with PAC, the floc was less and more stable. However, the treating results were poor and the cost was expensive.

3. At the constant temperature and dosage, the flocculation performance of PAC and CPAM decreased dramatically with the content of residual HPAM. At 37°C, when HPAM residue in wastewater increased from 100mg/L to 600mg/L, the light transmission decreased from 96.4% to 70% after treating with PAC at the dosage of 600mg/L and from 87.3% to 50.0% with CPAM at the dosage of 150mg/L.

1.2.2 Ultra-filtration membrane treatment technologies for the HPAM-containing wastewater

Unfortunately, none of these traditional separation techniques can meet complex demands for purifying the polymer-flooding wastewater of tertiary oil extraction. How to treat the oilfield polymer-flooding wastewater efficiently still remains unsettled.

Currently, ultra-filtration technology plays a more prominent role in the treatment of oily wastewater (Wu et al., 2008; Lu et al., 2009). However, the major problem arising from the membrane process is the decline in flux due to the concentration polarization and membrane fouling. The scientific practice suggests that the membrane fouling can be avoided (Field et al., 1995) when the operating flux is lower than a certain flux (critical flux). In contrast, when the operating flux exceeds this flux, the colloids initially present in the polarized layer will transform from the liquid phase into an irreversible cake layer (Chen et al., 1995; Bacchin et al., 2002a, 2002b). Since then, many studies have focused on the critical flux, including the effect of hydrodynamic factors such as cross-flow velocity (CFV) (Defrance & Jaffrin, 1995), sufficient shear stress (Li et al., 2000), sludge concentration (Le-clech, 2003) and particle size (Kwon et al., 1998; Kwon, 2000) on the critical flux for colloidal suspension, mineral suspension (Benkahla et al., 1995), protein or yeast suspension (Causserand et al., 1996) and activated sludge water(Bouhabila et al., 1998). As the study of the critical flux, the membrane resistance at sub-critical flux was investigated all along (Cho et al., 2002; Ognier et al., 2004). In addition, Chiu et al. (Chiu et al., 2005) reported that the gas could be used as a means of enhancing the critical flux in a non-circular multi-channeled (star-shaped) ceramic membrane module. Chong et al. (Chong et al., 2008) developed a sodium chloride tracer response technique to determine the critical flux of colloidal silica in reverse osmosis process.

In view of the characteristics of oilfield produced water quality, Wang (Wang et al., 2011) used the ultra-filtration membrane technique to treat synthetic oilfield polymer-flooding wastewater. In the experiment of fouling mechanism research, the first part and second part of filtration met the standard blocking and cake filtration model respectively. But the standard blocking period was very short at the beginning and the cake filtration period predominates in all filtration run. The critical flux was determined by transmembrane pressure (TMP)-step method in dead-end ultra-filtration test cell. The total fouling resistance increased with the increase of TMP. When the operating flux was below the critical flux, there was only concentration polarization phenomenon. The increasing trend of membrane resistance with the increase of TMP was unconspicuous. The intrinsic membrane resistance was the dominant resistance and the membrane fouling force was negative in this situation. And the filtration proceeds reached the biggest value. But once the operating flux exceeded the critical flux, the membrane pollution happened and the increasing rate of resistance accelerated.

The fouling resistance was the dominant resistance and the membrane fouling driving force became positive and higher. Moreover, the filtration proceeds were smaller and smaller. In the experiments of quantitative analysis of the critical flux, according to comparative results of the average rates of change of the critical flux for the concentration of HPAM, oil and suspended solid in the single solute solution, double solute solution and oilfield polymer-flooding wastewater, HPAM can decrease the average rate of change of the critical flux for oil and suspended solid. It has the crucial effect on the critical flux. The sequence of influence degree on the critical flux is the HPAM concentration$>$oil concentration$>$SS concentration and the percentage contribution is 84.58%, 14.36% and 1.06% respectively in the oilfield polymer-flooding wastewater. SEM images indicated that there was no membrane fouling formation at constant sub-critical flux. In contrast, the membrane was covered with a cake layer on the top surface of the membrane and the interaction between the particles and the membrane pore caused the pore narrowing, constriction and plugging at a constant supra-critical flux.

1.2.3 Photocatalysis degradation of HPAM in wastewater

Advanced oxidation processes (AOPs) are defined as oxidation processes in which hydroxyl radicals are the main oxidants involved. This radical is a very powerful oxidant (E^0: 2.80V versus SHE) which leads to a very effective oxidation process, such as Fenton and photo-Fenton catalytic reactions, H_2O_2/UV processes and TiO_2 photocatalysis (Faouzi et al.., 2006). Among the AOPs, heterogeneous photocatalysis oxidation using TiO_2 as photocatalyst has been extensively studied because of its low cost, high photoactivity, nontoxicity, photocorrosion resistance and other physical and chemical properties, and proved to be efficient and potentially advantageous. This semiconductor absorbs photons whose energy is higher than or equal to the band-gap energy. Thus, valence band electrons are promoted into the conduction band generating an electron-hole pair (e^-/h^+). These pairs are able to initiate oxidation and reduction reactions in the surface of TiO_2. The positive holes can oxidize the organic molecules adsorbed, through the formation of •OH radicals. Simultaneously, the photogenerated electron can produce radical species such as superoxide •O_2^- and hydroperoxide •HO_2. All these radicals initially oxidize the substrate in intermediate compounds which subsequently undergo a total mineralization in most of the cases (Serpone & Pellizzetti, 1989; David & Ollis, 2000).

The interest in photocatalysis is extensive, as shown by the number of publications on this subject which regularly appear in some journals, and thousands of papers have been published since the 1970s. Photocatalysis has come to describe the field of study and the technology in which irradiated semi-conductors generate photocharges that are ultimately poised at the surface. These photocharges undergo various processes, the most important of which are effectively separated and transferred to the contacting liquid, gas or solid for photooxidation of a large variety of organic substances to their complete mineralization. Wang et al. early works (Wang et al, 2006) discovered that TiO_2 particles can be self-potentially absorbed and orientedly arranged onto oil-water interface of emulsions. While the TiO_2 particles on the oil-water interface and in bulk water can be photocharged by UV radiation, reactive photoholes generated by photocatalytic process primarily oxidize the touched organic substances and damage the film between oil and water.

In their paper, Wang et al. (Wang et al, 2006) reported on a novel approach efficiently to achieve viscosity breaking of wastewater containing PAM. Initially by analyzing emulsification action and role of PAM in the wastewater, a process using photocatalytical technique was investigated by taking aim at viscosity breaking and degradation of PAM. Wang et al. refer to this process as "photocatalytical visbreaking". The experimental results show that viscosity of wastewater produced from polymer flooding was greatly decreased and the rate of PAM photodegradation is above 90% under short time of illumination by using photocatalysis over TiO_2 powders. The efficient breaking of viscosity favors treatments to feed the conventional system used in the water flooding with low viscosity of wastewater. The photocatalytic visbreaking can promisingly be used by efficient performance in oilfields. Chen et al. also think that it is a good method to treat PAM in water by photocatalytic oxidation (Chen et al., 2001).

Recently, the photocatalysts prepared by doping of rare earth oxides into anatase TiO_2 matrix have attracted much attention. TiO_2 is considered as a good host candidate for doping rare earth oxides due to its attractive properties such as mechanical, thermal, and anticorrosive properties. It is well known that the surface composition and structure of photocatalyst can greatly influence its activity. Some important results have been achieved by studies on the rare earth oxide-doped TiO_2 composites. For example, Saif et al. (Saif &

Abdel-Mottaleb, 2007) reported that TiO_2 nanocomposites doped with trivalent lanthanide ions (e.g. Tb^{3+}, Eu^{3+}, and Sm^{3+}) exhibited remarkable enhanced photocatalytic activity to textile dye degradation compared to pure TiO_2. Yan et al. reported that rare earth oxide-doped TiO_2 composites increased conversion of phenol and selectivity to CO_2 compared with pure TiO_2. Although some successful methods have been reported concerning about the preparation of the rare earth oxide-doped TiO_2 composites, new routes still need to be developed in order to lead to the composites with nanoscale, unique physicochemical properties, and interesting surface compositions.

In current Li's work(Li et al., 2009), a single step sol-gel-solvothermal method is applied to prepare rare earth oxide-doped titania nanocomposites, RE^{3+}/TiO_2, where $RE^{3+}=Eu^{3+}$, Pr^{3+}, Nd^{3+}, Gd^{3+}, and Y^{3+}. The morphology, phase structure, surface composition and structure, optical property, and textural property of the composites are well characterized.

As-prepared Eu^{3+}-, Pr^{3+}-, Gd^{3+}-, Nd^{3+}-, and Y^{3+}-doped TiO_2 composites with anatase phase, nanosize, and mesoporosity exhibited remarkably high UV-light photocatalytic activity to HPAM degradation. Moreover, Eu^{3+} (Pr^{3+}, Gd^{3+})/TiO_2-2.4 were the most photoactive among all tested materials including Degussa P25. This enhanced photocatalytic activity is attributed to the following properties of as-prepared RE^{3+}/TiO_2 composites: (a) quantum size effect; (b) unique textural properties (mosoporosity with larger BET surface areas and pore sizes); and (c) interesting surface compositions (more hydroxyl oxygen and adsorbed oxygen and some Ti^{3+} species existed at the surface of the products with respect to pure TiO_2). As-prepared photocatalysts are also essential for any practical application of photocatalytic oxidation process.

1.2.4 Photo-Fenton treatment of simulated HPAM in wastewater

Recently, some investigators have reported the successful application of advanced oxidation processes (AOPs) for PAM degradation (Vijayalakshmi & Giridhar, 2006; Ren & Chunk, 2006). One of advanced oxidation processes, Fenton's reagent, a mixture of H_2O_2 and Fe^{2+} (a powerful source of oxidative •OH generated from H_2O_2 in the presence of Fe^{2+} ions) or photo-Fenton reaction has been used in the degradation of many organic compounds because of its ease of operation (Murray & Parsons, 2004; Yardin & Chiron, 2006). The iron is the first most abundant metal and contained in many inexpensive natural minerals including tourmaline used in this study. Furthermore, hydrogen peroxide used as oxidant in these processes is cheaper than other oxidants.

In our study, we investigated the photo-assisted Fenton (photo-Fenton) reaction for its ability to oxidize HPAM. The photo-assisted Fenton can promisingly be used by efficient performance in oilfields. Effects of operating parameters such as initial hydrogen peroxide concentration, ratio of Fe^{2+}/H_2O_2 (mole ratio), HPAM concentration, and pH on the degradation rate of HPAM have been quantitatively discussed.

The experimental data demonstrated that photo-Fenton processes are promising techniques for the degradation of HPAM from aqueous solution. Based on the results, the following conclusions can be drawn.

1. Fenton and photo-Fenton processes lead to complete degradation of HPAM in relatively short time (~30min).

2. The optimal parameters for photo-Fenton process are: pH=3.0, the ratio of Fe^{2+}/H_2O_2=1:10 (mole ratio) and amount of H_2O_2=6mmol/L.

The employment of the UV lamp benefits the HPAM degradation. So, it is possible conclude that the UV lamp, though has little power, is very useful in Fenton process to increase the HPAM degradation. More results can be obtained in Li's study (Li, et al., 2006; Li, et al., 2007).

Even though these systems are considered as a very effective approach to remove organic compounds, it should be pointed out that there is a major drawback because the post-treatment of Fe sludge is an expensive process. This shortcoming can be overcome by using heterogeneous photo-Fenton reaction. Therefore, a lot of effort has been made in developing heterogeneous photo-Fenton catalysts. For example, Parra et al. prepared Nafion/Fe structured membrane catalyst and used it in the photo-assisted immobilized Fenton degradation of 4-chorophenol (Parra et al., 2004). However, Nafion/Fe structured membrane catalyst is much expensive for practical use. Thus, the low cost supports such as the C structured fabric (Parra et al., 2003; Yuranova, 2004), activated carbon (Ramirez, 2007), mesoporous silica SBA-15 (Calleja et al., 2005; Martinez et al., 2007; Martine et al., 2005), zeolite (Noorjaha et al., 2005; Kusic et al., 2006) and clay (Feng et al., 2006; Chen & Zhu, 2007), have been used for the immobilization of active iron species. Remirez et al. prepared the catalysts using four iron salts as precursors for the heterogeneous Fenton-like oxidation of Orange II solutions (Ramirez et al., 2007). The results showed that the nature of the iron salt had a significant effect on the process performance. So, it is necessary to discuss the photocatalytic activities of the catalysts by using different iron salts as precursors.

Liu(Liu et al., 2009) prepared a series of Fe(III)-SiO$_2$ catalysts at different OH$^-$/Fe mole ratio and by using two iron salts as precursors, namely Fe(NO$_3$)$_3$ and FeSO$_4$ and as-prepared catalysts were characterized by the BET, XRD and XPS method. The percentage of chemisorbed oxygen on the surface of catalysts prepared by FeSO$_4$ is higher than that prepared by Fe (NO$_3$)$_3$. The results confirm the formation of Fe (II)-SiO$_2$ when Fe (III)-SiO$_2$ was irradiated by photon. The photocatalytic activities of Fe (III)-SiO$_2$ catalysts were evaluated by the degradation of PAM from aqueous solution in the photo-Fenton reaction and all the catalysts exhibited a better photocatalytic activities. However, the precursor species and the OH$^-$/Fe mole ratio have influence on the photocatalytic activities of the catalysts. At the same OH$^-$/Fe mole ratio, the catalysts could present the better photocatalytic activities when using FeSO$_4$ as precursor. The best efficiency for the degradation of PAM in heterogeneous photo-Fenton reaction was 94.0% degradation in 90 min and 70.0% TOC removal in 180 min at an initial pH of 6.8. Moreover, it was observed that Fe leaching from Fe(III)-SiO$_2$ catalysts was negligible, indicating that the catalysts have a long-term stability and the degradation of PAM from aqueous solution are almost caused by the heterogeneous photo-Fenton reaction.

2. Sono-photocatalytic degradation of hydrating polyacrylamide in wastewater

Recently, some investigators have reported the successful application of advanced oxidation processes (AOPs) for HPAM degradation (Vijayalakshmi & Giridhar, 2006; Ren et al., 2006). One of advanced oxidation processes, photocatalytic degradation has been used in the degradation of many organic compounds and showed greatly obvious effects (Augugliaro et al., 1991). However, very few commercial applications of this technology are available at present due to low quantum efficiency and reuse of catalyst. Most of the photoinduced positive holes (h$^+$) and electrons (e$^-$) had recombined before they were trapped by hydroxyl or oxygen and quantum efficiency was usually less than 5%. In order to enhance the quantum efficiency, many measures were taken. Recently, the sonochemical method has been proven to be a useful technique (Hu et al., 2004). Sonolysis is a relatively innovative advanced oxidation processes based on the use of low to medium frequency (typically in the range 20-1000 kHz) and high energy ultrasound to catalyze the destruction of organic pollutants in waters. The chemical effects of ultrasound irradiation are the result of acoustic cavitation which is the formation and subsequent collapse of micro-bubbles in a liquid. At the extreme conditions generated inside the cavitation bubbles during collapse, vapor is homolytically cleaved leading to the formation of hydroxyl radicals that can oxidize the organic pollutants found in wastewaters (Vajnhandl & Marechal, 2005). Sonochemical treatment typically operates at ambient conditions and does not require the addition of extra chemicals or catalysts. The efficacy of AOPs to treat pollutants is eventually dictated by the rate of generation of free radicals and other reactive moieties and the degree of contact between the radicals and the contaminants both of which should be maximized. In this view, process integration is conceptually advantageous in water treatment since it can eliminate the disadvantages associated with each individual process. The simultaneous application of ultraviolet and ultrasound irradiation in the presence of TiO_2, i.e. sono-photocatalysis, represents an example of recent advances targeted at improving photocatalytic processes (Adewuyi, 2005; Gogate & Pandit, 2004). It is obvious that if the two irradiation modes are operated simultaneously, an additional source of free radicals will be available for the oxidative destruction of various pollutants.

The sono-photocatalytical degradation of a variety of organic substrates has attracted much attention. It might be also an efficient way to eliminate HPAM. However, there are only few studies investigating the sono-photocatalytic degradation of HPAM in aqueous system (Li et al., 2010a; 2011b). In this paper, HPAM was chosen as a model compound to obtain detailed information of the innovative photocatalysis. The photodegradation of the HPAM catalyzed by coupling system of ultraviolet irradiation (20 W UVA) over TiO_2 suspensions and ultrasound irradiation (42 kHz, 100 W) (US/UV/TiO_2) was investigated. Many factors are involved during the degradation of HPAM in US/UV/TiO_2 systems. The important operating parameters that affect the overall photocatalytic oxidation efficiency were investigated in detail, including amount of catalyst, initial concentration of reactant, and the dosage of hydrogen peroxide.

2.1 Materials and methods
2.1.1 Chemical
HPAM (Mw 500×10^4, hydrolysis degree 25%) was obtained from Shengli oil refining and chemical plant, Shengli, China and was used without further purification. Hydrogen peroxide (H_2O_2), hydrochloric acid (HCl) and sodium hydroxide (NaOH) were all of analytical grade and obtained from Tianjin Kermal Chemical Reagents Co. (Tianjin, China). TiO_2 photocatalyst (64.5% anatase, 35.5% rutile, specific surface area 40-45 m^2/g) was prepared by ultrasonic assisted method. The particle diameter of TiO_2 is 47.1-67.5 nm. The density is 690 kg/m^3. All experiments were carried out with use of deionized water.

2.1.2 Apparatus
Lambda-17 spectrophotometer (US Perkin-Elmer Company) was used to inspect the degradation processes of HPAM. Branson 2510E-DTH apparatus (Branson company, US)

was adopted to irradiate the solution of HPAM, operating at an ultrasonic frequency of 42 kHz and output power of 100 W.

2.1.3 Analysis

HPAM concentration was measured by the starch-cadmium iodine method (Li et al., 2009). The temporal concentration changes of HPAM during experimental processes were monitored by measuring characteristic absorption intensity of HPAM at 590 nm. The maximal absorbency of 0-100 mg/L HPAM solution abides by Lambert-Beers law and the calibration curve of standard HPAM solutions are used to estimate the degradation rate of HPAM. The degradation rate of HPAM was defined as follows Eq. (1):

$$\text{Degradation rate (\%)} = (1 - C_t/C_0) \times 100\% \tag{1}$$

Where C_0 is the concentration of HPAM after adsorption equilibrium in the dark ($t=0$), and C_t is the concentration of HPAM at reaction time t (min).

All these experiments were conducted in triplicates and the results were showed at the mean values.

2.1.4 Procedures

Experimental procedure was performed as follows: 100 mL of HPAM solution were introduced in the reaction vessel (bottom area=50 cm^2) and the appropriate amount of TiO$_2$ was added to achieve the desirable concentration. The resulting TiO$_2$ suspension was magnetically stirred for 30 min in the dark to ensure complete equilibration of adsorption/desorption of HPAM on the catalyst surface. After that period of time, UVA irradiation was provided by a 20 W ultraviolet lamp, which emit predominantly UV radiation at a wavelength of 254 nm, the lamp was turned on (this was taken as "time zero" for the reaction). In the experiment, the distance between the lamp and the interface irradiated by UV light is required to be 10 cm. Ultrasound irradiation was provided by an apparatus operating at 42 kHz frequency and 100 W of electric power output. Samples periodically drawn from the reaction vessel were centrifuged at 4000 rpm for 15 min to remove TiO$_2$ particles and then subjected to analysis.

2.2 Results and discussion

2.2.1 Photocatalytic and sono-photocatalytic degradation

In order to check the feasibility of sono-photocatalysis process for the degradation of HPAM, the following control experiments were performed and the results are presented in Fig.1, ultrasound alone (US), ultrasonic degradation of HPAM in the presence of TiO$_2$(US/TiO$_2$), photocatalysis (UV/TiO$_2$) and sono-photocatalysis (US/UV/TiO$_2$) in the presence of TiO$_2$. It is evident from Fig.1 that are relatively higher degradation of HPAM was achieved by combining sonolysis and photocatalysis than that observed during the individual processes. The preliminary experiments revealed no significant degradation of HPAM in the presence of US alone. However, an enhancement in the sonolytic reaction rate in the presence of semiconduct or particles was recently reported by Pandit et al. (Pandit et al., 2001; Shirgaonkar & Pandit, 1998) which is referred to as sonocatalysis. The suspended solids may also increase the extent of cavitation generated in solution and hence the sonochemical degradation rate by providing additional nuclei for bubble generation. Therefore, in order to study the effect of TiO$_2$ on the sonolytic degradation of HPAM in the absence of photo irradiation, the following experiment was conducted under the experimental conditions of [HPAM]=200 mg/L and TiO_2 amount of 600 mg/L, and the results obtained are shown in Fig. 1.

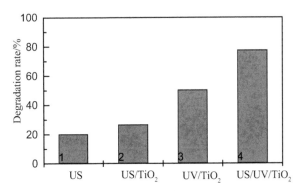

Fig. 1. Comparison of degradation rates of HPAM (200 mg/L) in the presence of TiO_2 (600 mg/L) using different processes

It was noted that about 26.7% degradation in 90 min under the sonocatalytic conditions was observed. An approximately 6.5% increase in the degradation amount observed in the presence of TiO_2 particles during sonolysis. This increase might be due to additional cavitation activity.

In the case of photocatalytic degradation, about 50.5% degradation (Fig. 1) was achieved in 90 min and can be explained using the general degradation mechanism available in the literature (Bhatkhande et al., 2002). The photoirradiation of the TiO_2 forms an electron in the conduction band and a hole in the valence band. As a consequence of such photoinduced charge separation on the semiconductor surface, electron exchange reactions occur at the water-semiconductor interface. The superoxide radical anion $O_2^-\bullet$ is formed by interaction of photo-generated conduction band electrons with adsorbed oxygen molecules, while $\bullet OH$ are formed via valence band hole oxidation of adsorbed water or hydroxyl anions (Carp et al., 2004) to generate $\bullet OH$ that subsequently oxidize the adsorbed pollutant molecules. Holes may also directly oxidize the adsorbed organic pollutant.

However, when both the US and UV are combined (sono-photocatalysis), a significant enhancement in the degradation (77.6% in 90 min) of HPAM was observed. About 27.2% increment in the degradation under the same processing time suggests that the hydroxyl radicals formed by both the advanced oxidation processes, viz., photocatalysis and sonolysis are involved in the sono-photocatalytic degradation of HPAM.

2.2.2 Effects of TiO_2 dosages on the degradation rate

The effect of TiO_2 powder concentration on the photocatalytic degradation of organics in the absence of ultrasound irradiation in aqueous solution has already been studied in the literatures (Ai et al., 2005). The necessity to optimize this factor was pointed out. Catalyst concentration has an optimum value, as using excess catalyst reduces the amount of photo-energy being transferred in the medium due to opacity offered by the catalyst particles. Fig. 2 shows the change of HPAM as a function of reaction time in $US/UV/TiO_2$ detected at different catalyst dosages 600, 800 and 1000 mg/L. On the whole, it was observed that along with the increase of catalyst dosage from 600 to 800mg/L, the degradation rate of HPAM in solution increased. However, the catalyst concentration amounting 800mg/L did not further enhance the degradation rate. The increased degradation is likely due to the increase of the total surface area (or number of active sites) of the photocatalysts available for photocatalytic reaction when increasing the dosage of TiO_2. When TiO_2 was overdosed, the intensity of light penetration attenuated and light scattering increased, which counteracted the positive effect coming from the dosage increment and therefore the overall performance reduced As the concentration of TiO_2 reaches a certain level, the suspension with a high concentration results in a lower light transmission of the system and then effective photons are decreased. As a result, it slows down the degradation of HPAM (Zhang et al., 2005). So, 800mg/L was the optimum amount of TiO_2 in terms of photodegradation under our experimental condition.

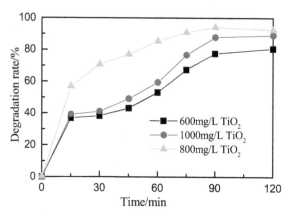

Fig. 2. Effect of added amount of TiO_2 on degradation rate of HPAM

2.2.3 Effect of initial concentration of HPAM on the degradation rate

The effect of initial concentration of HPAM on the sono-photocatalytic degradation rate was investigated over the concentration range of 80-200 mg/L, since the pollutant concentration is an important parameter in water treatment. Experimental results are presented in Fig. 3. It can be seen in Fig. 3 that degradation rate decreases with increasing initial concentration of the HPAM. The possible explanation for this behaviour is that as the initial concentration of the HPAM increases, the path length of photons entering the solution decreases and in low concentration the reverse effect is observed, thereby increasing the number of photon absorption by the catalyst in lower concentration. This suggests that as the initial concentration of the HPAM increases, the requirement of catalyst surface needed for the degradation also increases. Since illumination time and amount of catalyst are constant, the OH radical (primary oxidant) formed on the surface of TiO_2 is also constant. So the relative number of free radicals attacking the HPAM molecules decreases with increasing amount of the catalyst. The major portion of degradation occurs in the region (termed as reaction zone) near to the irradiated side, since the irradiation intensity in this region is much higher than that at the other side. Hence, at higher concentration, degradation decreases at sufficiently longer distances from the light source or reaction zone due to the retardation of penetration of light. Thus, the rate of degradation decreases with increase in concentration of HPAM (Neppolian et al., 2002).

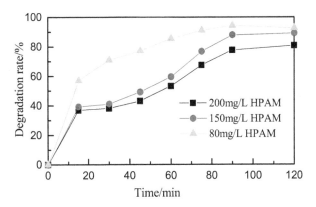

Fig. 3. Effect of initial concentration of HPAM on sono-photocatalytic degradation rate.

2.2.4 Effects of H_2O_2 dosages on the degradation rate

The use of inorganic peroxides has been demonstrated to enhance the rate of degradation because they trap the photo-generated electrons more efficiently than O_2. The effects of addition of the H_2O_2 into TiO_2 dispersion were examined (Fig. 4). The degradation rate of HPAM was increased in the presence of H_2O_2 and influenced by the dosages of H_2O_2. With the addition of H_2O_2, the degradation rate of HPAM ware increased abruptly at the low dosages of H_2O_2, but increasing H_2O_2 concentrations beyond 18 mmol/L had a negative effect on the process. This is because the addition of H_2O_2 can enhance the formation of ·OH. H_2O_2 would act as an electron donor to produce hydroxyl radicals by its reduction at the conduction band. The self-decomposition of H_2O_2 by UV light illumination or ultrasound irradiation would also produce hydroxyl radicals. Although more ·OH radical could be produced in the solution at higher oxidant concentrations, higher dosages of H_2O_2 may also act as an effective hydroxyl scavenger at higher H_2O_2 concentration (>18 mmol/L) as shown in the following reaction (Thandar et al., 2003): $H_2O_2 + ·OH \rightarrow H_2O + HO_2·$.

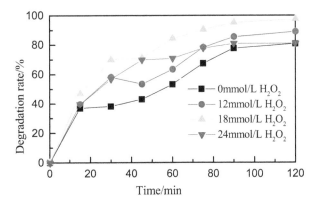

Fig. 4. Effect of dosage of H_2O_2 on degradation of HPAM in the solution containing 150 mg/L HPAM.

2.3 Degradation pathway of HPAM

During the process of TiO_2 sono-photocatalysis, numerous intermediates and products could be formed since $\cdot OH$ radicals did not exhibit a high degree of functional group selectively. UV-light irradiation of HPAM solution in the presence of TiO_2 was carried out under atmosphere, and three intermediates were identified with LC/MS (Fig. 5). The obtained peaks with m/z of 59.1, 59.7 (82.6), and 96.1, respectively, in the mass spectra corresponded to M-1 fragment of acetic acid (intermediate 1, M=60), M+1 (M+23) fragment of acetamide (intermediate 2, M=59), and M+23 fragment of propionamide (intermediate 3, M=73). An inorganic product, NO_3^- (M=62), was also detected with M+1/z=63. Moreover, acetic acid and NO_3^- were found to be dominant over the other two (Li & Mei, 2008). Therefore, a possible pathway for the degradation of HPAM had been proposed in Fig. 5.

To evaluate the conversion of the intermediates including acetic acid and acetamide and the formation of the product NO_3^- during the process of photocatalytic degradation of HPAM, a test was carried out in the suspension including TiO_2 and HPAM (200 mg/L, initial TOC 75 mg/L). The results demonstrated that TOC of the system decreased to 70% after 240 min UV-light irradiation, and further increasing the time of UV-light irradiation may result in total mineralization of HPAM (Fig. 5). As for the acetic acid, its concentration gradually increased as increasing UV-light irradiation time from 0 to 120 min, and then its concentration decreased slowly, indicating that it will continuously degrade into more small molecules. The changes of the concentrations of acetamide were slow during the process of UV-light irradiation. As for the final product, NO_3^-, its concentration increased from 0 to 7.4 mg/L after 240 min UV-light.

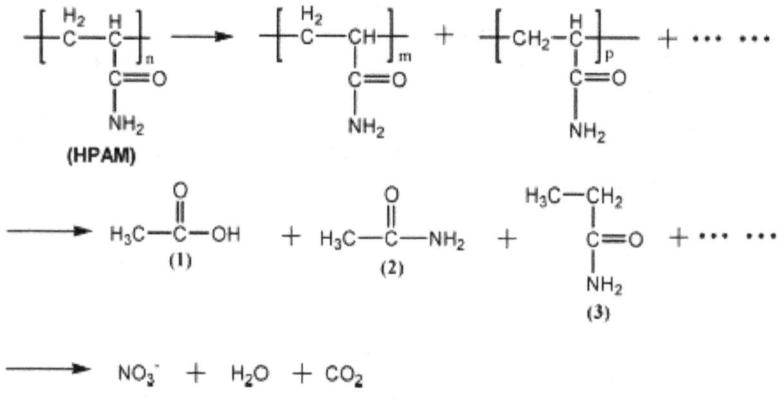

Fig. 5. Tentative pathway of sono-photocatalysis degradation of HPAM under UV-light irradiation of TiO_2

3. Conclusion

Among the treating technologies, AOPs constitute a promising technology for the treatment of wastewaters containing HPAM. UV/Fenton and Fenton-based reactions are capable of extensively degrading HPAM in a variety of aqueous solutions. Rigorous economic comparisons using an accepted standard measure of treatment efficiency are scarce. However, Fenton reactions can be performed at ambient temperature and do not require illumination, although they are usually enhanced by it. The reagents are readily available, easy to store, relatively safe to handle, and non-threatening to the environment. Drawbacks associated with the use of Fenton oxidation are the need to firstly reduce the pH, followed by a subsequent neutralization. This is not surprising since Fenton processes application requires strict pH control and sludges can be formed with related disposal problems. These drawbacks are intrinsic.

Photocatalytic or photochemical degradation processes are gaining importance in the area of wastewater treatment, since these processes result in complete organics mineralization with operation at mild conditions of temperature and pressure. However, there is still a problem that the photocatalytic efficiency of TiO_2 need to be improved since TiO_2 is photoactive only under near UV-light irradiation. On the other hand, recombination of photogenerated electron-hole pairs (e^--h^+) also results in low photo quantum efficiency of TiO_2. The effective separation of e^--h^+ pairs, therefore, is one of the most important subjects for broadening applications of TiO_2 photocatalyst.

In this study, the sono-photocatalytic oxidation of HPAM was examined in aqueous suspension. The evaluations of treatment performance were conducted for different amounts of TiO_2, initial concentration of HPAM, H_2O_2 dosage and reaction time. Based on the experimental findings, the following conclusions were drawn:

1. Coupling photocatalysis with ultrasound irradiation results in increased efficiency compared to the individual processes operating at common conditions. Interestingly, the overall sono-photocatalytic effect is greater than the additive effects of the two processes, thus indicating possible synergy.

2. TiO_2 had a good performance in $US/UV/TiO_2$ degradation HPAM. The concentration of 800 mg/L was the optimum dosage of TiO_2 under the experimental condition in terms of photocatalytic oxidation rate. Photocatalytic degradation increased with increasing TiO_2 loading (in the range 600-800 mg/L) and decreasing HPAM concentration (in the range 200-80 mg/L). Addition of H_2O_2 up to 18 mmol/L hindered degradation of HPAM, scavenging the photogenerated holes and hydroxyl radicals.

4. Acknowledgment

This research was supported in part by Educational Commission of Hubei Province of China (No. B200612013). Financial support for this work is gratefully acknowledged. I am also grateful to the anonymous referees for their insightful comments and suggestions, which clarified the presentation.

5. References

Augugliaro, V.; Palmisano, L.; Schiavello, M.; Sclafani, A.; Marchese, L.; Martra, G. & Miano, F. (1991). Photocatalytic degradation of nitrophenols in aqueous titanium dioxide dispersion. *Applied Catalysis*, Vol. 69, No. 1, 1991, pp. 323-340, ISSN 0926-860X

Adewuyi, Y. G. (2005). Sonochemistry in environmental remediation. II Heterogeneous sonophotocatalytic oxidation processes for the treatment of pollutants in water. *Environ. Sci. Technol.*, Vol. 39, No. 22, pp. 8557-8570. ISSN 0013-936X

Ai, Z. H.; Yang, P. & Lu, X. H. (2005). Degradation of 4-chlorophenol by a microwave assisted photocatalysis method. *Journal of Hazardous Materials*, Vol. B124 , No. 1-3, pp. 147-152, ISSN 0304-3894

Bao, M. T.; Chen, Q. G.; Li, Y. M. & Jiang, G. C. (2010). Biodegradation of partially hydrolyzed polyacrylamide production water after polymer flooding in an oil field. *Journal of Hazardous Materials*, Vol.184, No.1-3, pp.105-110, ISSN 0304-3894

Bilstad, T. & Espedal, E. (1996). Membrane separation of produced water. *Water Sci. Technol.* Vol. 34, No. 9, pp.239-246, ISSN 0273-1223

Bacchin, P.; Meireles, M. & Aimar, P. (2002). Modeling of filtration: from the polarised layer to deposit formation and compaction, *Desalination*, Vol.145, pp.139-146, ISSN 0011-9164

Bacchin, P.; Si-Hassen, D.; Starov, V.; Clifton, M. J. & Aimar, P. (2002). A unifying model for concentration polarization, gel-layer formation and particle deposition in crossflow membrane filtration of colloidal suspension, *Chemical Engineering Science*, Vol.57, No.1 , pp.77-91, ISSN 0009-2509

Benkahla, Y. K.; Ould-Dris, A.; Jaffrin, M. Y. & Si-Hassen, D. (1995). Cake growth mechanism in cross-flow microfiltration of mineral suspensions. *J. Membr. Sci.*, Vol.98, pp.107-117, ISSN 0376-7388

Bouhabila, E. H.; Ben Aim, R. & Buisson, H. (1998). Microfiltration of activated sludge using submerged membrane with air bubbling (application to wastewater treatment). *Desalination*, Vol.118, pp.315-322, ISSN 0011-9164

Bhatkhande, D. S.; Pangarkar, V. G. & Beenackers, A. C. M. (2002). Photocatalytic degradation using TiO_2 for environmental applications-A review. *J. Chem. Technol. Biotechnol.*, Vol.77, pp. 102-116, ISSN 1097-4660

Causserand, C.; Rouaix, S.; Akbari, A. & Aimar, P. (2004). Improvement of a method for the characterization of ultrafiltration membranes by measurements of tracers retention. *Journal of Membrane Science*, Vol. 238, No, 1-2, pp. 177-190, ISSN 0376-7388

Cho, B. D. & Fane, A. G. (2002). Fouling transients in nominally sub-critical flux operation of a membrane bioreactor. *J. Membr. Sci.*, Vol.209, pp.391-403, ISSN 0376-7388

Chiu, T. Y. & James, A. E. (2006). Critical flux enhancement in gas assisted microfiltration. *J. Membr. Sci.*, Vol.281, pp.274-280, ISSN 0376-7388

Chong, T. H.; Wong, F. S. & Fane, A. G. (2008). Implications of critical flux and cake enhanced osmotic pressure (CEOP) on colloidal fouling in reverse osmosis: Experimental observations. *J. Membr. Sci.*, Vol.314, pp.101-111, ISSN 0376-7388

Cheryan, M. & Rajagopalan, N. (1998). Membrane processing of oily streams. Wastewater treatment and waste reduction. *J. Membr. Sci.*, Vol.151, No.1, pp.13-28, ISSN 0376-7388

Chen, Y.; Cui, J. M.; Wang, B. H.; Li, S. Q. & Liu, J. (2001). Feasibility of photocatalytic oxidation of degrading polyacrylamide in water. *Journal of Daqing Petroleum Institute*, Vol.25, No.2, pp.82-83, ISSN 1000-1891

Calleja, J.; Melero, G. A.; Martinez, F. & Molina, R. (2005). Activity and resistance of iron containing amorphous, zeolitic and mesostructured materials for wet peroxide oxidation of phenol. *Water Research*, Vol. 39, No.9, pp.1741-1750, ISSN 0043-1354

Chen, J. & Zhu, L. (2007). Heterogeneous UV-Fenton catalytic degradation of dyestuff in water with hydroxyl-Fe pillared bentonite. *Catal. Today*, Vol.126, pp.463-470, ISSN 0920-5861

Chen, V.; Fane, A. G.; Madaeni, S. & Wenten, I. G. (1995). Particle deposition during membrane filtration of colloids: transition between concentration polarization and cake formation. *J. Membr. Sci.*, Vol.125, pp.109-122, ISSN 0376-7388

Carp, O.; Huisman, C. L. & Reller, A. (2004). Photoinduced reactivity of titanium dioxide. *Progress in Solid State Chemistry*, Vol. 32, No. 1-2, pp. 33-37, ISSN 0079-6786

David, F. & Ollis, D. F. (2000). Photocatalytic purification and remediation of contaminated air and water. *Chimie/Chemistry*, Vol. 3, pp. 405-411

Defrance, L. & Jaffrin, M. Y. (1999). Comparison between filtrations at fixed transmembrane pressure and fixed permeate flux: application to a membrane bioreactor used for wastewater treatment. *J. Membr. Sci.*, Vol.152, pp.203-210, ISSN 0376-7388

Espinasse, B.; Bacchin, P. & Aimar, P. (2002). On an experimental method to measure critical flux in ultrafiltration. *Desalination*, Vol.146, No1-3, pp. 91-96, ISSN 0011-9164

Faouzi, M; Ca~nizares, P.; Gadri, A.; Lobato, J.; Nasr, B.; Paz, R.; Rodrigo, M. A. & Saez, C. (2006). Advanced oxidation processes for the treatment of wastes polluted with azoic dyes. *Electrochimica Acta*, Vol.52, No.1, pp. 325-331, ISSN 0013-4686

Feng, J. Y.; Hu, X. J. & Yue, P. L. (2006). Effect of initial solution pH on the degradation of Orange II using clay-based Fe nanocomposites as heterogeneous photo-Fenton catalyst. *Water Res.*, Vol.40, pp. 641-646, ISSN 0043-1354

Field, R. W.; Wu, D.; Howell, J. A. & Gupta, B. B. Critical flux concept for microfiltration fouling. *J. Membr. Sci.*, Vol.100, pp.259-272, ISSN 0376-7388

Gogate, P. R. & Pandit, A. B. (2004). A review of imperative technologies for wastewater treatment II: hybrid methods. *Advances in Environmental Research*, Vol.8, No. 3-4, pp. 553-597, ISSN 1093-0191

Han, D. K.; Yang, C. Z.; Zhang, Z. Q.; Lou, Z. H. & Chang, Y. I. (1999). Recent development of enhanced oil recovery in China. *J. Petrol. Sci. Eng.*, Vol.22, No.1-3, pp.181-188, ISSN 0920-4105

Hu, X. L.; Zhu, Y. J. & Wang, S. W. (2004). Sonochemical and microwave-assisted synthesis of linked single-crystalline ZnO rods. *Materials Chemistry and Physics*, Vol. 88, pp. 421-426, ISSN 0254-0584

Kwon, D. Y.; Vigneswaran, S.; Fane, A. G. & Aim, R. B. (2000). Experimental determination of critical flux in cross-flow microfiltration. *Sep. Purif. Technol.*, Vol.19, No.3, pp.169-181, ISSN 1383-5866

Kwon, D. Y. & Vigneswaran, S. (1998). Influence of particle size and surface charge on critical flux of crossflow microfiltration. *Water Sci. Technol.*, vol. 38, No. 4-5, pp.481-488, ISSN 0273-1223

Kong, J. & Li, K. (1999). Oil removal from oil-in-water emulsions using PVDF membranes. *Sep. Purif. Technol.*, Vol.16, No.1, pp.83-93, ISSN 1383-5866

Kusic, K.; Koprivanac, N. & Selanec, I. (2006). Fe-exchanged zeolite as the effective hetenogeneous Fenton-type catalytic for the organic pollutant minimization: UV irradiation assistance. *Chemosphere*, Vol. 65, pp.65-73. ISSN 0045-6535

Li, Q. X.; Kang, C. B. & Zhang, C. K. (2005). Waste water produced from an oilfield and continuous treatment with an oil-degrading bacterium. *Process Biochem.*, Vol.40, pp.873-877, ISSN 0006-291X

Li, J. H.; Yang, X.; Yu, X. D.; Xu, L. L.; Kang, W. L.; Yan, W. H.; Gao, H. F.; Liu, Z. H. & Guo, Y. H. (2009). Rare earth oxide-doped titania nanocomposites with enhanced photocatalytic activity towards the degradation of partially hydrolysis polyacrylamide. *Applied Surface Science*, Vol. 255, pp.3731-3738, ISSN 0169-4332

Lu, Y.; Sun, H.; Meng, L. L. & Yu, S. L. (2009). Application of the Al_2O_3–PVDF nanocomposite tubular ultrafiltration (UF) membrane for oily wastewater treatment and its antifouling research. *Sep. Purif. Technol.*, Vol. 66, pp.347-352, ISSN 1383-5866

Liu, T.; You, H. & Chen, Q. W. (2009). Heterogeneous photo-Fenton degradation of polyacrylamide in aqueous solution over Fe (III)-SiO_2 catalyst. *Journal of Hazardous Materials*, Vol.162, pp.860-865, ISSN 0304-3894

Li, H.; Fane, A. G.; Coster, H. G. L. & Vigneswaran, S. (2000). An assessment of depolarization models of cross-flow microfiltration by direct observation through the membrane. *J. Membr. Sci.*, Vol. 172, pp. 135-147, ISSN 0376-7388

Le-clech, P.; Jefferson, B. & Judd, J. S. (2003). Impact of aeration, solids concentration and membrane characteristics on the hydraulic performance of a membrane bioreactor. *J. Membr. Sci.*, Vol.218, pp.117-129, ISSN 0376-7388

Li, F. X. & Mei P. (2008). Study on sono-photocatalytic degradation of hydrating polyacrylamide in waste water. *Journal of Oil and Gas Technology,,* vol.30, No. 3, pp. 157-160, ISSN 1000-9752

Li, F. X. & Xie, J. H. (2010). Study on photocatalytic degrading hydrolyzed polyacrylamide solution in wastewater treatment. *Journal of Oil and Gas Technology,* vol.32, No. 4, pp. 153-156, ISSN 1000-9752

Li, F. X. & Huang, Y. (2011). Study on the photodegradation of hexachlorobenzene catalyzed by TiO_2. *Journal of Anhui Agricultural Sciences,* vol.39, No. 4, pp. 2185-2188, ISSN 0517-6611

Li, F. X.; Lu, X. H.; Li, X. B. & Mei, P. (2006). Recent developments in researches of advanced oxidation processes for treating oilfield wastewaters. *Oilfield Chemistry,* Vol.23, No. 2, pp.188-192, ISSN 1000-4092

Li, J. L.; Li, H.; Chen, Y. & Wang, B. H. (2007). Influence of various ions on the degradation of hydrolyzed polyacrylamide by Fenton technique. *Petroleum Processing and Petrochemicals,* Vol.38, No. 11, pp.29-31, ISSN 1005-2399

Murray, C. A. & Parsons, S. A. (2004). Removal of NOM from drinking water: Fenton's and photo-Fenton's processes. *Chemosphere,* Vol. 54, pp.1017-1023, ISSN 0045-6535

Martinez, F.; Calleja, G.; Melero, J. A. & Molina, R. (2007). Iron species incorporated over different silica supports for the heterogeneous photo-Fenton oxidation of phenol. *Appl. Catal. B: Environ.,* Vol.70, pp.452-460, ISSN 0926-3373

Martinez, F.; Calleja, G.; Melero, J. A. & Molina, R. (2005). Heterogeneous photo-Fenton degradation of phenolic aqueous solutions over iron-containing SBA-15 catalyst. *Appl. Catal. B: Environ.,* Vol.60, pp.181-190, ISSN 0926-3373

Noorjaha, M. V.; Kumari, D.; Subrahmanyam, M. & Panda, L. (2005). Immobilized Fe(III)-HY: an efficient and stable photo-Fenton catalyst. *Appl. Catal. B: Environ.,* Vol.57, pp.291-298, ISSN 0926-3373

Neppolian, B.; Choi, H. C. & Sakthivel, S. (2002). Solar/UV-induced photocatalytic degradation of three commercial textile dyes. *Journal of Hazardous Materials B,* vol. 89, pp. 303-317. ISSN 0304-3894

Ognier, S.; Wsiniewski, C. & Grasmick, A. (2004). Membrane bioreactor fouling in sub-critical filtration conditions: a local critical flux concept. *J. Membr. Sci.,* Vol.229, pp.171-177, ISSN 0376-7388

Pandit, A. B.; Gogate, P. R. & Mujumdar, S. (2001). Ultrasonic degradation of 2, 4, 6-trichlorophenol in presence of TiO₂ catalyst. *Ultrason. Sonochem.*, Vol.8, pp. 227-231, ISSN 1350-4177

Parra, S.; Henao, L. & Mielczarski, E (2004). Synthesis, testing, and characterization of a novel Nafion membrane with superior performance in photoassisted immobilized Fenton catalysis, *Langmuir*, Vol. 20, pp. 5621-5629, ISSN 0743-7463

Parra, S.; Guasaquillo, I.; Enea, O. & Melczarski, E. (2003). Abatement of an azo dye on structured C-Nafion/Fe-ion surfaces by photo-Fenton reactions leading to carboxylate intermediates with a remarkable biodegradability increase of the treated solution, *The Journal of Physical Chemistry B*, Vol. 107, pp. 7026-7035, ISSN 1520-6106

Ren, G.; Sun, D. & Chunk, J. S. (2006). Advanced treatment of oil recovery wastewater from polymer flooding by UV/H₂O₂/O₃ and fine filtration. *J. Environ. Sci.*, Vol.18, pp.29-32, ISSN 1001-0742

Ramirez, J. H.; Maldonado-Hodar, F. J. & Perez-Cadenas, A. F. (2007). Azo-dye Orange II degradation by heterogeneous Fenton-like reaction using carbon-Fe catalysts. *Appl. Catal. B: Environ.*, Vol. 75, pp.312-323, ISSN 0926-3373

Ramirez, J. H.; Costa, C. A. & Madeira, L. M. (2007). Fenton-like oxidation of Orange II solutions using heterogeneous catalysts based on saponite clay. *Appl. Catal. B: Environ.*, Vol.71, pp.44-56, ISSN 0926-3373

Scholz, W. & Fuchs, W. (2000). Treatment of oil contaminated wastewater in a membrane bioreactor. *Water Res.*, Vol.34, No.14, pp.3621-3629, ISSN 0043-1354

Su, D. L. (2007). Kinetic performance of oil-field produced water treatment by biological aerated filter. *Chinese Journal of Chemical Engineering*, Vol.15, No.4, pp.591-594, ISSN 1004-9541

Serpone, N. & Pellizzetti, E. (1989). *Photocatalysis, Fundamentals and Applications*, Wiley, New York, ISBN 0471626031

Saif, M. & Abdel-Mottaleb, M. S. A. (2007). Titanium dioxide nanomaterial doped with trivalent lanthanide ions of Tb, Eu and Sm: Preparation, characterization and potential applications. *Inorganica Chimica Acta*, Vol.360, No.9, pp.2863-2874, ISSN 0020-1693

Shirgaonkar, I. Z. & Pandit, A. B. (1998). Sonophotochemical destruction of aqueous solution of 2, 4, 6-trichlorophenol. *Ultrason. Sonochem.*, Vol. 5, pp. 53-61, ISSN 1350-4177

Taylor, K. C., Burke, R. A., Nasr-El-Din, H. A. & Schramm, L. L. (1998). Development of a flow injection analysis method for the determination of acrylamide copolymers in brines. *J. Petrol. Sci. Eng.*, Vol.21, No.1-2, pp.129-139, ISSN 0920-4105

Thoma, G. J.; Bowen, M. L. & Hollensworth, D. (1999). Dissolved air precipitation/solvent sublation for oil-field produced water treatment. *Sep. Purif. Technol.*, Vol.16, No.2, pp.101-107, ISSN 1383-5866

Thandar, A.; William, A. A. & Mehrab, M. (2003). Photocatalytic treatment of cibacron brilliant yellow 3G-P. *Journal of environmental science and health, Part A-Toxic/Hazardous Substances and Environmental Engineering*, Vol. A38, No. 9, pp. 1903-1914, ISSN 0360-1234

Vijayalakshmi, S. P. & Giridhar, M. (2006). Photocatalytic degradation of poly (ethylene oxide) and polyacrylamide. *Journal of Applied Polymer Science*, Vol.100, pp. 3997-4003, ISSN 1097-4628

Vajnhandl, S. & Marechal, L. (2005). Ultrasound in textile dyeing and the decoloration / mineralization of textile dyes. *Dyes Pigments,* Vol. 65, No. 2, pp.89-101, ISSN 0143-7208

Wang, B. H.; Chen, Y.; Liu, S. Z.; Wu, H. J. & Song, H. (2006). Photocatalytical visbreaking of wastewater produced from polymer flooding in oilfields. *Colloids and Surfaces A: Physicochem. Eng. Aspects,* Vol.287, pp.170-174, ISSN 0927-7757

Wu, C. J.; Li, A. M.; Li, L. & Zhang, L. (2008). Treatment of oily water by a poly (vinyl alcohol) ultrafiltration membrane, *Desalination,* Vol.225, pp.312–321, ISSN 0011-9164

Wang, X. Y.; Wang, Z.; Zhou, Y. N.; Xi, X. J.; Li, W. J. & Yang, L. Y. (2011). Study of the contribution of the main pollutants in the oilfield polymer-flooding wastewater to the critical flux. *Desalination,* Vol.273, No.2-3, pp.375-385, ISSN 0011-9164

Yuranova, T.; Enea, O.; Mielczarski, E. & Mielczarski, J. (2004). Fenton immobilized photoassisted catalysis through a Fe/C structured fabric. *Appl. Catal. B: Environ.,* Vol. 49, pp. 39-50, ISSN 0926-3373

Yardin, C. & Chiron, S. (2006). Photo-Fenton treatment of TNT contaminated soil extract solutions obtained by soil flushing with cyclodextrin. *Chemosphere,* Vol. 62, pp. 1395-1402, ISSN 0045-6535

Zhang, Y. Q; Gao, B. Y.; Lu, L.; Yue, Q. Y.; Wang, Q. & Jia, Y. Y. (2010). Treatment of produced water from polymer flooding in oil production by the combined method of hydrolysis acidification-dynamic membrane bioreactor-coagulation process. *Journal of Petroleum Science and Engineering,* Vol.74, No.1-2, pp.14-19, ISSN 0920-4105

Zhang, X. W.; Wang, Y. Z. & Li, G. T. (2005). Effect of operating parameters on microwave assisted photocatalytic degradation of azo dye X-3B with grain TiO_2 catalyst. *Journal of Molecular Catalysis A: Chemical,* Vol.237, No. 5, pp. 199-205, ISSN 1381-1169

Zhao, X.; Wang, Y. M.; Ye, Z. F. & Alistair, G. L. (2006). Oil field wastewater treatment in biological aerated filter by immobilized microorganisms. *Process Biochemistry,* Vol.41, No.7, pp.1475-1483, ISSN 1359-5113

Zhao, X. F.; Liu, L. X.; Wang, Y. C.; Dai, H. X.; Wang, D. & Cai, H. (2008). Influences of partially hydrolyzed polyacrylamide (HPAM) residue on the flocculation behavior of oily wastewater produced from polymer flooding. *Separation and Purification Technology,* Vol.62, pp.199-204, ISSN 1383-5866

5

Chemical Degradation of Chlorinated Organic Pollutants for *In Situ* Remediation and Evaluation of Natural Attenuation

Junko Hara
Institute for Geo-resources and Environment,
National Institute of Advanced Industrial Science and Technology,
Japan

1. Introduction

Chlorinated organic compounds, prevalent contaminants found in the geo-environment, pose an ecological risk even at trace concentrations. More volatile chlorinated compounds such as VOCs (volatile organic compounds) have been detected in urban areas and industrial zones because of the use of these compounds as components of industrial solvents and both raw and intermediate synthetic products. Chlorinated organic compounds quickly evaporate from surface water but remain in groundwater and soil for a long time. Recently, several remediation techniques have been developed that can entirely remediate chlorinated organic compounds to non-toxic materials. However, high molecular weight chlorinated organic compounds (e.g., polychlorinated biphenyls (PCBs) and dichloro-diphenyl-trichloroethane (DDT)) are highly toxic chemicals that persist for long periods of time in the environment and bioaccumulate. They are categorized as persistent organic pollutants (POPs). Although the amount and date of use vary by country, POPs were widely used for pesticides and disease control in crop production and industrial processes during the period of industrial production after World War II around the globe. DDT, PCBs and dioxins are the best known POPs. DDT is used to control mosquitoes, which carry malaria, and PCBs were useful in electrical transformers and large capacitors. Among these POPs, PCBs, hexachlorobenzene (HCB), chlordanes, dichloro-diphenyl-dichloroethane (DDE) and dieldrin show significant ecological accumulation, such as in human milk, human blood and other biological media.

The Stockholm Convention is intended to protect human health and the environment, starting with the reduction or elimination of the production, use, and/or release of 12 species of POPs (PCBs, HCB, aldrin, dieldrin, endrin, DDT, chlordane, heptachlor, toxaphene, mirex, polychlorinated dibenzo-*p*-dioxins (dioxins), and polychlorinated dibenzofurans (furans). The 9 additional chemicals adopted in amendments to the Stockholm convention as new POPs are α-hexachlorocyclohexane, β-hexachlorocyclohexane, chlordecone, hexabromobiphenyl, hexabromobiphenyl ether and heptabromobiphenyl ether, Lindane, pentachlorobenzene, perfluorooctane sulfonic acid and its salts and perfluorooctasulfonyl fluoride, tetrabromobiphenyl ether and pentabromobiphenyl ether in 2009. The Stockholm Convention has led to a general global decline in the concentration of these chemicals in the environment. However, some individual POPs still persist and accumulate in fatty tissue and are present in higher concentrations at higher levels in the food chain, with long-range mobility through natural processes, because complete removal of POPs from the environment is difficult. Experimental degradation methods for POPs at room temperature have been reported, using chemical catalysis, bacteria, UV, photocatalysis, Fenton reagent and other methods. Among these techniques, reductive dechlorination processes encounter difficulty in achieving complete dechlorination and degradation of the chemical structure. Powerful oxidative processes are assumed to show some possibility for complete degradation of POPs. Reports of complete degradation of POPs are few, even when oxidative processes are applied for on-site remediation over a very extensive polluted area.

This chapter reviews remediation methodology for chlorinated organic pollutants and chemical remediation methods using ferric sulphide compared with reaction with zero-valent iron, which widely used as a practical in-situ remediation method for soil and groundwater pollution by VOCs. Ferric sulphides also have outstanding ability to degrade chlorinated organic pollutants, but the dechlorination processes differs from that of zero-valent iron. The natural remediation capability, reaction products, and reaction mechanisms using ferric sulphide for trichloroethylene, dieldrin and chlorinated benzenes are reported in this chapter.

2. Reported remediation methods for chlorinated organic pollutants

Several remediation methodologies for chlorinated organic pollutants such as bioremediation, bioaugmentation, and chemical or physical remediation have been reported by many researchers. In these remediation methodologies, chlorinated organic contaminants can be transformed chemically, photochemically, or biochemically by oxidation or reduction in the soil and the groundwater environments.

Bioremediation technology has developed rapidly in the last 15 years, and the application of this technology has been extended to several species of contaminants, including volatile organic compounds (VOCs), polycyclic aromatic hydrocarbons (PAHs) and polychlorinated biphenyls (PCBs). Reductive dechlorination by microorganisms under anaerobic conditions is an advantageous process because reductive dechlorination allows the reoxidation of metabolic intermediates. For VOCs, PCE (tetrachloroethylene) or TCE (trichloroethylene), which is most widely reported chemicals, are capable of reductive dechlorination by dehalococcoides, dehalobacter and desulfuromonas (Hollinger et al., 1993; Gerritse et al., 1996, 1999; Loffler et al., 1999; Krumholz et al., 1996; Maymo-Gatell, 1997). These microorganisms dechlorinated the PCE/TCE to cis-dichloroethylene (cis-DCE) or vinyl chloride (VC). Dehalococcoides especially (including the culture) is reported to have achieved complete dechlorination of PCE/TCE to ethylene. In aerobic or oxidizing environments, chlorinated ethylenes are oxidized to CO_2 and the chlorinated ethylenes are co-metabolized to CO_2 via trichloroacetate, trichloroethanol, trichloroethylene epoxide, dichloroacetate, glyoxylate, formate, oxalate, etc. The reaction pathway and the number of intermediates differ according to the species of bacteria in the local environment (Newman and Wackett, 1991, 1995, 1997; Weightman et al., 1985, 1993; Li and Wackett, 1992; Kim et al., 2009; Rosenzweig et al., 1993; Oldenhuis et al., 1989; Motosugi et al., 1982). Oxygen served as the electron acceptor in the aerobic oxidation, and non-specific microbial oxygenase enzymes produced by aerobic microorganisms participated in co-metabolism reactions.

Anaerobic microbial communities in sediments also dechlorinate PCBs. The ease of dechlorination of positions on aromatic rings is usually meta > para > ortho, and on biphenyl rings, mono–ortho- and ortho-chlorobiphenyls were not degraded after one year of reaction (Teidje *et al.*, 1993). PCE was also oxidatively degraded by Pseudomonas strain sp. P2. In this reaction, PCE is metabolized to chlorobenzoic acids with one to three chlorine atoms. At the actual reaction site, reductive and oxidative biological reaction is estimated to occur because of conditions at the site. Each reaction is therefore assumed to proceed partly under anaerobic conditions and partly under aerobic conditions at one polluted site, and this collaboration of reactions under different sets of conditions leads to complete remediation at the natural site.

Many chemical remediation methods have been investigated for rapid in-situ or off-site remediation of soil and groundwater. Among the chemical and photochemical dechlorination methods, zero-valent iron (ZVI) has frequently been reported as the remediation technique for reductive dechlorination of chlorinated ethenes and ethanes. Among actual chemical remediation techniques, the ZVI methodology is widely used for in situ remediation for soil (e.g., injection by direct drilling) and groundwater (application in a reactive permeable barrier). The chlorinated organic materials degrade under anaerobic (reducing) conditions using ZVI. This reductive dechlorination is generally divided into hydrogenolysis and reductive elimination. Both reaction mechanisms are accompanied by a net transfer of two electrons. Although reductive dechlorination is occasionally referred to as hydrolytic reduction, this term is misleading because hydrolysis is incidental to the actual reduction. Some other zero-valent metals such as zinc and platinum are also reported as reductive remediation catalysts instead of ZVI. The reaction pathway for TCE using ZVI and other metals is the same as the reductive dechlorination process using bioremediation.

Although the application of ZVI for remediation of high molecular weight persistent organic compounds is difficult, zero-valent metals such as Pd, Pt, Ni and Cu are also used as catalysts in the ZVI methodology for dechlorination of PCBs. Metal catalysis enhances the reductive dechlorination capability of ZVI. In the case of persistent organic compounds, electrochemical methods using metal electrodes are used for reductive dechlorination.

Fenton reaction and photochemical reaction are the predominant oxidative dechlorination processes and can remediate contaminants rapidly. The Fenton reaction is the oxidation of organic substrates by iron(II) and hydrogen peroxides. The Fenton reagent is effective in treating various industrial wastewaters polluted by chlorinated organic compounds, aromatic amines, pesticides, and surfactants. This oxidation system is based on the formation of reactive oxidizing species able to degrade the contaminants effectively in wastewater. In the Fenton chemistry, a two-reaction pathway is advanced as the first step. Production of hydroxyl radicals and a non-radical pathway using ferric ion production have been reported (Barbusinski, 2009). The nature of the oxidizing species is still controversial. Some researchers showed that the hydroxyl radical is the major species in the Fenton mechanism, and other groups showed that this reaction includes the formation of reactive oxidizing species such as ferric ion. Considering that the Fenton reaction occurs chemically and biologically as well as in the natural environment, there is a possibility that both mechanisms coexist in the Fenton reaction. Several hazardous pollutants can be oxidized by the Fenton reaction; for example, chlorophenol is degraded to hydroxyacetic intermediates (Barbeni *et al.*, 1987a) and perchloroethylene is transformed to dichloroacetic acid, formic acid and CO_2 (Leung *et al.*, 1992). More detailed information about Fenton reaction is discussed in the chapter "Fenton's Process for the Treatment of Mixed Waste Chemicals".

In natural waters exposed to solar radiation, the Fenton reaction is often perceived as a possible source of hydroxyl radical in sunlit waters. The photolysis of nitrates, metal-to-ligand charge transfer reactions, and photoFenton reactions are included as other possibilities for the Fenton reaction, and H_2O_2 and Fe(II) are photochemically produced in these sunlit waters. Photochemical dechlorination is reported for persistent high molecular weight organic compounds such as DDT and PCBs (Mochizuki 1977; Van Beek et al, 1982; Shimakoshi et al., 2004). Photochemical PCBs dechlorination in alkaline isopropyl alcohol effectively degrades PCBs to biphenyl, chloride ions, acetone and water (Mochizuki, 1977). Dechlorination of DDT was catalyzed by hydrophobic vitamin B12 (heptamethyl cobyrinate perchlorate) with irradiation by visible light (Shimakoshi et al., 2004). The Co(I) species of cobalamin and the related cobalt complex are supernucleophiles and react with an alkyl halide to form alkylated complexes with dehalogenation (Shimakoshi H. et al., 2004). In this reaction, DDT is mono-dechlorinated to DDD. Photocatalytic treatment assisted by TiO_2 is also widely reported for TCE, DDT, chlorinated phenol, chloroform, etc. (Ahmed and Ollis, 1984; Barbeni et al., 1986, 1987b; Dible and Raupp, 1990, 1992; D'Oliveria et al., 1990; Kondo and Jardim, 1991; Borello et al., 1989) These oxidative degradations proceed by radicals arising from the photoreaction of other compounds present in the system.

In a similar reaction, UV/H_2O_2 and UV/Fenton are used for the remediation of chlorinated ethylenes and ethanes, chlorobenzenes, benzenes, and chlorinated phenols (Froclich, 1992; Moza et al., 1988; Sundstrom et al., 1986, 1989; Weie et al., 1987). UV/H_2O_2 oxidation involves the dissociation of H_2O_2 to form the hydroxyl radical. The hydroxyl radical oxidizes toxic organic materials by abstraction of protons to produce organic radicals. Some toxic materials degrade to lower molecular weight acids and finally transform to CO_2. However, many reports only estimate degradation ability and reaction kinetics and do not mention reaction products, especially when high molecular weight persistent organic compounds are involved.

These strong chemical oxidation technologies such as Fenton reagent, UV radiation, catalysis and photochemical treatment have the prospect of degradation of persistent organic pollutants to non-toxic compounds.

3. Remediation ability of ferric sulphides

The metal sulphides show a great diversity in electrical and magnetic properties. The sulphides of the transition metals can be considered as intermediate between the transition metal oxides. The small particles of metal sulphide are often superior in electrical and chemical properties, as well as in catalysis.

Pyrite is the most abundant metal sulphide mineral and is treated as industrial waste at many mining sites. An outcrop including pyrite leads to oxidation in aerobic weathering processes and causes acidification of environmental water. Pyrite is also distributed in acidic coastal sulphate soils, which are naturally formed under waterlogged anaerobic conditions. These acidic sulphide soils are located in tropical coastal areas in West Africa, South and Southeast Asia, and northeast South America. These acidic conditions are mainly developed in recent or semi-recent sediments close to the sea. In these environments, the sulphur in pyrite is derived from the sulphate in seawater, which is biologically reduced to sulphide in the anaerobic mud. The organic matter which serves as the energy source for the sulphate-reductive bacteria is usually abundant in plants in the coastal area. Ferrous iron is also derived from reduction of insoluble ferric compounds resulting from the weathering of clay.

The reactivity of pyrite when the surface is exposed to H_2O, O_2, and a mixture of H_2O and O_2, has been studied experimentally by Guevremont et al. (1997, 1998a, b, c, d). The pyrite surface exposed to H_2O vapour up to 1 bar was oxidized, although the reaction site is limited to the defect site on pyrite surface. The surface reaction with O_2 vapour showed no oxidation. Substantial surface reaction is observed in the reaction with the H_2O/O_2 mixture. The intermediate oxidation products, sulphur oxoanion and zero-valent sulphur, are also identified and removed by use of either the O_2 or H_2O (Kendelewicz et al., 2004). The oxidation mechanism of pyrite with the H_2O/O_2 mixture involves competitive adsorption of O_2 and H_2O at the surface Fe site, oxidation of surface Fe sites by O_2, dissociation of co-adsorbed H_2O at Fe sites, and charge redistribution in surface S atoms (Rosso et al., 1999). This pathway allows for the production of hydroxyl radicals from dissociated water and subsequent nucleophilic attack of these hydroxyl radicals at surface S sites. The oxygen in the final product sulphate arises from water molecules (Taylor et al., 1984; Usher et al., 2004, 2005). The hydroxyl radical from the pyrite oxidation process causes the degradation of chlorinated organic compounds.

3.1 Degradation ability for major volatile organic compounds

Chlorinated ethylenes are widespread groundwater and soil contaminants. Due to the prevalent pollution and the efforts to treat these compounds, substantial research has been conducted to identify the mechanism of reaction under various environmental conditions. As mentioned above, the chemical dechlorination method using ZVI is in wide use as a practical in situ remediation technique for soil and groundwater pollution with TCE. This reductive dechlorination reaction process includes hydrogenolysis, β-elimination and a hydrogen addition reaction, and degrades chlorinated ethylenes and ethanes to ethylene and ethane. The dechlorination of TCE by pyrite (ferric sulphide) under anaerobic conditions is also reported to be reductive dechlorination like ZVI (Weerasooriya R., 2001; Lee and Batchelor, 2002, 2003). Other chlorinated ethylenes and ethanes are also reductively dechlorinated under anaerobic conditions (Kriegman-King and Reinhard, 1994; Lee and Batchelor, 2003). The reductive dechlorination of chlorinated compounds occurs by the transfer of electrons from the mineral surface.

Fig. 1 shows the dechlorination of TCE by metallic sulphides and zero- valent iron. In this reaction, 100 ppm of TCE is dechlorinated with powdered pyrite and Milli-Q-grade-ultrapure water under closed aerobic conditions. TCE is detected by a headspace method using gas chromatography (GC) (GC17A, Shimadzu Co. Ltd.). Pyrite and chalcopyrite showed an outstanding dechlorination rate, and complete dechlorination was confirmed by the mass balance of chloride ion. According to the recent pyrite oxidation work, the most reactive surface component is S^{2-}, and the second most reactive surface component is the surface atom of the first disulphide layer (S_2^{2-}) with sulphur atoms of the disulphide groups beneath the surface layer being least reactive. Oxidized iron (Fe^{3+}) states are proposed to arise after fracturing of S-S bonds by electron transfer from Fe^{2+} to this S^- state, which then reacts rapidly to sulphate (Nesbitt et al., 1998; Schaufuss et al., 1998). If this oxidation reaction is similar with metallic sulphide, the S-S bonds in metallic sulphide oxidize chlorinated contaminants, as assumed from this oxidation model. Disulphide metallic minerals involving S-S bonds are therefore assumed to have greater degradation ability than the mono-sulphide metals. In our experimental estimation in Figure 1, the disulphide metallic minerals are only pyrite and chalcopyrite, which have superior degradation ability relative to the other mono-sulphide metals and ZVI.

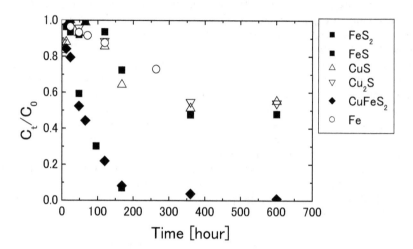

Fig. 1. Dechlorination of trichloroethylene (TCE) by metallic sulphide and zero-valent iron under aerobic conditions. C_0 is the initial concentration of TCE (100 mg/L), C_t is the concentration of TCE at time t. The reaction was represented as the evaluation of the normalized remaining percentage of TCE with time.

TCE is also degraded by pyrite under aerobic conditions (Hoa *et al.*, 2008). TCE dechlorination in anaerobic and aerobic pyrite suspensions is observed with time, but there is no outstanding degradation under anaerobic conditions and the dechlorination rate of TCE is proportional to the increase in oxygen. Fig. 2 shows the disappearance of TCE, the reaction intermediates and the final product in the reaction of TCE with a pyrite suspension under aerobic conditions. Under aerobic conditions, 98 % of TCE was degraded after about 2 weeks. The TCE degrades to dichloroacetic acid, glyoxylic acid, formic acid, oxalic acid and CO_2 by oxidative processes (Hoa *et al.*, 2009). This degradation process is similar to the oxidative metabolic pathway of TCE (Kim *et al.*, 2009; Li *et al*, 1992). Figure 3 shows the expected dechlorination pathway of TCE by pyrite under aerobic conditions. The dechlorination pathway of TCE is divided into three pathways. All pathways show direct dechlorination of three chlorine atoms in TCE. The main pathway is degradation of TCE to formic acid (Eq. (1)).

$$C_2HCl_3(TCE) + 4^\bullet OH + H_2O \rightarrow 2CH_2O_2(formic \quad acid) + 3HCl + \frac{1}{2}O_2 \qquad (1)$$

$$CH_2O_2(formic \quad acid) + 2^\bullet OH \rightarrow CO_2(g) + 2H_2O \qquad (2)$$

Formic acid has toxicological properties toward aquatic organisms, but the degradation rate of formic acid by pyrite suspensions is also high, similar to the dechlorination of TCE to formic acid, and formic acid continuously transforms to $CO_2(g)$ without accumulation (Eq. (2)).

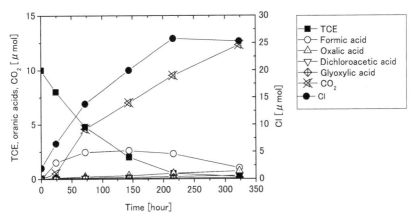

Fig. 2. Transformation of trichloroethylene (TCE) and detected reaction products in pyrite suspension (Edited from Hoa *et al.* (2008)). The initial TCE concentration is 10 µmol, the oxygen volume is 0.268 mmol/L, and pyrite is 10 m²/L. The left side of the Y-axis shows the molecular weight of TCE, CO_2 and organic acids (formic acid, oxalic acid, dichloroacetic acid, glyoxylic acid). The right side of the Y-axis shows the chloride molecular weight as degraded from TCE.

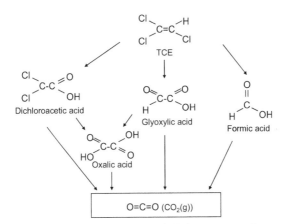

Fig. 3. Degradation pathway of trichloroethylene (TCE) using a pyrite suspension under aerobic conditions (Edited from Hoa *et al.* (2009)).

3.2 Degradation capability for dieldrin

Dieldrin is a cyclodiene pesticide which has persisted in the soil over decades in some agricultural fields. Some agricultural crops and animals accumulate dieldrin through the food chain. From the previous report of TCE dechlorination by pyrite, we see that TCE is able to be reductively degraded under anaerobic conditions and oxidatively degraded under aerobic conditions. The reductive dechlorination under anaerobic conditions is not observed using pyrite suspensions. Dieldrin has a higher persistence than TCE and is less amenable to reductive dechlorination under anaerobic conditions. Remediation techniques for dieldrin (UV/Fenton reagent, UV/chemical reaction, Pd/C catalyst, solar photocatalysis, and bioremediation) have been reported (Books, 1980; Maule et al., 1987; Bandala et al., 2002; Kusvuran and Erbatur, 2004; Chiu et al., 2005; Zinovyev et al., 2005; Dureja et al., 1987; Baczynski et al., 2004), but dieldrin transforms to mono- or di- dechlorinated intermediates still having a bicyclic ring structure. A study of the use of Fenton reagent for aldrin reports the transformation from aldrin to oxalic acid, acetic acid, chlorohexanone, cyclohexanol, cis-2-hydroxy-cyclohexanone, cis-2-methyl-cyclohexanol, 4-hydroxycyclohexanone, cis-4-methylcyclohexanol, 1-cyclopropyl-1-hydroxyethylene, and trans-dihydroxycyclohexane (Kusvuran and Erbatur, 2004).

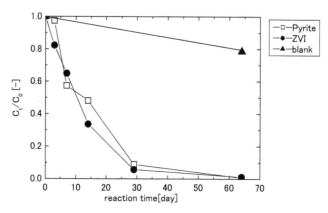

Fig. 4. Degradation profile of dieldrin using pyrite and zero-valent iron under aerobic conditions. The blank denotes the non-metallic catalysis condition. C_0 is the initial concentration of dieldrin (0.131 μmol), C_t is the concentration of dieldrin at time t. (Edited from Hara et al., (2009)

The degradation of dieldrin using ZVI and pyrite under anaerobic conditions is shown in Fig. 4. ZVI and pyrite have similar dechlorination capabilities for dieldrin, but the reaction process differs completely for the two reagents (Hara et al., 2009). In the case of ZVI, dieldrin partially transforms to nono-dechlorinated products ($C_{12}H_9Cl_5$, $C_{12}H_9Cl_5O$) and aldrin ($C_{12}H_8Cl_5$). This reaction is a reductive dechlorination proceeded by the generation of H+ and an electron arising from oxidation of ZVI. This reaction is stopped only this pathway. In contrast, the mono-dichlorinated reaction intermediates and aldrin are not detected in the reaction using pyrite. Dieldrin is oxidatively degraded to water-soluble reaction intermediates by pyrite.

Fig. 5 shows the dechlorination rate of dieldrin using pyrite under anaerobic to aerobic conditions, changing with oxygen concentration (O_2 = 0 ~ 833 μmol). Dieldrin was gradually degraded under every condition of oxygen concentration, except for the O_2 = 0 μmol condition. In Fig. 5, 5 ppm (0.131 μmol) of dieldrin is used with a powdered pyrite suspension under each oxygen condition. Dieldrin is detected by GC/MS (Shimadzu Co. Ltd.) after solvent/solvent extraction using acetone and hexane. The water-soluble reaction intermediates are detected by IC-TOF/MS (ICS-3000 (Dionex) and JMS-T 100LP (JEOL). A little oxygen (O_2 = 10 μmol) accelerates the dechlorination of dieldrin in comparison to the anaerobic conditions. Approximately 99 % of the dieldrin was degraded after about one month under the most enhanced reaction conditions (O_2 = 10 μmol), and approximately 43 % of dieldrin remained under O_2 = 300 μmol in the same reaction time, which is the slowest dechlorination condition. In these reactions, chloride ion resulting from the dechlorination was detected in the aqueous phase. The mass balance of dechlorinated chloride and degraded dieldrin is 90 % in the most degraded condition (O_2 = 10 μmol). This discrepancy in mass balance is obvious in the anaerobic conditions, but the mass balance agrees well under aerobic conditions (O_2 = 300 and 833 μmol).

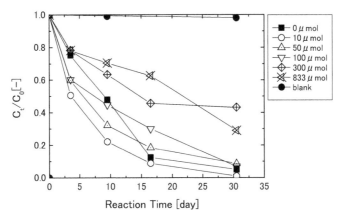

Fig. 5. Oxygen dependence on degradation of dieldrin in pyrite suspension (Hara, 2011). C_0 is the initial concentration of dieldrin (0.131 μmol), C_t is the concentration of dieldrin at time t. O_2 = 833 μmol denotes the air volume conditions in this experiment. The blank is without pyrite.

As reaction products, formic acid, oxalic acid, malonic acid, succinic acid, acetoxyacetyl, lactic acid, and pyruvic acid are detected as the main reaction products in the case of aerobic conditions (O_2 = 833 μmol). Acetic acid, glycolic acid, hydroxybutyric acid, glyoxylic acid, propionic acid, glutaric acid, levulinic acid and sulphur-containing organic acids such as methanesulphonic acid, sulphopropionic acid, etc. are also detected as minor reaction products. Formic acid is predominantly generated, as shown in (Eq. (3)).

$$C_{12}H_8Cl_6O\,(dieldrin) + 24\,{}^{\bullet}OH + H_2O \rightarrow 12CH_2O_2\,(formic\quad acid) + 6HCl + O_2 \qquad (3)$$

Formic acid is also detected in TCE dechlorination by a pyrite suspension under aerobic conditions, and the formic acid continuously degrades to CO_2 (Eq. (2)). Dieldrin is therefore assumed to finally transform to CO_2 as a main reaction pathway.

In the case of a low O_2 volume or anaerobic conditions, these low molecular weight organic acids generated under oxygen-rich conditions are much lower in abundance and some organic acids are not detected. Instead of the generation of organic acids, 3-chloro-4-methyl-2-pentanol ($C_6H_{13}ClO$) and dibutyl phthalate ($C_{16}H_{22}O_4$) are generated. One part of these intermediates is continuously degraded to a low molecular weight organic acid, but the residual volume is also higher than the volume observed under aerobic conditions.

The reaction pathway of dieldrin proceeds mostly by oxidative degradation under anaerobic and aerobic conditions. Although the oxidative ability is obviously different from the content of oxygen, the active oxidant to degrade the dieldrin is promoted from both the pyrite/H_2O and pyrite/O_2 interface reaction, due to oxidants produced under either aerobic or anaerobic condition. The difference of degradation ability in oxygen volume is assumed to depend on the difference of radical species arising form pyrite/H_2O and pyrite/O_2 interface.

Considering the reaction products, the degradation pathway of dieldrin results mainly in the direct production of organic acids, which are easy to produce under oxygen-rich conditions. Under anaerobic or micro-aerobic conditions, the pathway of ring opening and the addition reaction of low molecular weight organic acids arising from $C_{16}H_{22}O_4$ and the pathway of generation of chlorinated hydrocarbons arising from $C_6H_{13}ClO$ (which continuously transforms to formic acid and malonic acid) become predominant.

3.3 Degradation ability for chlorinated benzenes

Dieldrin having a bicyclic ring could be decomposed by pyrite. This section discusses the ability to dechlorinate chlorobenzenes whose main structure is the benzene ring.

Chlorobenzenes are divided into 12 species based on the number and configuration of chlorine in the molecule: monochlorobenzene (mono-CB), 3 types of dichlorobenzene (1,2-, 1,3-, 1,4- di-CB), 3 types of trichlorobenzene (1,2,3-, 1,2,4-, 1,3,5- tri-CB), 3 types of tetrachlorobenzene (1,2,3,4-, 1,2,3,5-, 1,2,4,5- tetraCB), pentachlorobenzene (penta-CB), and hexachlrobenzene (hexa-CB).The chlorobenzenes are used in pesticides, deodorants, or as intermediates in a chemical synthesis process. Some tri- or tetra-chlorinated benzenes are extensively used as insulating materials. The risk associated with chlorinated benzenes increases relative to the increase in the chlorine number, because of their increasing lack of volatility with increasing chlorine number, the highest level of risk is associated with the misuse or accidental release of the mono- to trichlorinated benzenes. These compounds spread into the atmosphere are reported to be photolyzed or chemically reacted, and that in groundwater and soils these compounds are mainly remediated by microbial degradation. However, the residence time is increased because of the organic constituents present in the soil and groundwater, which results in adsorption and accumulation in the soil ecosystems. The major remediation method for more the accumulation of chlorobenzenes with a higher chlorination level in soils is incineration at high temperatures for digging out of soils.

In the previous reports of electrochemical dehalogenation of chlorinated benzenes (Miyoshi et al., 2004; Mohammad and Dennis, 1997; Farwell et al., 1975; Kargina et al., 1997; Guena et al., 2000), the chlorine is eliminated step by step from the highly chlorinated benzenes to yield less-chlorinated benzenes and finally transform to benzene. Farwell et al. (1975) reported on chlorobenzenes and the main cathodic reaction pathway for hexachlorobenzenes as follows: hexachlorobenzene → pentachlorobenzene → 1,2,3,5-tetrachlorobenzene → 1,2,4-trichlorobenzene → 1,4-dichlorobenzene → monochlorobenzene → benzene. This dechlorination pathway is promoted by the electrochemical reductive dechlorination. Benzene is detected as the final reaction product.

Oxidative dechlorination of chlorobenzenes has been reported using bacteria, Fenton reagent, UV/H_2O_2, and TiO_2-assisted photocatalysis. In the microbial metabolism reported by Reineke and Knackmuss (1984), chlorobenzene was gradually degraded to 3-chloro-cis -1,2-dihydroxycylohexa-3,5-diene, 3-chlorocatechol, 2-chloro-cis,cis-muconate, trans-4-carboxymethylenebuten-4-olide, maleylacetate, and 3-oxoadipate. Fenton reagent effectively degrades chlorobenzene to chlorohydroxycyclohexadienyl radical in the first step, and the radical dimerizes to produce dichlorobiphenyls, with bimolecular disproportionation to produce chlorophenol and chlorobenzene under non-oxygen conditions or in the absence of other strong oxidants (Reineke W. and Knackmuss H-J, 1984). In the presence of oxygen or other strong oxidants, reactions of the oxidant (O_2) with chlorohydroxycyclohexadienyl radical is predominant and results in lower concentrations of dichlorobiphenyl, which decreases remarkably, and chlorobenzoquinone is formed. Chlorophenol isomers were further oxidized by hydroxyl radical and formed chlorinated and non-chlorinated diols. Chlorobenzenes were also dechlorinated by the UV/H_2O_2 treatment method (Sundstrom *et al.*, 1989), and chlorobenzene was degraded to phenol, biphenyl, chlorobiphenyl isomers, and benzaldehyde with only UV, and chlorobenzenes also transform to chlorophenol and various isomers of chlorobiphenyl and dichlorobiphenyl in the UV/H_2O_2 system.

Degradation of chlorobenzenes in pyrite suspensions is shown in Fig. 6 and in Table 1, with four volatile species: mono-chlorobenzene, 1,2-dichlorobenzene, 1,3-dichlorobenzene, and 1,2,4-trichlorobenzene. The concentrations of these compounds were determined by headspace methods. The experiments were conducted at 25 °C at 200 rpm in the pyrite suspension under anaerobic conditions. The initial concentrations of chlorobenzenes were set to 10 mg/L using the each pure chlorobenzenes dissolved in hexane water. The initial rate of degradation is 1,2,4-trichlorobenzene > 1,2-dichlorobenzene > monochlorobenzene > 1,3-dichlorobenzene, and more than 90 % of these chlorobenzenes were dechlorinated after merely 10 days. The degraded amount of chlorobenzenes is not sensitive to differences in the distribution of their isomers. However, the amount of chloride ion arising from dechlorination of chlorobenzenes is less than the entire chlorine content from each of the chlorobenzenes. This disagreement in mass balance also shows less than one chloride ion degraded from each chlorobenzene. The order of total dechlorination ability is 1,2-dichlorobenzene > monochlorobenzene > 1,2,4-trichlorobenzene > 1,3-dichlorobenzene. Among the 4 species of chlorobenzenes estimated here, 1,3-dichlorobenzene is the hardest to degrade because it has a meta-site chloride configuration. The meta-site dechlorination is not easier than 1,2-(ortho) and 1,4-(para-) dichlorobenzenes.

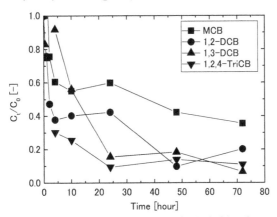

Fig. 6. Degradation of monochlorobenzene, 1,2- and 1,3-dichlorobenzene, and 1,2,4-trichlorobenzene in pyrite suspension under aerobic conditions. C_0 is the initial concentration of chlorobenzenes (10 mg/L), C_t is the concentration of chlorobenzenes at time t.

species	$-\Delta C^t_{CBs}/C^0_{CBs}$ [mol%]	$+\Delta C^t_{Cl}/C^0_{CBs}$ [mol%]
MonoCB	88.1	76.5
1,2-DiCB	96.4	88.4
1,3-DiCB	89.1	56.6
1,2,4-TriCB	95.6	68.7

Table 1. The fraction of degraded chlorinated benzenes and chloride ion arising from chlorinated benzenes after 10 days.

Fig. 7 shows the ratio of the residual chlorinated chlorobenzenes relative to the degradation of all species of chlorinated benzenes by pyrite suspension. Table 2 is the fraction of dechlorinated chloride ion from each chlorobenzene and the initial chloride content of the chlorobenzenes. The volatile chlorinated benzenes, mono-, 1,2-di-, 1,3-di- and 1,2,4-trichlorobenzenes were analyzed by the headspace method in addition to solvent extraction analysis to determine their concentrations. The non-degraded volatile chlorobenzenes adsorbed on the pyrite surface are correctly estimated here. The reaction products were also analyzed by GC/MS after eluting in the organic solutions along with the other chlorinated benzenes.

The ratio of complete dechlorination is higher for low molecular weight chlorobenzenes, and 1,2-dichlorobenzene is easier to degrade than 1,3-dichlorobenzene. The greater electron deviation due to chlorine configuration allows the degradation of chlorobenzenes. More than 80 to 90 % of mono-, di-, tri-chlorobenzenes, and 1,2,3,4- and 1,2,3,5-tetrachlorobenzene are degraded from the original concentrations. These compounds are easy to degrade, but there is a small variation depending on chlorine configuration. The residual ratio is 1,2- < 1,3- < 1,4- among the dichlorobenzenes, and 1,2,3- < 1,2,4- < 1,3,5- among the trichlorobenzenes. There is a significant difference in dechlorination ability for the tetrachlorobenzene isomers. The symmetrical chlorine configuration, 1,2,4,5-, on the benzene ring is stable and hard to degrade, so 1,2,4,5-tetrachlorobenzene, tetra- and hexa-chlorobenzenes with the 1,2,4,5-chlorine configuration are extremely difficult to degrade. Considering the mass balance of persistent chlorobenzenes and chloride ions, di-, tri-, and tetra- (limited to 1,2,3,4- and 1,2,3,5-) chlorobenzenes are transformed into chlorinated intermediates because the detected concentration of chloride ion is only one mole of chloride ion per one mole of chlorobenzene and that corresponds to a decline of the initial concentration of each of the chlorobenzenes.

The GC/ MS analysis of the solvent-extracted samples after 10 days indicates the presence of one or more dechlorinated chlorobenzenes from the initial materials for each chlorobenzene, except for hexachlorobenzene. In the experimental dechlorination of hexachlorobenzene, pentachlorobenzene is detected after 10 days. Hexachlorobenzene is assumed to have a slow dechlorination rate so that a little of the compound could be detected. As an example of reaction products, Fig. 8 shows the analytical result of GC/MS and NMR analysis for the solvent extracted from the reaction system of monochlorobenzenes. $C_9H_{14}(OH)Cl$, $C_{13}H_{17}(OH)_3$, $C_{16}H_{23}SO_2$ are detected by MS together with cyclic sulphur molecules such as S_6, S_7, S_8. The other saturated or unsaturated straight-chain hydrocarbons, such as 3-decyne ($C_{10}H_{18}$), 8,10-dodecadienial ($C_{12}H_{20}O$) and 2,2,6-trimethyl-1,4-cyclohexanedione ($C_9H_{14}O_2$) are detected as one of the reaction products of tri- to hexachlorobenzenes. The NMR spectrum also shows the alcohol, ether, and aromatic methyl peaks ($-CH_2OH$, -O-, -COH, C_6H_6-CH_3) with the benzene and chlorobenzene peaks. One part of the benzene ring structure is assumed to be opening in this reaction process.

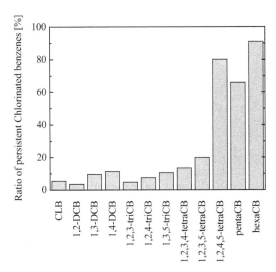

Fig. 7. Residual ratio of 12 species of chlorinated benzenes estimated under aerobic conditions in pyrite suspension after 10 days (Hara *et al.* (2006).

species	$+\Delta C^t_{Cl-} /C^0_{Cl-}$ [mol%]
monochlorobenzene	89
1,2-dichlorobenzene	47
1,3-dichlorobenzene	45
1,4-dichlorobenzene	45
1,2,3-trichlorobenzene	30
1,2,4-trichlorobenzene	29
1,3,5-trichlorobenzene	30
1,2,3,4-tetrachlorobenzene	20
1,2,3,5-tetrachlorobenzene	18
1,2,4,5-tetrachlorobenzene	11
pentachlorobenzene	7.5
hexachlorobenzene	5.5

Table 2. The fraction of dechlorinated chloride ion from each chlorobenzene and initial chloride content of chlorobenzenes.

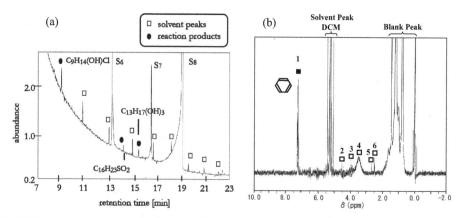

Fig. 8. The spectrum of reaction intermediates arising from transformation of monochlorobenzene in pyrite suspension. (a) MS spectrum. (b) MNR spectrum. In the NMR spectrum, 1. aromatic peak (benzene and monochlorobenzene), 2-4. alcohol and ether peak ($-CH_2OH$, -O-, -COH), 5. aromatic methyl (methylbenzene).

From the analytical results, the benzene ring could have been opened by the dechlorination process. One part of the benzene ring opened to straight chain unsaturated hydrocarbons and these hydrocarbons polymerized with other straight chain unsaturated hydrocarbons. Some polymerized straight chain hydrocarbons remain, and some hydrocarbons are combined with unsaturated ring structures. A ring-opening reaction of benzene is assumed to be a common reaction for every chlorobenzene under our experimental conditions, not only turning into cyclohexanes. These reactions are initiated when one chlorine site is dechlorinated by pyrite. Therefore, at least one mole of chlorine ions was dissolved in solution for every mole of chlorinated benzene molecules that reacted with pyrite. Furthermore, non-detected chlorine ions may still bind the carbon atoms that are produced by this reaction.

The reactions are proceeded by the oxidant arising on the pyrite interface. The polymerization and ring-opening reactions are similar to the dechlorination of Dieldrin under anaerobic conditions.

4. Discussion

Pyrite suspension shows degradation ability predominantly for TCE, Dieldrin and chlorobenzenes as shown in Section 3. These degradations of chlorinated organic compounds are initiated by the oxidant arising from the pyrite interface reaction.

The surface reaction of pyrite/H_2O and pyrite/O_2 produces the radical as follows (Cohn *et al.*, 2006). In the pyrite/H_2O interface under non-oxygen conditions, H_2O is hydrolysed to •OH and H^+ (Eq. (4) and Eq. (5))

$$Fe^{3+}(pyrite) + H_2O \rightarrow Fe^{2+}(pyrite) + {}^{\bullet}OH + H^+ \qquad (4)$$

$$2({}^{\bullet}OH) \rightarrow H_2O_2 \qquad (5)$$

Commonly, the radical is apt to arise under aerobic conditions, but the hydroxyl radical can be generated under anaerobic conditions and it acts as an oxidant.

In the presence of oxygen (aerobic conditions), superoxide $(O_2 \bullet)$- was detected as an intermediate from H_2O_2 formation. The superoxide $(O_2 \bullet)$- is generated by the oxidation of iron on the pyrite surface as follows:

$$Fe^{2+}(pyrite) + O_2 \rightarrow Fe^{3+}(pyrite) + (O_2^{\bullet})^- \tag{6}$$

$$Fe^{2+}(pyrite) + (O_2^{\bullet})^- + 2H^+ \rightarrow Fe^{3+}(pyrite) + H_2O_2 \tag{7}$$

The reactions of the pyrite/O_2 interface generate hydroxyl radicals from hydrogen peroxide by the oxidation of surface iron (Eq. (6) and Eq. (7)), and hydrogen peroxide is also used for the reduction of ferric to ferrous ion in the reaction shown in Eq. (8). However, the hydroxyl radicals are not noticeably produced under aerobic condition.

$$Fe^{2+}(aq) + H_2O_2 \rightarrow {}^{\bullet}OH + OH^- + Fe^{3+}(aq) \tag{8}$$

Under aerobic conditions, superoxide and hydrogen peroxide are dominant oxidants used for degradation of chlorinated organic compounds.

Although there is a difference in oxidative ability among oxidants, oxidants arise from the pyrite interface under both aerobic and anaerobic conditions. The hydroxyl radical has an oxidant ability superior to superoxide and hydrogen peroxide, but the total amount of ferrous ion on the pyrite surface is limited under anaerobic conditions; therefore, the total volume of the hydroxyl radical is also limited under anaerobic conditions as opposed to aerobic conditions. The difference between reaction products under aerobic and anaerobic conditions is caused by thus radical character of the pyrite interface.

The oxidation and polymerization of chlorinated organic compounds are caused by several radicals arising from the pyrite/H_2O, pyrite/O_2 interface reaction.

5. Conclusion

This chapter reviews several remediation methods for chlorinated organic pollutants, and emphasizes the degradation ability of natural ferric sulphide (pyrite) for TCE, Dieldrin and chlorobenzenes. The transformation of TCE and Dieldrin under aerobic conditions becomes clear and the main reaction product is formic acid, which continuously degrades to CO_2. Although the 1,2,4,5-chloride configuration in the chlorobenzene and benzene ring structure is hard to degrade, one part of the benzene ring is transformed into several hydrocarbons.

The oxidative reaction of the pyrite suspension has great degradative ability not only for low molecular weight organic pollutants such as VOCs but also for high molecular weight persistent organic compounds with a stable benzene ring and bicyclic rings. Natural ferric sulphide has a remediation ability predominantly for chlorinated organic pollutants.

Iron and ferric sulphides (e.g., pyrite) are widely distributed in the subsurface layer in the global environment. The reductive degradation ability of iron and the oxidative degradation ability of ferric sulphides assist the innovative remediation techniques that are conducted at in situ normal temperature and pressure. The ferric ion and ferric sulphide latent in natural systems have a potential for natural attenuation of contamination.

Pyrite is also widely distributed at mining sites and soils in coastal regions of the Asian and the Pacific areas located in tropical, subtropical and temperate climate areas. For instance, the Mekong Delta in Vietnam is a vast acidic sulphate soil region that includes metallic sulphide. Garvalho *et al.* (2008) reported the agrochemical and polychlorobiphenyl residues in the Mekong River Delta. Several chlorinated compounds, such as DDT, HCH (hexacychlochlorohexane), PCBs and endosulfan were detected in sediment and biota. However, the concentrations of PCBs and pesticide residues in the aquatic environment in the Mekong River Delta are lower than the values reported in other regions of Vietnam and Asia. The aquatic environment of the Mekong Delta is endowed with natural ferric sulphide and water, and consequently the natural environment in this region is assumed to enhance PCB degradation. The acidic sulphate soils distributed in coastal area assumes to actually attenuate the contamination of chlorinated organic compound and prevent the expansion of pollution.

6. Acknowledgments

The author would like to thank Dr. Tsuji (Yamagata Environmental Research Institute) for assistance with the analysis and detection of reaction intermediates.
This work was made possible by a grant-in-aid for scientific research from the Japan Society for the Promotion of Science (B-17760656).

7. References

Ahmed S. and Ollis D.F. (1984) Solar photoassisted catalytic decomposition of the chlorinated hydrocarbons trichloroethylene and trichloromethane. *Solar Energy*, Vol.32(5), 597-601.

Baczynski T.P., Grotenhuis T. and Knipscheer P. (2004) The dechlorination of cyclodiene pesticide by metanogenic granular sludge. *Chemosphere*, Vol.55, 653-659.

Bandala E.R., Gelover S. and Leal M.T. (2002) Solar Photocatalytic degradation of aldrin. *Catalysis Today*, Vol.76, 189-199.

Barbeni M., Minero C. and Pelizzetti E. (1987a) Chemical degradation of chlorophenols with Fenton's Reagent. *Chemosphere*, Vol.16(10-12), 2225-2237.

Barbeni M., Morello M., Pramauro E. and Pilizzetti E. (1987b) Sunlight Photodegradation of 2,4,5-trichlorophenoxy-acetic acid and 2,4,5-trichlorophenol on TiO$_2$. Identification of itermediates and degradation pathway. *Chemosphere*, Vol.16(6), 1165-1179.

Barbeni M., Pramauro E. and Pilizzetti E. (1986) Photochemical degradation of chlorinated dioxins, biphenyls phenols and benzene on semiconductor dispersion. *Chemosphere*, Vol.15(9), 1913-1916.

Barbusinski K. (2009) Fenton reaction controversy concerning the chemistry. *Ecological chemistry and engineering*, Vol.16(3), 347-258.

Borello R., Minero C., Pramauru E., Pelizzetti E., Serpone N. and Hidaka H. (1989) Photocatalytic degradation of DDT mediated in aqueous semiconductor slurries by simulated sunlight. *Environmental Toxicology and chemistry*, Vol.8, 997-1002.

Brooks G.T. (1980) The preparation of some reductively dechlorinated analogues of deildrin, endosulfan and isobenzan. *Journal of Pesticide Science*, Vol.5, 565-574.

Chiu T., Yen J., Hsieh Y. and Wang Y. (2005) Reductive transformation of dieldrin under anaerobic sediment culture. *Chemosphere*, Vol.60, 1182-1189.

Cohn C.A., Mueller S. Wimmer E., Leifer N., Greenbaum S., Strongin S.R. and Schoonen M.A.A. (2006) Pyrite-induced hydroxyl radical formation and its effect on nucleic acids. *Geochemical Transactions*, Vol.7(3), 1-11.

Dibble L.A. and Raupp, G. B. (1990) Kinetic of the gas-solid heterogeneous photocatalytic oxidation of trichloroethylene by near UV illuminated titanium dioxide. *Catalysis Letters*, Vol.4, 345-354.

Dibble L.A. and Raupp, G. B. (1992) Fluidized-Bed photocatalytic oxidation of trichloroethylene in contaminated airstreams. *Environmental Science and Technology*, Vol.26, 492-495.

Dureja P., Walia S. and Mukerjee S. K. (1987) Superoxide mediated dehydrohalogenation of photodieldrin and photoaldrin. *Tetrahedron letters*, Vol.28(8), 895-896.

D'Ollvelra J.-C., Al-Sayyed G. and Plchat P. (1990) Photodegradation of 2- and 3-chlorophenol in TiO_2 Aqueous Suspensions. *Environmental Science and Technology*, Vol.24, 990-996.

Farwell O., Beland F. A., and Geer R. D. (1975) Reduction pathways of organohologen compounds. Part 1. Chlorinated benzenes. *Electroanalytical Chemistry and Interfacial Electrochemistry*, Vol. 61, 303-313.

Froelich E.M. (1992) Chemical Oxidation. *Technomic Publishing Company, Inc.* Eckenfelder W. W., Nowers A.R. and Roth J. A. (eds.) Lancaster, PA, 104-113.

Garvallho F.P., Villeneuve J.P, Cattini C., Tolosa I., Dao Dinh Thuan and Dang Duc Nhan (2008) Agrochemical and polychlorobyphenyl (PCB) residues in the Mekong River delta, Vietnam. *Marine Pllution Bullutin*, Vol.56, 1476-1485.

Gerritse J., Drzyzga O., Kloetstra G., Keijmel M., Wiersum L.P., Hutson R., Collins M.D. and Gottschal J.C. (1999) Influence of different electron donors and acceptors on dehalorespiration of tetrachloroethene by Desulfitobacterium frappieri TCE1. *Applied and Environmental Microbiology*, Vol.65(12), 5212-5221.

Gerritse J., Renard V., Pedro Gomes T.M., Lawson P.A., Collins M.D. and Gottschal J.C. (1996) Desulfitobacterium sp. Strain PCE1, an anaerobic bacterium that can grow by reductive dechlorination of tetrachloroethene or ortho-chlorinated phenols. *Archives of Microbiology*, Vol.165(2), 132-140.

Guena T., Wang L., Gattrell M. and MacDougall (2000) Mediated Approach for the electrochemical reduction of chlorobenzens in Nonaqueous media. *Journal of the Electrochemical Society*, Vol. 147, 248-255.

Guevremont J.M., Elsetinow A.R., Strongin D.R., Bebie J.M. and Schoonen M.A.A. (1998b) Structure Sensitivity of pyrite oxidation: Comparison of the (100) and (111) planes. *American Mineralogist*, Vol.83, 1353-1356.

Guevremont J.M., Strongin D.R. and Schoonen M.A.A. (1998c) Photoemission of adsorbed Xenon, X-ray photoemission spectroscopy, and temperature-programmed desorption studies of H_2O on FeS_2 (100). *Langmuir* 14: 1361-1366.

Guevremont J.M., Strongin D.R. and Schoonen M.A.A. (1998d) Thermal chemistry of H_2S and H_2O on the (100) plane of pyrite: Unique reactivity of defect sites. *American mineralogist*, Vol.83, 1246-1255.

Guevremont J.M., Bebie J., Elsetinow A.R., Strongin D.R. and Schoonen M.A.A. (1998a) Reactivity of the (100) plane of pyrite in oxidizing gaseous and aqueous environments: Effect of Surface imperfections. *Environmental Science and Technology*, Vol.32, 3743-3748.

Guevremont J.M., Strongin D.R. and Schoonen M.A.A. (1997) Effects of surface imperfections on the binding of CH_3OH and H_2O on FeS_2 (100): Using adsorbed Xe as a probe of mineral surface structure. *Surface Science*, Vol.391, 109-124.

Hara J. (2001) The effect of oxygen on chemical dechlorination of dieldrin using iron sulfides. *Chemosphere*, Vol.82, 1308-1313.

Hara J., Kawabe Y., Komai T. and Inoue C. (2009) Chemical degradation of dieldrin using ferric sulfide and iron powder. *International Journal of Environmental Science and Engineering*, Vol.1-2, 91-96.

Hara J., Inoue C., Chida T., Kawabe Y., Komai T. (2006) Dehalogenation of Chlorinated Benzenes by iron sulfide. *International Journal of Power and Energy Systems*, Vol.1(1), 239-243.

Hoa T. P., Kitsuneduka M., Hara J., Suto K., Inoue C. (2008) Trichloroethylene Transformation by Natural Mineral Pyrite: The deciding role of oxygen. *Environmental Science and Technology*, Vol.42. 7470-7475.

Hoa T.P., Suto K. and Inoue C. (2009) Trichloroethylene transformation in Aerobic pyrite suspension: pathways and Kinetic modeling. *Environmental Science and Technology*, Vol.43, 6744-6749.

Holliger C., Schraa G., Stams A.J.M. and Zehnder A.J.B. (1993) A highly purified enrichment culture couples the reductive dechlorination of tetrachloroethene to growth. *Applied and Environmental Microbiology*, Vol.59 (9), 2991-2997.

Kargina O., MacDougall B., Kargin Y. M. and Wang L. (1997) Dechlorination of monochlorobenzene using organic mediators. *Journal of the Electrochemical Society*, Vol.144, 3715-3721.

Kendelewiz T., Dolyle C.S., Bostick B.C. and Brown G.E. (2004) Initial oxidation of fractured surface of pyrite (FeS_2(100)) by molecular oxygen, water vapor, and air. *Surface Science*, Vol.558, 80-88.

Kim S., Kim D., Pollack G. M., Collins L. B. and Rusyn I. (2009) Pharmacokinetic analysis of trichloroethylene metabolism in male B6C3F1 mice: Formation and disposition of trichloroacetic acid, dichloroacetic acid, S-(1,2-dichlorovinyl) glutathione and S-(1,2-dichlorovinyl)-L-Cysteine. *Toxicology and applied pharmacology*, Vol.238, 90-99.

Kondo M.M. and Jardim W.F. (1991) Photodegradation of chloroform and urea using Ag-loaded titanium dioxide as catalyst. *Water Research*, Vol.25(7), 823-827.

Kriegman-King M.R. and Reinhard M. (1994) Degradation of carbon tetrachloride by pyrite in aqueous solution. *Environmental Science and Technology*, Vol.28, 692-700.

Krumholz L.R., Sharp R. and Fishbain S.S. (1996) A freshwater anaerobe coupling acetate oxidation to tetrachloroethylene dehalogenation. *Applied and Environmental Microbiology*, Vol.62(11), 4108-4113.

Kusveran E. and Erbatur O. (2004) Degradation of aldrin in adsorbed system using advanced oxidation processes: comparison of the treatment methods. *Journal of Hazardous Materials*, 106B, 115-125.

Lee W. and Batchelor B. (2002) Abiotic reductive dechlorination of chlorinated ethyelens by iron-bearing soil minerals. 1. pyrite and magnetite. *Environmental Science and Technology*, Vol.36, 5147-5154.

Lee W. and Batchelor B. (2003) Reductive capacity of natural reductants. *Environmental Science and Technology*, Vol.37, 535-541.

Leung S.W., Watts R. J. and Miller G.C. (1992) Degradation of Perchloroethylene by Fenton's Reagent: Speciation and Pathway. *Journal of Environmental Quality*, Vol.21(3), 377-381.

Li S. and Wackett L.P. (1992) Trichloroethylene oxidation by toluene dioxygenase. *Biochemical and Biophysical research communications*, Vol.185(1), 443-451.

Löffler F.E., Sanford R.A. and Tiedje J.M. (1996) Initial characterization of a reductive dehalogenase from Desulfitobacterium chlororespirans Co23. *Applied and Environmental Microbiology*, Vol.62(10), 3809–3813.

Maule A., Plyte S. and Quirk A.V. (1987) Dehalogenation of organochlorine insecticides by mixed anaerobic microbial populations. *Pesticide Biochemistry and Physiology*, Vol.27, 229-236.

Maymó-Gatell X., Anguish T. and Zinder S.H. (1999) Reductive dechlorination of chlorinated ethenes and 1,2-dichloroethane by Dehalococcoides ethenogenes 195. *Applied and Environmental Microbiology*, Vol.65(7), 3108–3113.

Miyoshi K., Kameyama Y. and Matsumura M. (2004) Electrochemical reduction of organohalogen compound by noble metal sintered electrode. *Chemosphere*, Vol.56, 187-193.

Mochizuki S. (1977) Photochemical dechlorination of PCBs. *Chemical Engineering Science*, Vol.32, 1205-1210.

Mohammad S. M. and Dennis G. P. (1997) Electrochemical reduction of di-, tri- and tetrahalobenzenes at carbon cathodes in dimethylformamide. Evidence for a halogen dance during the electrolysis of 1,2,4,5-tetrabromobenzene. *Journal of Electroanalytical Chemistry*, Vol.435, 47-53.

Motosugi K., Esaki N. and Soda K. (1982) Purification and properties of a new enzyme, DL-2-haloacid dehalogenase, from Psurdomonas sp. *Journal of Bacteriology*, Vol.150, No.2, 522-527.

Moza P. N., Fytianos K., Samanidou V. and Korte F. (1988) Photodecomposition of chlrophenols in aqueous medium in presence of hydrogen peroxide. *Bulletin of Environmental Contamination and toxicology*, Vol.41, 678-682.

Nesbitt H.W. and Muir I.J. (1998) Oxidation states and speciation of secondary products on pyrite and arsenopyrite reacted with mine waste waters and air. *Mineralogy and petrology*, Vol.62, 123-144.

Newman L.M. and Wackett L.P. (1991) Fate of 2,2,2-trichloroacetaldehyde (chloral hydrate) produced during trichloroethylene oxidation by methanotrophs. *Applied Environmental Microbiology*, vol.57 (8), 2399-2402.

Newman L.M. and Wackett L.P. (1995) Purification and characterization of toluene 2-monooxygenase from Burkhokderia cepacia G4. *Biochemistry*, Vol.34(43), 14066-14076.

Newman L.M. and Wackett L.P. (1997) Journal of Bacteriology, No.179(1), 90-96.

Oldenhuis R., Vink R.L., Janssen D.B., Witholt B. (1989) Degradation of chlorinated aliphatic hydrocarbons by methylosius trichosporium OB3b expressing soluble methane monooxygenase. *Applied Environmental Microbiology*, Vol.55(11), 2819-2826.

Reineke W. and Knackmuss H.J. (1984) Microbial Metabolism of Haloaromatics: Isolation and properties of Chlorobenzene-Degrading Bacterium. *Applied and Environmental Microbiology*, Vol.47(2), 395-402.

Rosenzweig A.C., Frederick C.A., Lippard S.J., Nordlund P. (1993) Crystal structure of bacterial non-haem iron hydroxylase that catalyses the biological oxidation of methane. *Nature*, Vol.366(6455), 537-43

Rosso K.M., Becker U. and Hochella M. F. J. (1999) The interaction of pyrite{100} surface with O_2 and H_2O: Fundamental oxidation mechanisms. *American Mineralogist*, Vol.84, 1549-1561.

Schaufuss A.G., Nesbitt H.W., Kartio I., Laajalehto K., Bancroft G.M. and Szargan R. (2000) Reactivity of surface chemical states on fractured pyrite. *Surface Science*, Vol.411, 321-328.

Shimakoshi H., Tokunaga M. Baba T. and Hisaeda Y. (2004) Photochemical dechlorination of DDT catalyzed by a hydrophobic vitamin B12 and a photosensitizer under irradiation with visible light. *Chemical Communications*, 1806-1807.

Sundstrom D.W., Klei H.E., Nalette T.A., Reidy D. J., Weir B. A. (1986) Destruction of Halogenated Aliphatics by Ultraviolet Catalyzed Oxidation with Hydrogen Peroxide. *Hazardous Waste and Hazardous Materials*, 3(1), 101-110.

Sundstrom D.W., Weir B.A. and Klei H.E. (1989) Destruction of aromatic pollutions by UV light catalyzed oxidation with hydrogen peroxide. *Environmental Progress*, Vol.8 (1), 6-11.

Taylor B. E., Wheeler M.C. and Nordstorm D.K. (1984) Stable isotope geochemistry of acid mine drainage: Experimental oxidation of pyrite. *Geochimica et Cosmochimica Acta*, Vol.48, 2669-2678.

Tiedje, J. M., Quensen III J. F., Chee-Sanford, J., Schimel J. P. and Boud S. A. (1993) Microbial reductive dechlorination of PCBs. *Biodegradation*, Vol.4, 231-240.

Usher C.R., Cleveland C.A., Strongin D.R. and Schoonen M.A. (2004) Origin of oxygen in sulfate during pyrite oxidation with water and dissolved oxygen: An in situ horizontal attenuated total reflectance infrared spectroscopy isotope study. *Environmental Science and Technology*, Vol.38, 5604-5606.

Usher C.R., Paul K.W., Narayansamy J., Kubicki J.D., Sparks D.L., Schoonen M.A.A. and Strongin D.R. (2005) Mechanistic aspects of pyrite oxidation in an oxidizing gaseous environment: An in situ HATR-IR isotope study. *Environmental Science and Technology*, Vol.39, 7576-7584.

Van Beek H.C.A., Van der Stoep H.J., Van Oort H. and Van Leene J. (1982) Photochemical Radical Chain Dechlorination of DDT in 2-Propanol. *Industrial & Engineering Chemistry Product Research and Development*, Vol.12, 123-125.

Weerasooriya R. and Dharmasena B. (2001) Pyrite-assisted degradation of trichloroethene (TCE). *Chemosphere*, Vol.42, 389-396.

Weie B.A., Sundstrom D.W. and Klei H.E. (1987) Destruction of Benzene by Ultraviolet Light-Catalyzed oxidation with Hydrogen Peroxide. *Hazardous Waste and Hazardous Materials*, Vol.4(2), 165-176.

Weightman A.J., Weightman A.L., Slater J.H. (1985) Toxic effects of chlorinated and brominated akanoic acids on Pseudomonas putida PP3: selection at high frequencies of mutations in genes encoding dehalogenases. *Applied Environmental microbiology*, Vol.49(6), 1494-1501.

Weightman A.L., Weightman A.J. and Slater J.H. (1992) Microbial dehalogenation of trichloroacetic acid. *World journal of microbiology and biotechnology*, Vol.8, 512-518.

Zinovyev S.S., Shinkova N.A., Perosa A. and Tundo P. (2005) Liquid phase hydrodechlorination of dieldrin and DDT over Pd/C and Raney-Ni. *Applied Catalysis B: Environemtal*, Vol.55, 39-8.

6

Research on Pressure Swing Adsorption of Resin for Treating Gas Containing Toluene

Ruixia Wei and Shuguo Zhao

Hebei Polytechnic University, Tangshan, Hebei
China

1. Introduction

1.1 Introduction of volatile organic compounds

Volatile organic compounds (VOC) means any compound of carbon, excluding carbon monoxide, carbon dioxide, carbonic acid, metallic carbides or carbonates, and ammonium carbonate, which participates in atmospheric photochemical reactions.

Volatile organic compounds or VOCs are organic chemical compounds whose composition makes it possible for them to evaporate under normal indoor atmospheric conditions of temperature and pressure. This is the general definition of VOCs that is used in the scientific literature, and is consistent with the definition used for indoor air quality.

Since the volatility of a compound is generally higher the lower its boiling point temperature, the volatility of organic compounds are sometimes defined and classified by their boiling points.For example, the European Union uses the boiling point, rather than its volatility in its definition of VOCs.

A VOC is any organic compound having an initial boiling point less than or equal to 250° C measured at a standard atmospheric pressure of 101.3 kPa. VOCs are sometimes categorized by the ease they will be emitted. For example, the World Health Organization (WHO) categorizes indoor organic pollutants as very volatile, volatile, and semi-volatile. The higher the volatility (lower the boiling point), the more likely the compound will be emitted from a product or surface into the air. Very volatile organic compounds (VVOCs) are so volatile that they are difficult to measure and are found almost entirely as gases in the air rather than in materials or on surfaces. The least volatile compounds (SVOCs) found in air constitute a far smaller fraction of the total present indoors while the majority will be in solids or liquids that contain them or on surfaces including dust, furnishings, and building materials.

Many VOCs are dangerous to human health or cause harm to the environment. VOCs are numerous, varied, and ubiquitous. They include both man-made and naturally occurring chemical compounds. Anthropogenic VOCs are regulated by law, especially indoors, where concentrations are the highest. VOCs are typically not acutely toxic, but instead have compounding long-term health effects. Because the concentrations are usually low and the symptoms slow to develop. The main hazard is as the following aspects:

1. The ability of organic chemicals to cause health effects varies greatly from those that are highly toxic, to those with no known health effect. As with other pollutants, the extent and nature of the health effect will depend on many factors including level of exposure

and length of time exposed. Eye and respiratory tract irritation, headaches, dizziness, visual disorders, and memory impairment are among the immediate symptoms that some people have experienced soon after exposure to some organics. At present, not much is known about what health effects occur from the levels of organics usually found in homes. Many organic compounds are known to cause cancer in animals; some are suspected of causing, or are known to cause, cancer in humans. Eye, nose, and throat irritation; headaches, loss of coordination, nausea; damage to liver, kidney, and central nervous system. Some organics can cause cancer in animals; some are suspected or known to cause cancer in humans. Key signs or symptoms associated with exposure to VOCs include conjunctival irritation, nose and throat discomfort, headache, allergic skin reaction, dyspnea, declines in serum cholinesterase levels, nausea, emesis, epistaxis, fatigue, dizziness.

2. The reaction of photochemical smog in the sunlight will occur among nitrogen oxides, hydrocarbons and photochemical oxidants of the atmosphere.The main component of photochemical smog is ozone, peroxy acetyl nitrate Cool (PAN), aldehydes and ketones and so on. They stimulate people's eyes and respiratory system, endangering people's health and even harm plant growth.

3. Halogenated hydrocarbons VOCs may destroy the ozone layer and change the Earth's heat balance. According to the Indian National Academy of Sciences report, the emissions of chlorofluorocarbons into the atmosphere have increased the atmospheric methane and chloride absorption of infrared radiation and heat hinder the discharge of the Earth which will make the Earth's temperature,climate change.

VOCs	environmental acceptable concentration (mg/m^3)	health hazards
Toluene	100	Headache, dizziness, nausea, pulmonary emphysema
Benzene	5	Cancer, leukemia, respiratory paralysis
Xylene	100	Anemia, leukemia, red blood cells reduced, skin and mucous membrane irritation
Methanol	200	Neurological disorders, vomiting, insomnia, headaches,
Acetone	750	Cramps
Ethyl acetate	150	Irritate the eye, skin, numbness, headache, cough, nausea
Carbontetrachloride	5	Eye irritation, paralysis Abdominal pain, nausea, vomiting, cancer
Acetaldehyde	10	Mucosal erosion, blurred vision, pulmonary edema
Ether	400	Paralysis, nervous system damage, liver and kidney damage
Acetonitrile	40	Headache, dizziness, breathing difficulties, damage the central nervous
Acrylonitrile	20	Nausea, vomiting, difficulty breathing

Table 1. Common VOCs environmental acceptable concentration and human health hazards

As the VOCs harmful environmental effects, many countries have developed a corresponding law to limit emissions of VOCs. "Clean Air Act 1990" of the United States requires 90% reduction in emissions of the 189 kinds of toxic chemicals which of about 70% belongs to VOCs. In 1996 Japan adopted legislative restrictions of 53 kinds of VOCs emissions and limited 149 kinds of VOCs emissions in 2002. China also enacted in 1997 and implemented the "Integrated emission standard of air pollutants"which limits 33 pollutant emission limits, including benzene, toluene, xylene and other volatile organic compounds. The VOCs harmful environmental effects and human health can not avoid in terms of current technology, so there is an urgent need for effective technology to control VOCs.

1.2 Vocs treatment technology

VOCs treatment technology is divided into two categories: Destruction processes and Recuperation processes. VOC controls include all technologies which either collect the VOCs for recovery and reuse, or destroy the VOCs. If the VOCs have recovery value, which typically implies single-VOC exhaust streams, and if the cost of recovery is less than the cost of purchasing new VOC, which typically implies relatively concentrated exhaust streams, then recovery makes sense. Carbon adsorption, scrubbing, and condensation are typical recovery techniques. Note that the installation and operation of a recovery technology may more than pay for itself if the recovery value of the VOC is high enough. If the VOC stream has no recovery value, if, for example, it is a mixture, or if there are disposal concerns, such as for toxic compounds, then destruction probably makes the most sense. Thermal and catalytic oxidation and biofiltration would be useful in this case.

1.2.1 Destruction processes

1.2.1.1 Thermal oxidation processes

Thermal oxidation is the process of oxidizing combustible materials by raising the temperature of the material above its auto-ignition point in the presence of oxygen, and maintaining it at high temperature for sufficient time to complete combustion to carbon dioxide and water. Time, temperature, turbulence (for mixing), and the availability of oxygen all affect the rate and efficiency of the combustion process. These factors provide the basic design parameters for VOC oxidation systems.

There are three basic types of thermal oxidation systems: direct flame, recuperative, and regenerative.

Direct flame systems or flares rely on contact of the waste stream with a flame to achieve oxidation of the VOCs. These systems are the simplest thermal oxidizers and the least expensive to install, but require the greatest amount of auxiliary fuel to maintain the oxidation temperature, thus entailing the highest operating cost. Flares are useful for destruction of intermittent streams.

Recuperative thermal oxidation systems use a tube or plate heat exchanger to preheat the effluent stream prior to oxidation in the combustion chamber. Thermal recovery efficiencies typically are limited to 40-70% to prevent auto-ignition in the heat exchange package, which could damage the package. Supplemental fuel therefore is usually required to maintain a high enough temperature for the desired destruction efficiency. Recuperative systems are more expensive to install than flares, but have lower operating costs.

Regenerative thermal oxidation systems typically incorporate multiple ceramic heat exchanger beds to produce heat recovery efficiencies as high as 95%. An incoming gas stream passes through a hot bed of ceramic or other material, which simultaneously cools the bed and heats the stream to temperatures above the auto-ignition points of its organic constituents. Oxidation thus begins in the bed, and is completed in a central combustion chamber, after which the clean gas stream is cooled by passage through another ceramic heat exchanger. Periodically the flow through the beds is reversed, while continuous flow through the unit is maintained. Regenerative thermal oxidation systems are the most expensive thermal oxidizers to build, but the added capital expense is offset by savings in auxiliary fuel.

ESOCOV is a regenerative thermal oxidation process on ceramic beds. The process is especially well adapted to a mixture of gases with concentrations between 1 and 10 g/Nm3 and flows from 1.000 to 100.000 Nm3/h.The gas passes over a ceramic bed in which air is progressively heated and the VOC's are destroyed by oxidation above 800 C. The direction of the airflow is changed on a regularly basis in order to charge and discharge the calories in the bed(s). The thermal efficiency amounts to 90 - 98 %. In this way the regenerating systems are autothermal, so, without additional energy for concentrations higher than 1,5 g/Nm3. Addition of a catalyst to have the oxidation at a lower temperature (between 200 and 400 C).

1.2.1.2 Catalytic oxidation

Catalytic oxidation converts volatile organic compounds (VOC) into carbon dioxide and water, as do other oxidation processes, with no byproducts requiring disposal. Catalytic oxidation is well suited to applications with VOC concentrations ranging up to 25% of the lower explosion limit. With proper selection of catalyst, operating conditions, and equipment design, catalytic oxidation can attain VOC conversions of up to 99%. Advantages of this technology are low fuel usage, particularly with the proper choice of heat exchanger, little nitrogen oxide formation, given low operating temperatures, and little formation of partial oxidation products, such as carbon monoxide and aldehydes. Disadvantages include susceptibility to catalyst poisons, and the sensitivity of the catalysts to high temperatures.

Catalysts for VOC oxidation typically are either precious metals supported on ceramic or metal monoliths (honeycombs) or on ceramic pellets, or base metals supported on ceramic pellets. Catalyst life exceeds five years with the proper choice of catalyst, and may be extended with catalyst washing and regeneration techniques. Recent generations of catalysts have much longer lives and greater poison resistance than their forebears, and have greater capabilities, including the destruction of chlorinated organics.

As with any process, proper equipment design is essential to performance and operating cost. Typical catalytic oxidizer components include the catalyst housing, blower, burner, heat exchanger, controls, and stack. Small units are often skid-mounted and delivered to the site ready for installation.

As vent streams are often below the temperature at which catalytic oxidation is effective, most oxidizers use burners to preheat these streams to reaction temperatures, often from 400-800 °F. Heat is recovered using either recuperative or regenerative heat exchangers. As the latter can provide 95% heat recovery, streams with low VOC levels can be processed with minimal fuel usage.

1.2.1.3 Biological treatment processes

Biological method is essentially the use of microbial life activities to the emissions of VOCs into simple inorganic (such as CO_2 and H_2O) and microbial composition of the material itself. Common processes are biological filtration, biological washing. The biggest difference of biological filtration from biological wastewater treatment process is: in the exhaust gas through the organic material must first transfer to the liquid (or solid surface of the film) in the mass transfer process, and then in the liquid (or solid surface of the biological layer) adsorption by microbial degradation. Biological method is particularly suitable for processing gas is greater than 17000m3 / h, the gas concentration is less than 1000ppm.

Biological method has many advantages compared to other technologies, it's simple, low operation cost, low investment relative to other methods, wide range of applications, while not easy to produce secondary pollution. Especially for low concentrations of VOCs (for example, when only a few ppm) the results of treatment is good. However, biological method is related to gas and liquid (or solid) mass transfer and chemical and biological degradation processes. The influencing factors are many and complex. Now, biodegradable technology is not enough in-depth theoretical study, so the biological treatment of VOCs present the design and operation is still basically remain in the level of experience. At the same time it also requires a larger footprint, which is limiting the biological treatment of extensive use of VOCs.

1.2.2 Recuperation processes

(1) Cryogenic Condensation processes

Low temperature or cryogenic condensation is a process that can be used as an effective means for VOC emissions control. Cryogenic condensation technology is based on lowering the vapor pressure of a component by reducing the temperature of the process stream thus increasing the recovery of the components in the liquid phase.

Since nitrogen gas is widely used in the chemical process industry, is inert and is typically transported and stored in it's liquid state at low temperature and high pressure, it is a convenient media to use. The low temperature capability of liquid nitrogen allows for the design of highly efficient condensation systems. At temperatures below –120°F, the vapor pressure of most organic compounds is depressed sufficiently to condense 95 to 99+% of the compounds from a typical emissions stream. In addition, the vented nitrogen can be recycled for reuse within the plant.

Cryogenic condensation is well suited for VOC emission control because of its ability to respond instantly to changes in VOC flow rate and solvent loading. It can recover virtually any VOC species even under varying conditions. Cryogenic condensation can deal with all organics (even in the presence of water) and can function when the concentration and composition are changing over time. This flexibility makes it particularly suitable for VOC control in multi-product, multi-purpose plants where batch or continuous processes are employed.

(2) membrane

Volatile organic compounds (VOCs) are involved in atmospheric pollution and green house effect.Some of these compounds might be recovered, instead of being released to the atmosphere, by several methods such as condensation, absorption, adsorption, etc. Among these processes, vapor permeation has several advantages since it requires compact equipment, it is non destructive and it is notenergy-intensive. Over the past ten years, vapor permeation has been proven to be a feasible alternative to conventional processes in the recovery of several halogenated VOCs and monomers. In recent years, this process has found other applications such as in the recovery of hydrocarbon VOCs from the petroleum industry facilities; these applications are still under development.

Within the membranes used for the recovery of volatile organic compounds, composite membranes offer several advantages over other kinds. They are composed of a selective, defect-free layer that performs the vapor separation while another porous layer gives mechanical strength. Poly dimethyl siloxane (PDMS) is one of the most used polymers as selective permeation layer. It can be easily fabricated and thus is readily available for its use on large scales. The use of dimensionless solubility parameters showed that PDMS has good selectivities towards a wide variety of VOCs (e.g., hydrocarbons).

The recovery of toluene, propylene and 1,3-butadiene, which are compounds of particular concern in the petroleum industry, is focused on in this study. Since several petroleum activities such as oil storage or distribution, emit pollutants at low flow rates and variable concentrations, vapor permeation appears to provide a flexible recovery solution.

(3) absorption processes : ESOLAV

ESOLAV is a range of absorption processes or washers which transfers the VOC's into a liquid phase in order to be solubilized, oxidized or separated. It is especially well suited for VOC's soluble in the washing solution with weak and low concentrations (< 1 g/Nm3) and for gasflows between 500 and 50.000 m3/h.

(4)adsorption with regeneration processes : ESOSORB

ESOSORB is a solvent adsorption process on beds of activated carbon with recuperation by desorption through steam. The process is very well adapted to gases containing 1 or 2 solvents in a concentration lower than 30 g/Nm3 and gas flows between 200 and 200.000 Nm3/h.These units are built out of several adsorbent beds containing activated carbon. The air to be treated is guided over the activated carbon where the VOC's ared adsorbed until the activated carbon is saturated. If one of the adsorbing beds is saturated, a regeneration process is set in motion. After passage of the steam for the desorption a fan dries and cools the activated carbon beds, which makes them ready for a new adsorption cycle.

Adsorption on activated carbon is useful for recovery of VOCs with intermediate molecular weights (typically about 45-130): smaller compounds do not adsorb well, and larger compounds cannot be removed during regeneration, which typically is by steam stripping. Adsorption is most effective at lower temperatures, so that cooling of hot exhaust gas streams may be necessary. Further, dehumidification of very humid streams may be necessary for the carbon to have the greatest capacity. While carbon is the dominant adsorbent used, alumina, zeolites, and polymers have been used in some processes. Carbon can also be used to remove compounds in a once-through process with off-site regeneration.

1.3 Adsorption with regeneration processes and its principles
1.3.1 Adsorption

Adsorption refers to the binding of a dissolved solute onto the surface of a solid adsorbing material. It is a surface phenomenon and should not be confused with absorption, which is a term with a much broader meaning and generally deals with penetration of species into material. Adsorption can be used to separate a solute from a mixture of solutes, or a solute from a solvent. This is achieved by contacting the solution with the adsorbing material which is also called the adsorbent. The solute/s, which adsorb on the adsorbent, is/are referred to as adsorbate/s. The release of adsorbed material from an adsorbent is called desorption which is the reverse of adsorption.

Adsorption is a selective process and this selectivity is due to differences in the following: Molecular weight or size; Solute shape; Polarity; Electrostatic charge.

The physical binding of an adsorbate onto an adsorbent takes place due to non-covalent interactions such as: van der Waals forces; Electrostatic interactions; Hydrophobic interactions; Hydrogen bonding.

This type of adsorption takes place at ordinary temperature and is called physical adsorption or simply adsorption. Certain types of adsorption take place at much elevated temperatures when activation energy is available to break chemical bonds and facilitate chemical changes. Such processes are referred to as chemical adsorption or chemisorption.

The adsorbent material can be natural or synthetic. These generally have amorphous or microcrystalline structure and thus have very high specific surface area (surface area per unit amount of adsorbent). Commonly used adsorbents include clays like kaolin and bentonite, activated carbon, silica gel, activated alumina, zeolite (molecular sieves), etc.

Some of the advantages of adsorption over competing separation technologies are:High selectivity (e.g. affinity adsorption);Ability to handle very dilute solute concentrations.Major disadvantages are:It is a surface phenomenon,therefore the interior of the adsorbent material is not involved ;Batch or semi-batch operations generally have to be used;In certain cases adsorbents have to be regenerated;

In certain cases adsorption results in loss of product quality (e.g. with certain bioproducts)

Some of the common applications of adsorption are:Gas separation using molecular sieves by pressure swing adsorption;Removal of toxic gases from air (e.g. gas masks);Fractionation of industrial chemicals using gas or liquid chromatography;Removal of trace amounts of CS2 and H2S;Removal of phenolic and other toxic chemical from waste water;Removal of chlorinated hydrocarbons from waste gas streams;De-hydration or de-humidification of gases;Water purification by deionization and ion-exchange;Fractionation and recovery of protein bio-products;Affinity separations of bio-products.

1.3.2 PSA

Pressure swing adsorption (PSA) is a technology used to separate some gas species from a mixture of gases under pressure according to the species' molecular characteristics and affinity for an adsorbent material. It operates at near-ambient temperatures and so differs from cryogenic distillation techniques of gas separation. Special adsorptive materials (e.g., zeolites) are used as a molecular sieve, preferentially adsorbing the target gas species at high pressure. The process then swings to low pressure to desorb the adsorbent material.

Pressure swing adsorption principle can be illustrated in Figure 1. The gas component in a defined adsorption on the adsorbent is a function of temperature and pressure, usually available as shown below those adsorption isotherms. The figure shows the A, B two gases at the same temperature in a certain adsorbent adsorption isotherms on. Obviously, the same pressure A is more easily adsorbed than B. If A and B mixture through the adsorption column filled with the adsorbent, under relatively high pressure PH adsorption, at relatively low pressure PL desorption. The partial pressure of component A easily adsorbed are respectively PAH, and PAL, and the partial pressure of component B hard adsorbed are respectively PBH and PBL. The figure shows that under the relatively high-pressure component A is preferentially adsorbed, while the component B-rich gas stream in the outflow set. It is due to the equilibrium adsorption amount of qAH of component A is much higher than the equilibrium adsorption amount of qBH of component B. To make the adsorbent regeneration, the bed pressure is reduced to PL, the equilibrium adsorption capacity of the component A and B are respectively qAL and qBL. In the process which a new equilibrium is reached, the amount of desorption are qAH-qAL and qBH-qBL. This change in bed pressure periodically, the A, B mixture can be separated.

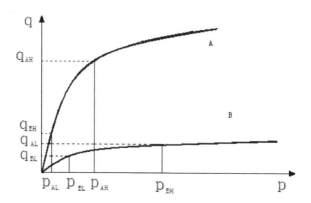

Fig. 1. The basic principle of pressure swing adsorption

Pressure swing adsorption processes rely on the fact that under pressure, gases tend to be attracted to solid surfaces, or "adsorbed". The higher the pressure, the more gas is adsorbed; when the pressure is reduced, the gas is released, or desorbed. PSA processes can be used to separate gases in a mixture because different gases tend to be attracted to different solid surfaces more or less strongly. If a gas mixture such as air, for example, is passed under pressure through a vessel containing an adsorbent bed that attracts nitrogen more strongly than it does oxygen, part or all of the nitrogen will stay in the bed, and the gas coming out of the vessel will be enriched in oxygen. When the bed reaches the end of its capacity to adsorb nitrogen, it can be regenerated by reducing the pressure, thereby releasing the adsorbed nitrogen. It is then ready for another cycle of producing oxygen enriched air.This is exactly the process used in portable oxygen concentrators used by emphysema patients and others who require oxygen enriched air to breathe.

Using two adsorbent vessels allows near-continuous production of the target gas. It also permits so-called pressure equalisation, where the gas leaving the vessel being depressured is used to partially pressurise the second vessel. This results in significant energy savings, and is common industrial practice.

Aside from their ability to discriminate between different gases, adsorbents for PSA systems are usually very porous materials chosen because of their large surface areas. Typical adsorbents are activated carbon, silica gel, alumina and zeolite. Though the gas adsorbed on these surfaces may consist of a layer only one or at most a few molecules thick, surface areas of several hundred square meters per gram enable the adsorption of a significant portion of the adsorbent's weight in gas. In addition to their selectivity for different gases, zeolites and some types of activated carbon called carbon molecular sieves may utilize their molecular sieve characteristics to exclude some gas molecules from their structure based on the size of the molecules, thereby restricting the ability of the larger molecules to be adsorbed.

One of the primary applications of PSA is in the removal of carbon dioxide (CO_2) as the final step in the large-scale commercial synthesis of hydrogen (H_2) for use in oil refineries and in the production of ammonia (NH3). Refineries often use PSA technology in the removal of hydrogen sulfide (H2S) from hydrogen feed and recycle streams of hydrotreating and hydrocracking units. Another application of PSA is the separation of carbon dioxide from biogas to increase the methane (CH4) content. Through PSA the biogas can be upgraded to a quality similar to natural gas. Nitrogen generator units employ the PSA technique to produce high purity nitrogen gas (99.5% or greater) from a supply of compressed air.

2. The part of experiment

2.1 Experimental materials and equipments
Granular activated carbon was from Takeda Pharmaceutical Chemistry Kabushiki Kaisha, Environmental Company). XAD-4, NDA-150 and ND-90 resin were made by Nan Da Ge De Environmental Protection Technology Co Ltd, and their characters were listed in Table 2. High performance liquid chromatography (Waters 600) was manufactured by USA Waters Company. Experimental device of pressure swing adsorption was made by Shanghai Tonguang Technology & Education Equipment Co Ltd..

2.2 Chemical properties of toluene
Toluene, formerly known as toluol, is a clear, water-insoluble liquid with the typical smell of paint thinners. It is a mono-substituted benzene derivative, i.e., one in which a single hydrogen atom from the benzene molecule has been replaced by a univalent group, in this case CH3.It is an aromatic hydrocarbon that is widely used as an industrial feedstock and as a solvent. Like other solvents, toluene is sometimes also used as an inhalant drug for its intoxicating properties; however, inhaling toluene has potential to cause severe neurological harm. Toluene is an important organic solvent, but is also capable of dissolving a number of notable inorganic chemicals such as sulfur.

Toluene reacts as a normal aromatic hydrocarbon towards electrophilic aromatic substitution. The methyl group makes it around 25 times more reactive than benzene in such reactions. It undergoes smooth sulfonation to give p-toluenesulfonic acid, and chlorination by Cl2 in the presence of FeCl3 to give ortho and para isomers of chlorotoluene. It undergoes nitration to give ortho and para nitrotoluene isomers, but if heated it can give dinitrotoluene and ultimately the explosive trinitrotoluene (TNT).With other reagents the methyl side chain in toluene may react, undergoing oxidation. Reaction with basify potassium permanganate and diluted acid (e.g., sulfuric acid) or potassium permanganate with concentrated sulfuric acid, leads to benzoic acid, whereas reaction with chromyl chloride leads to benzaldehyde (Étard reaction). Halogenation can be performed under free radical conditions. For example, N-bromosuccinimide (NBS) heated with toluene in the presence of AIBN leads to benzyl bromide. Toluene can also be treated with elemental bromine in the presence of UV light (direct sunlight) to yield benzyl bromide. Toluene may also be brominated by treating it with HBr and H2O2 in the presence of light.

Toluene is a common solvent, able to dissolve paints, paint thinners, silicone sealants, many chemical reactants, rubber, printing ink, adhesives (glues), lacquers, leather tanners, and disinfectants. It can also be used as a fullerene indicator, and is a raw material for toluene diisocyanate (used in the manufacture of polyurethane foam) and TNT. In addition, it is used as a solvent to create a solution of carbon nanotubes. It is also used as a cement for fine polystyrene kits (by dissolving and then fusing surfaces) as it can be applied very precisely by brush and contains none of the bulk of an adhesive.

Industrial uses of toluene include dealkylation to benzene, and the disproportionation to a mixture of benzene and xylene in the BTX process. When oxidized it yields benzaldehyde and benzoic acid, two important intermediates in chemistry. It is also used as a carbon source for making Multi-Wall Carbon Nanotubes. Toluene can be used to break open red blood cells in order to extract hemoglobin in biochemistry experiments.

2.3 Experimental setup process

Air cames out of the air compressor filter which provides pressure for the experimental device, it filter out the oil, then into the organic gas generating device.The reaction had occurred after installation of benzene into the vapor mixing with air. Organic gases from the device comes into the adsorption column of the pipe with bypass for the determination of adsorption column inlet concentration. Adsorption column filled with resin and adsorption of gas access to outdoors by pipeline in the trachea on the road ,passing through the bypass outlet to determine the concentration after adsorption. Experimental setup is as follows:

2.4 Experimental methods

1. The pretreatment of resin

The three kinds of resins were extracted by Soxhlet extractor with absolute alcohol to get rid of the porogen, catalyst, reaction solvent and other impurities. The process was stoped when the circumfluence liquid was colorless and transparent. Then the resins were washed with distilled water, filtered and dried in the ovens at 60°C after dried in the air. At last they were put into the dryer for a backup

2. Comparison of toluene adsorption between resin and activated carbon

XAD-4, NDA-150, ND-90 resin and activated carbon of same quality were taken as adsorbent, adsorbing 10 min at 0.1MPa pressure and room temperature, with flux of 3L/min, and selected the optimum adsorbing resin by comparing adsorbing effects of 3 types of resin and activated carbon.

3. Experiment of the optimum adsorption flux

The resin and activated carbon were taken as adsorbent of toluene gas, under 0.1MPa pressure and room temperature, in-gas flux of 3, 4, 5, 6, 7L/min. After they adsorbed 10 min, removal rate of toluene in air were tested to confirm the optimum adsorbing flux of resin and activated carbon

4. Experiment on variance of desorption effect under different pressures

Desorbed resin and activated carbon after adsorption by the optimum adsorption flux and time under the pressure of -0.05Mpa, -0.04Mpa, -0.03Mpa, -0.02Mpa, -0.01Mpa, continued stable desorption for 12 min, tested desorption rate of toluene at 4, 6, 8, 9, 10, 11, 12 min, to confirm the optimum desorbing pressure and time

5. Experiment on stability

Under the optimum adsorption and desorption conditions of pressure swing confirmed by the above experiment, 100 batches of experiment on stability were conducted with resin and activated carbon as adsorbent, and removal rate of toluene each time and observed characters of resin and activated carbon were tested.

Gases treated in experiments of 1)-5) were all air containing toluene of 0.5178mg/L.

2.5 Analysis method

Toluene concentration was tested by Waters 600 High Performance Liquid Chromatograph (HPLC), and chromatographic column was C18 reversed phase column. Flowing phase was carbinol: water (80:20). Flow speed: 0.8ml/min, ultraviolet detector, 254nm in wavelength.

To investigate surface area and aperture distribution of resin and activated carbon, we adopted BET method, nitrogen gas as adsorbent, equipment type be Micromeritics ASAP2010 (USA). Infrared Spectroscopy was tested by Fourier transform infrared spectroscopy by using potassium bromide and pressed film method of resin powder. Element analysis was tested by Perkin-Elmer240c (USA) elemental analyzer.

resin	XAD-4	NDA-150	ND-90	granular activated carbon
structure	macroporous adsorption resin	hypercrosslinked	amino-modified hypercrosslinked	carbon build-up
geopolarity	non-polar	weak-polar	mid-polar	weak-polar
BET surface area (m²/g)	850	906	819.1	880
average pore diameter (nm)	5.83	1.7	1.5	4.0
microporous surface area (m²/g)	3.1	561.3	463.3	231.7
micropore volume (mL/g)	0.0051	0.2256	0.2186	0.1048
porosity (%)	40	53	52	42
granularity (mm)	0.4-0.6	0.4-0.6	0.4-0.6	0.4-0.5
oxygen content (%)	0	2.9	1.5	0.2
amino content (mmol/g)	0	0	1.51	0

Table 2. The nature of resin ND-900, NDA-101 and NDA-99

3. Results and discussion

3.1 The comparison of toluene adsorption of resin with activated carbon

Seen from Fig.2 evidently, the adsorption quantity of resin NDA-150 is the highest under the same condition, mainly relating to aperture, specific surface area and polarity of every adsorbent. Toluene is weak-polar molecule, so the adsorbability is minor on the non-polar resin XAD-4. In addition, from table1 we know that the aperture of resin XAD-4 is mainly big pore, but the other three kinds of sorbent have a certain amount of micropores.

Adsorbent with. micropores adsorbs molecule not only depending on the high specific surface area, but also filling function of micro-aperture and capillarity, which both play significant role [8-10], so the adsorption quantity of resin XAD-4 for toluene is the lowest. Resin ND-90 is amino-modifing hyper-cross-linked resin, is mid-polar resin in spite of a lot of micropores, be adverse for the adsorption of toluene for higher polarity, so the adsorption quantity of toluene is less. The polarity of active carbon and NDA-150 matches with the polarity of toluene, the discrepancy of specific surface area between two adsorbents is not big, and the main distinguish is micropore volume. The micropore volume of NDA-150 is more than twice as activated carbon, so the adsorption quantity of resin NDA-150 for toluene is the highest. Therefore, resin NDA-150 and activated carbon were taken as adsorbent in the following experiment.

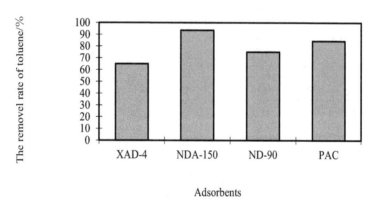

Adsorbents

Fig. 2. The comparison of every adsorbent in toluene adsorption

3.2 The confirm of the best adsorption flow

According to the fig.3, we know that the removal of toluene reduced with the increase of intake flow. Adsorption process is very complicated, generally, can be divided into three phases: ①out-diffusion, gas molecules get on the surface of adsorbent form outside space; ②in-diffusion, gas molecules go deeply into adsorption surface along absorbent channel; ③adsorption on the internal adsorption surface. Toluene molecules have no time to contact with adsorbent fully and be adsorbed when adsorption flux increased, and effuse from the adsorption column with airflow. Thereby, only by controlling the flow can toluene molecular absorbed by adsorbent. From fig.3, we learn that the resin NDA-150 has optimal adsorption effect when the adsorption flow is 3L/min, but the absorption effect at 5L/min has very small difference with that of 3L/min, however, adsorption efficiency reduced evidently when the gas flow increase from 5L/min to 6L/min. There are certain requests for adsorption volume in actual adsorption process, and the adsorption volume of exhaust gas was not great in the adsorption process when the adsorption efficiency is the highest, hence, confirmed the optimal adsorption flow was 5L/min considering the two factors, the absorption efficiency and the adsorption volume disposal gas. For activated carbon, adsorption efficiency declined evidently when the gas flow increased from 4L/min to 5L/min, so the optimal adsorption flow is 4L/min.

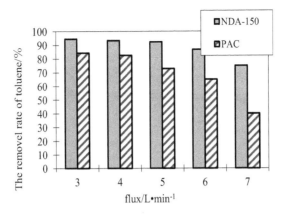

Fig. 3. The adsorption results of different adsorption flows

3.3 The confirm of the best adsorption time

As can be seen from the Fig4, the adsorption efficiency of adsorbent declined as time increased. It will take some contact time for toluene gas molecules entering into micropore of adsorbent; therefore, the longer the time, the much more amount of toluene entering micropore, tends to saturation in the end. Seen from Fig.4, the maximum absorption efficiency of NDA-150 resin and activated carbon are absorption for 5 minutes. There is slight difference of absorption effect between 10 min and 5 min for absorption time. At the same time, absorption efficiency descended significantly when the absorption time varying from 10 min to 12 min. Resin had the maximum adsorption efficiency when the absorption time was 5 min, but the volume of gas that resin dealt with was too small to satisfy the demand for the volume of gas that resin dealt with in actual process. Allowing for the two factors of absorption efficiency and absorption capacity, we determined the best adsorption time of NDA-150 resin and activated carbon were in 10 minutes.

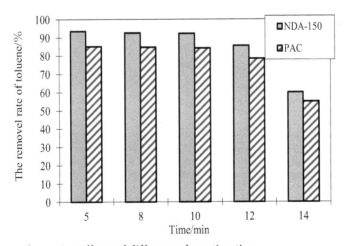

Fig. 4. Different absorption effects of different adsorption time

3.4 The confirm of desorption pressure and desorption time

From Fig. 5 and Fig. 6, we know that desorption efficiency of activated carbon increases with the time in the same condition of desorption negative pressure, and tends to balance gradually. With a certain desorption time, increasing desorption negative pressure makes desorption efficiency increased, and acquires the best desorption efficiency in -0.05MPa.

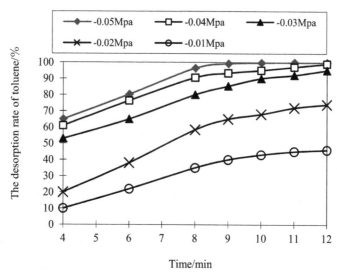

Fig. 5. The variation of desorption result of NDA-150 resin at under different desorption pressure with the time

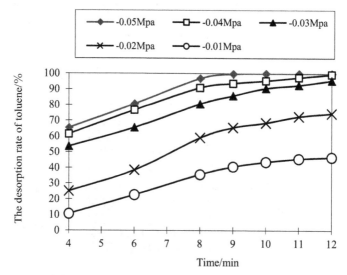

Fig. 6. The variation of desorption result of activated carbon at under different desorption pressure with the time

In theory, the greater desorption negative pressure, the higher desorption efficiency. However, in actual project, the greater negative pressure, the greater requirements of equipment and energy. As a result, we determined the best desorption negative pressure of resin and activated carbon is 0.05Mpa from this experiment. In the desorption pressure of - 0.05Mpa, desorption rate could reach 99% after 9 minutes, while desorption efficiency didn't increase significantly when desorption time varied from 10 minutes to 12 minutes, closing to balance. Accordingly, chose the optimum desorption time for 10 minutes, and desorption efficiency were 99.8 percent at this point.

3.5 The stability test

It can be seen from the results of stability (listed in table 3): because of the weak mechanical strength of activated carbon, the more the number of repeated experiments, the greater impact under pressure, much seriously damaged, seriously affected its treatment. It is only by adding new activated carbon in time in order to guarantee treatment results. Conversely, the resin is synthetic organic polymer, has powerful mechanical strength. It didn't have damage within 100 batches of experiments and effect on the absorption efficiency of resin for toluene. Therefore, it is feasible to treat organic waste gas using resin as absorbent.

The number of experiments	Toluene removal (%)		Character	
	activated carbon	NDA-150 resin	activated carbon	NDA-150 resin
20	83.5	92.3	intact	intact
40	82	92.5	A portion of activated carbon had crack, and small particles of activated carbon emerged	intact
60	79.5	92	The amount of activated carbon that had cracking increased, walls of adsorption column are stuck by activated carbon powder.	intact
80	75	92	A small amount of active carbon powder appears at the bottom of adsorption column	intact
100	70	92.4	The quantity of activated carbon powder increased	intact

Table 3. The results of stability

4. Conclusions

Having studied pressure swing adsorption for purifying toluene gas of low concentration by using activated carbon and resin as absorbent respectively, the experimental results indicate:

NDA-150 resin has the best adsorption effect on adsorption of air containing toluene.

The optimum adsorption flow of resin and activated carbon is 5L/min and 4L/min respectively, the optimum adsorption time both are 10min, and the best adsorption efficiency respectively is 92.3% and 83.5%.

With desorption negative pressure be -0.05Mpa, the optimal desorption time of the resin and activated carbon both are 10min, and desorption rates can reach 99%.

It can be seen from the results of the stability test that resin as absorbent superior to activated carbon in treating air including toluene.

5. Acknowledgment

This work was funded by the Hebei Province Natural Science Foundation (grant no. E2009000581). The authors wish to express their appreciation to the Analytical Center at Changzhou Petro-Chemical College for the measurement of specific surface area and pore diameter.

6. References

Faisal I. Khan, Aloke Kr. Ghoshal, Removal of Volatile Organic Compounds from Polluted Air,*Journal of Loss Prevention in the Process Industries*, vol. 13, pp. 527-545, May(2000)

Zhang Lin, Chen Huanlin, Cai Hong,Study Progress of Membrane Method Process of Volatile Organic Compounds Emissions,*Environmental Protection of Chemical Industry*, vol. 22, pp. 75-80, February 2002.

You Yongyan, Chen Fanzhi, Huang Shujie,Research Development of Activated Carbon Fiber Desorbing Volatile Organic Compounds, *Industry Catalysis*, vol. 14, pp. 63-66, April 2006

Wang Yanfang, Sha Haolei, Yu Jianming,Technology Progress of VOCs Waste Gas of Low Concentration, *Energy environmental protection*, vol. 21, pp. 8-12, March 2007

Sun Yue, Zhu Zhaolian, Resin Adsorption sulfa deal with the production of intermediate waste, *Chemical Environmental Protection*, vol. 23, pp.9-14, January 2003

Wang Xuejiang, Zhang Quanxing, Li Aimin, Research on the Adsorption Behavior of NDA-100 Macroporous Resin to Salicylic Acid in Water Solution, *Journal of Environmental Science*, vol. 22, pp. 658-660, May 2002

Karan.l, T., Mehmet, K., Kildu., J.E., Wigton, A., Role of granular activated carbon surface chemistry on the adsorption of organic compounds: 2. NOM,*Envrion. Sci. Technol.*, vol. 33, pp. 3225-3233, October 1999

Wei Ruixia, Chen Jinlong, Chen Lianlong, Fei Zhenghao, Li Aimin, Zhang Quanxing, Study of adsorption of lipoic acid on three types of resin,*Reactive and Functional Polymers*, vol. 59, pp. 243-252, November 2004

7

Electrochemical Incineration of Organic Pollutants for Wastewater Treatment: Past, Present and Prospect

Songsak Klamklang[1], Hugues Vergnes[2],
Kejvalee Pruksathorn[3] and Somsak Damronglerd[3]
[1]Technology Center, SCG Chemicals Co., Ltd., Siam Cement Group (SCG), Bangkok
[2]Laboratoire de Génie Chimique, UMR CNRS 5503, BP 84234, INP-ENSIACET
[3]Department of Chemical Technology, Faculty of Science,
Chulalongkorn University Bangkok
[1,3]Thailand
[2]France

1. Introduction

Water is a combination of two parts, hydrogen and oxygen as H_2O. However, pure water is only produced in a laboratory, water in general is not pure composition of hydrogen and oxygen. Eventhough, distilled water still has measurable quantities of various substances such as ions, mineral or organic compounds (http://www.environmental-center.com). These substances should be considered as the impurities that dissolved into water during flow through hydraulic pathway. Nowadays, there are some increasing on both population and consumption of natural resources to serve endless needs. Water is most important resource and becomes limited of use due to contamination from discharge of both domestic and industry. The discharge of domestic wastewater contains a large amount of organic pollutants. Industry also contributes substantial amounts of organic pollutants. However, some organic substrates discharged from industry contain a high toxicity and refractory organic pollutants.

Figure 1 presents the example of a partially closed water cycle. In the cycle the organic pollutants are neither removed by sorption nor biodegradation. Nevertheless, there are some organic pollutants pass all barriers such as wastewater treatment or underground passage and appear in raw waters used for drinking water production. The other group of organic pollutants may originate from consumer products used in household, pesticides applied in agriculture or chemicals used in industry (Reemtsma & Jekel,2006).

Wastewater treatment consists of applying known technology to improve or upgrade the quality of a wastewater. Usually wastewater treatment will involve collecting the wastewater in a central, segregated location and subjecting the wastewater to various treatment processes (Hanze et al., 1995). Wastewater treatment can be organized or categorized by the nature of the treatment process operation being used such as physical, chemical or biological treatment. Biological treatment of polluted water is the most economical process and commonly used for the elimination of degradable organic pollutants present in wastewater. However, situation fully differs when the wastewater contains toxic and refractory substrates, biological treatment may not suitable tool for treatment of contaminated wastewater (Grimmet al., 1998).

This chapter review related technologies and case studies of application of electrochemical incineration in industrial and restaurant wastewater treatment.

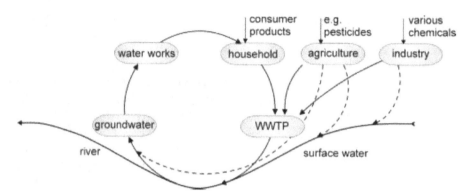

Fig. 1. Schematic of water cycle (Reemtsma & Jekel,2006).

2. Electrochemical incineration

Electrochemistry is a clean, versatile and powerful tool for the destruction of organic pollutants in wastewater. Electrochemical oxidation of organic compounds in aqueous solution is an anodic process occurring in the potential region of water discharge to produce oxygen (Kapalka et al., 2009).

Two different pathways are described in the literatures for the anode oxidation of undesired organic pollutants (Grimm et al., 1998). Electrochemical conversion transforms only the toxic pollutants refractory to biological treatments into biocompatible organics, so that biological treatment is still required after the electrochemical oxidation (Comninellis., 1994). The ideal electrode material which can be used in the electrochemical conversion method must have high electrochemical activity for aromatic ring opening and low electrochemical activity for further oxidation of the aliphatic carboxylic acids which are in general biocompatible (Comninellis., 1994).

Electrochemical incineration or combustion is method completely oxidizes the organic pollutants to CO_2 by physisorbed hydroxyl radicals. In this case, the electrode material must have high electrocatalytic activity towards the electrochemical oxidation of organics to CO_2 and H_2O (Comninellis., 1994).

2.1 Mechanism of electrochemical incineration

Study on electrochemical oxidation for wastewater treatment goes back to the 19th century, when electrochemical decomposition of cyanide was investigated (Kuhn., 1971). Extensive investigation of this technology commenced since the late 1970s (Chen., 2004). During the last two decades, research works have been focused on the efficiency in oxidizing various pollutants on different electrodes, improvement of the electrocatalytic activity and electrochemical stability of electrode materials, investigation of factors affecting the process performance, and exploration of the mechanisms and kinetics of pollutant degradation.

Experimental investigations focus mostly on the behaviors of anodic materials, the effect of cathodic materials was not investigated extensively although Azzam et al. (Azzam et al, 1999) have found a considerable influence of the counter electrode material in the anodic destruction of 4-chlorophenol.

Comninellis (Comninellis., 1994) has presented the mechanism of electrochemical oxidation and it was used as fundamental of electrochemical wastewater treatment. According to the electrochemical conversion and combustion of organics is presented on metal oxide anode (MO_x). H_2O in acid or OH^- in alkali solution is discharged at the anode to produce adsorbed hydroxyl radical according to the equation (1).

$$MO_x + H_2O \xrightarrow{\ k_1\ } MO_x(^\bullet OH) + H^+ + e^- \tag{1}$$

k_1 = Electrochemical rate constant for H_2O discharge

Secondly, the adsorbed hydroxyl radicals may interact with the oxygen already present in the metal oxide anode with possible transition of oxygen from the adsorbed hydroxyl radical to the lattice of the metal oxide anode forming the so-called higher oxide (MO_{x+1}) as represented in equation (2).

$$MO_x(^\bullet OH) \xrightarrow{\ k_2\ } MO_{x+1} + H^+ + e^- \tag{2}$$

k_2 = Electrochemical rate constant for transition of oxygen into oxide lattice

Thus, It could be considered that at the anode surface, two states of active oxygen can be presented in physisorbed active oxygen (adsorbed hydroxyl radicals, $^\bullet OH$) and chemisorbed active oxygen (oxygen in the oxide lattice, MO_{x+1}).

In the absence of any oxidizable organics, the physisorbed and chemisorbed active oxygen produce dioxygen according to the equation (3) and (4).

$$MO_x(^\bullet OH) \xrightarrow{\ k_0\ } \frac{1}{2}O_2 + H^+ + e^- + MO_x \tag{3}$$

k_0 = Electrochemical rate constant for O_2 evolution

$$MO_{x+1} \xrightarrow{\ k_d\ } MO_x + \frac{1}{2}O_2 \tag{4}$$

k_d = Electrochemical rate constant for O_2 evolution

In the presence of oxidizable organics, it is speculated that the physisorbed active oxygen ($^\bullet OH$) should cause predominantly the complete combustion of organics according to equation (5), and chemisorbed active oxygen (MO_{x+1}) participate in the formation of selective oxidation products as represented in equation (6).

Complete combustion:

$$R + MO_x(^\bullet OH) \xrightarrow{\ k_c\ } CO_2 + zH^+ + ze^- + MO_x \tag{5}$$

k_c = Electrochemical rate constant for the combustion of organics

Selective oxidation:

$$R + MO_{x+1} \xrightarrow{\ k_s\ } RO + MO_x \tag{6}$$

k_s = Electrochemical rate constant for the selective oxidation of organics

2.2 Electrochemical incineration performance

There are several researchers' work on the parameters that affect on the electrochemical oxidation process efficiency. In this topic, some important parameters for electrochemical treatment by electrochemical incineration have been reviewed.

Support materials

Vercesi (Vercesi et al, 1991) worked on searching for a good Dimensionally Stable Electrode (DSA) that has long service life for O_2-evolution based on the effects of support materials. Titanium, tantalum, zirconium, niobium and some of their alloys were used as support materials on the performance of IrO_2-Ta_2O_5 coated electrodes. The thermal behavior and oxygen affinity sequence of the metals, in relation to the electrode preparation procedure, were determined thermogravimetrically. The electrochemical corrosion of base metals was represented in Figure 2, the tantalum-based materials presented the highest stability with the minimum corrosion rate. The chemical and electrochemical stability of the base metals was found to be directly related to the service life of the electrode, measured in 30% H_2SO_4 at 80°C and 750 mA/cm². Tantalum-based electrodes represented the highest service life, 1700 h and 120 h for titanium-based electrodes.

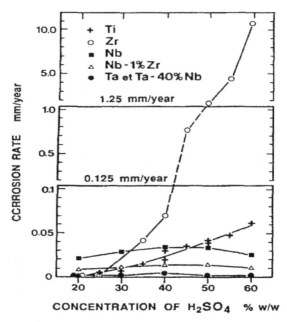

Fig. 2. Electrochemical corrosion rate of base metals as a function of H_2SO_4 concentration at anode potential of 2 V/SCE (Vercesi et al, 1991).

Coating materials

The active coating materials of electrodes are very important for pollutants degradation in electrochemical oxidation process. It could point the project to be benefit or insolvent. The coating enables the electrical charge transport between the base metal and the electrode/electrolyte interface. It must have high chemical and electrochemical stability and able to catalyse the desired electrochemical reaction.

There are some works focusing on the active coating materials for electrochemical degradation of organic pollutants in wastewater. Comninellis (Comninellis & Vercesil, 1991) found that the Ta_2O_5-doped IrO_2 represented the highest service life. However, the same author (Comninellis., 1994) proposed in 1994 that the mechanism of electrochemical oxidation of organic pollutant on IrO_2 electrode was the electrochemical selective oxidation while that on SnO_2 electrode was the electrochemical combustion. On SnO_2 electrode, the organic pollutants were oxidized to CO_2 and water. The SnO_2 electrode presents the highest current efficiency. Because of its highest oxidation state which contains excess oxygen in the oxide lattice as possible. At SnO_2 surface, the hydroxyl radicals are accumulated and favor the combustion of organics while the other oxides having a high concentration of oxygen vacancies and can favor selective oxidation of organics more than combustion.

Not only SnO_2 electrode has the good attractive to be used as electrode for wastewater treatment in electrochemical oxidation, the diamond thin films have been widely used in electrochemical studies due to the unique properties like chemical stability, large potential range and mechanical resistance. Their applications embrace the electroanalysis, electrosynthesis, fuel cell and the organic pollutant degradation in wastewater (Diniz et al, 2003). Some studies showed that conducting diamond electrodes could be grown by energy-assisted (plasma or hot-filament) chemical vapor deposition on several substrates, such as silicon, titanium, niobium, tantalum, molybdenum, glassy carbon. Although all these substrates are currently used, they still have some drawbacks. In fact, a silicon substrate is very brittle, Nb, Ta, W are too expensive and the stability of the diamond layer deposited on the Ti substrate is still not satisfactory, because cracks may appear and cause the detachment of the diamond film during long-term electrolysis (Panizza & Cerisola, 2005). The organic compound electrochemical oxidation efficiency strongly depends on the used anode material and diamond is very interesting due to its superior properties (Diniz et al, 2003). In recent years, there are some publications on the application of diamond electrode for wastewater treatment (Diniz et al, 2003, Lissens., 2003). However, the application of diamond electrode for organic pollutant degradation still limited due to the cost of diamond electrode is too expensive.

pH
Normally, the efficiency of the oxidation of organics tends to be superior in alkaline solution. That also holds for the anodic treatment using standard electrode materials. The wastewater accessible to treatment process has any pH and pH adjustment before treatment to the more favorable value above 7 will be too expensive. Moreover, Stuki (Stucki et al, 1991) proved that on SnO_2 electrodes, the degradation of benzoic acid was pH independent. Cañizares (Cañizares et al, 2004) proposed that the pH does not influence in the global oxidation rate even if initial oxidation rate was higher in alkaline media. However, after the galvanostatic method development, the oxidation rate in acidic media surpasses those in the alkaline media because the accumulation of oxalic acid in alkaline media was higher than that in acidic media, due to its lower oxidazability at alkaline conditions.

Current density
The effect of current density was studied many times, Comninellis (Comninellis & Nerini, 1995) found that the degradation of phenol was independent from the current density and that the phenol elimination depends only on the specific electrical charge. Almost complete phenol elimination can be archived after the passage of an electrical charge of 17-20 Ah/dm^3.

The higher current densities increase the initial reaction rate ($dCOD/dt$) (COD : Chemical Oxygen Demand) but decrease lightly the initial efficiencies of the process ($dCOD/dQ$) (Q : charge). Nevertheless, both treatments end nearly at the same charge passed. This behavior is a characteristic of electrochemical systems in which both direct and mediated oxidation reactions play an important role. But the low current density experiment achieves initially a higher mineralization rate. This fact can be easily explained that the amount of oxalic acid, that is an end of chain product and difficult to destroy, accumulated in the high current density experiment is greater than those obtained in the low current density experiments (Cañizares et al, 2004).

Although, the current density might not affect on the kinetic of surface electrochemical oxidation of organic pollutant, it also enhances the production of bulk chemicals, which may contribute to the degradation process through parallel reaction schemes.

Klamklang (Klamklang, 2006, Klamklang, 2007 and Klamklang et al, 2010) has found that increasing current density from 5 to 10 mA/cm2, leads to less degradation rate of oxalic acid by electrochemical oxidation. This behavior is characteristic of mass transfer-controlled processes (Kesselman, 1997). In such systems, the increase of current density cannot increase the organic removal efficiency at the electrode, but only favors oxygen evolution as the anodic side reaction which hides the electrode and prevents contact between hydroxyl radicals and organic pollutants. When the system does not generate only adsorbed hydroxyl radicals or other active oxygen, the decrease of organic pollutant removal efficiency is observed.

Temperature

An increase in the temperature leads to more efficient processes by global oxidation. While direct oxidation processes remain almost unaffected by temperature, this fact may be explained in terms of the presence of inorganic electrogenerated reagents. The oxidation carried out by these redox reagents is a chemical reaction. Consequently, its rate normally increases with temperature. But the oxidation process can be carried out either at the electrode surface and by electrogenerated reagents-mainly hypochlorite and peroxodisulphates. However, the new organic intermediates are not formed with the increase of temperature, indicating that the process mechanisms do not vary with temperature (Cañizares et al, 2004).

2.3 Electrocatalytic electrodes

Electrochemical degradation of organic pollutants in the wastewater needs specific electrodes. Couper (Couper et al, 1990) has reviewed some properties of typical substrates used as electrocatalytic electrodes and report that electrode must have low voltage drop through the substrates and substrate-solvent interface. Many metallic electrodes could answer these criteria and many alloys can also be used as good composite electrodes covered with an active layer and a long service life of electrodes.

The complexity of electrode behaviors and our lack of detailed insights make it impossible to select the optimum electrode for a given process on a theoretical basis. Instead, an empirical approach must be used. The initial selection is based on process experience, and this is then tested and refined during an extensive development program. Indeed, it is very difficult to predict the success of an electrode material or to define its lifetime without extended studies under realistic process conditions. Accelerated testing is rarely satisfactory except to indicate catastrophic failure.

There are some general guidelines to assist the choice of an electrode material;

Physical stability
The electrode material must have adequate mechanical strength, must not be prone to erosion by the electrolyte, reactants, or products, and must be resistant to cracking.

Chemical stability
The electrode material must be resistant to corrosion, unwanted oxide or hydride formation, and the deposition of inhibiting organic films under all conditions (e.g., potential and temperature) experienced by the electrode.

Suitable physical form
It must be possible to fabricate the material into the form demanded by the reactor design, to facilitate sound electrical connections, and to permit easy installation and replacement at a variety of scales. The shape and design of the electrode may take into account the separation of products, including the disengagement of gases or solids.

Rate and product selectivity
The electrode material must support the desired reaction and, in some cases, significant electrocatalytic properties are essential. The electrode material must promote the desired chemical change while inhibiting all competing chemical changes.

Electrical conductivity
This must be reasonably high throughout the electrode system including the current feeder, electrode connections, and the entire electrode surface exposed to the electrolyte. Only in this fashion it is possible to obtain a uniform current and potential distribution as well as to avoid voltage losses leading to energy inefficiencies.

Cost and lifetime
A reasonable and reproducible performance including a lifetime probably extending over several years must be achieved for an acceptable initial investment.

It is important to note that the choice of working and counter electrodes cannot be made independently since the chemistry at each has consequences to the solution composition throughout the cell. Indeed, the selection of electrode material and its form must be an integrated decision within the prospective of the cell and process design. In some cases such as the manufacture of pharmaceutical products, the electrodes and their compounds must have a low toxicity.

This section has reviewed the ways in which the choice of electrode material influences the design of electrochemical reactors and process performance.

Energy Consumption
The specific energy consumption should be minimized in order to minimize the power costs. In general, the total power requirement has contributions for both electrolysis and movement of either the solution or the electrode. The design of electrodes and cell has an important role in reducing each of these components. Thus, a very open flow-through porous electrode will have a low pressure drop associated with it, giving rise to modest pumping costs and facilitating reactor sealing. A high surface area electrode which itself a turbulence promoter in bed electrode, will give rise to a moderately high mass transfer coefficient and active area without the need for high flow rates through the cell; the pumping cost will again be moderately low.

The direct electrolytic power could be minimized by
- Obtaining a current efficiency approaching 1.0
- Minimizing the cell voltage.

It is therefore important to select the electrode material and operating conditions so as to maintain a high current efficiency. This also assists the operation of the process by reducing the amount of product purification that is necessary and/or byproducts that must be handled.

The cell voltage is a function of the reversible cell voltage, the over-potentials at the two electrodes, and ohmic drops in the electrolyte, the electrodes, busbars etc., and any separator in the cell. Again, the maintenance of a low cell voltage demands attention to the design of both electrodes and cell. Where possible, the following features should be included:

-   The counter electrode reaction should be chosen so as to minimize the reversible cell voltage. This requires the availability of a suitably stable electrode material.
-   The over-potentials at both electrodes should be minimized by the use of electrocatalysts.
-   The electrodes, current feeders, and connectors should be made from highly conducting materials to lower ohmic drops.
-   The electrodes should facilitate low IR drop in the electrolyte by, for example, allowing efficient gas disengagement and passage out of the interelectrode gap. Meshes as well as louvred and lantern blade electrodes can be used.
-   Electrode and cell design should allow a small interelectrode or electrode membrane gap. In the limit, the electrode may touch the membrane as in zero-gap or solid polymer electrolyte cells.
-   A separator should be avoided by suitable selection of the counter electrode chemistry or, if essential, a thin conductive membrane should be used.

Current efficiency

Current efficiency is the fraction of the total charge passed that is used in the formation of the desired product. This can be a strong function of electrode material, e.g., because of differences in the rate of hydrogen evolution as a competing reaction. Competing reactions can also lead to the corrosion and/or erosion of the electrode material as well changes to the electrode (e.g., by formation of a hydroxide or oxide or the deposition of another metal onto the surface).

Material Yield

This is the fraction of the starting material that is converted into the desired product. This is also dependent on electrode material in many cases. Values less than one indicate byproducts and hence perhaps the need to introduce additional purification steps that inevitably increases the complexity of the overall process and costs.

Space-Time Yield

One of the most valuable statements of reactor performance is the space-time yield or weight of product per unit time per unit volume of reactor. It is determined by the current density, the current efficiency, and the area of electrode per unit volume of cell, all dependent on the electrode material and its form. Commonly, the cell is operated in conditions where the electrode reaction is mass-transport controlled (especially when a high fractional conversion is desirable or when the concentration of reactant is limited by solubility or process considerations. Then, the current density is determined by the concentration of reactant and the mass transport condition. The latter is therefore frequently enhanced by the use of high flow rates, turbulence promoters, and/or electrode movement.

The current is proportional to the active electrode area in the cell. A compact cell design requires a high area per unit cell volume. This suggests the use of a three-dimensional electrode but such electrodes make it difficult to maintain a uniform fluid flow and electrode potential, i.e., to control the reaction environment. Hence, the use of porous, flow-through electrodes often involves a trade-off between enhanced electrode area and material yield and/or current efficiency.

Other Factors
Of course, other factors are important in the design of electrodes and cells. These include cost, safety, ease of maintenance, and convenience to use. It is also essential that the performance of the electrodes is maintained throughout the projected operating life of the cell, maybe several years. Examples of problems that frequently arise include (a) deposition onto cathode surfaces of hydrogen evolution catalysts due to trace transitional metal ions in the electrolyte and (b) poisoning of PbO_2 anodes by organic molecules leading to enhanced corrosion as well as oxygen evolution.

3. Application of electrochemical incineration in wastewater treatment

3.1 Electrochemical incineration for restaurant wastewater treatment
There are a lot of restaurants, cafeterias and food centers in the big cities, which everyday make large amounts of wastewater. Generally, there is unavailable of on-site treatment for each restaurant. The direct discharge of wastewater from these restaurants and food shops to the drainage system is a huge problem to the municipal wastewater collection and treatment works. The oil and grease contained in the wastewater aggregate and foul the sewer system and generate an unpleasant odor (Chen et al, 2000).

Basically, restaurant wastewater treatment facilities must be highly efficient in removing oil and grease, cause no food contamination and be compact size. Low capital and operating costs are important because profit margins of most restaurants are small. In addition, the technology has to be simple so that it can be operated easily either by a chef or a waiter (Chen et al, 2000).

Conventional biological processes are therefore ruled out due to the requirement of large space, long residence time and skilled technicians. Chemical coagulation/settlement is not practicable because of the low efficiency in removing light and finely dispersed oil particles and possible contamination of foods by chemicals. The G-bag approach, which uses a bag of absorbent to capture the pollutants and degrade the pollutants with the immobilized microorganisms on the absorbent, seems to be a good alternative only if the system can be designed as simple and free from fouling (Chen et al, 2000).

In this work, the treatment of restaurant wastewater by electrochemical incineration is obtained with continuous electrochemical oxidation system described elsewhere Klamklang (Klamklang, 2006, Klamklang, 2007 and Klamklang et al, 2010). A simple three-electrode electrochemical reactor with 18 ml of capacity with using Ti/SnO_2 electrodes as anode, the 316L stainless steel was used as cathode. The feed solution was fed to the reactor by peristaltic pump and the effluent was collected at sample trap. The apparatus is presented in Figure 3. (Klamklang, 2006). The operating conditions were represented in Table 1.

Parameter	Operating condition
Current density	5-10 mA/cm^2
SnO$_2$ film thickness	1.8-3.6 micron
Residence time	2-3 hr
Elapse time	24 hr
Stirring	300 rpm

Table 1. Operating conditions for continuous electrochemical oxidation.

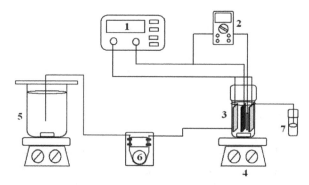

1) Power supply, 2) Voltmeter, 3) Electrochemical reactor set, 4) Magnetic stirrer, 5) Feed reservoir, 6) Peristaltic pump, 7) Sample trap

Fig. 3. Schematic diagram of continuous electrochemical oxidation apparatus.

The experiments performed in a continuous mixed flow reactor were carried out for the determination of the effects of the current density, residence time and SnO_2 film thickness on organic pollutant degradation. Due to the very small electrode area and easy to observe the change of Total Organic Carbon (TOC), the wastewater, which feed to the system, was diluted to around 140 mg TOC/L. The investigated current densities were 5 and 10 mA/cm^2 and residence times were 2 and 3 hr.

Influence of current density

The influence of current density in continuous mixed flow experiments is presented in part A of the Figure 4. The electrochemical degradation of organic pollutants presented in actual restaurant wastewater takes place slowly and its TOC removal efficiency presented in part B of this figure is higher at lower current density. The gain in efficiency being overwhelmed by the lower current values applied. This result may not be surprising on the basis of the previously discussed influence of current density in batch experiments, which indicated to a weak behavior for the characteristic of diffusion-controlled processes. Increase in current density cannot increase the organic removal efficiency at the electrode, but only favours the anodic side reaction which decreased the organic pollutant removal efficiency. It agrees with Figures 5 (A) and (B) that the destruction of organic pollutants in term of Chemical Oxygen Demand (COD) was decreased with increasing of the current density from 5 to 10 mA/cm^2. The equilibrium efficiencies of both TOC and COD removal were 62% when current density was 5 mA/cm^2, while their removal efficiencies were 47% when the current density was 10 mA/cm^2.

Influence of residence time

The presented results in batch experiments show that the increasing of residence time after first 2 hr was not greatly affect on the organic pollutant degradation efficiency due to the change reaction order from zero-order to first-order reaction with reduction of TOC. However, it would be of practical interest to test how much an increase or decrease in the wastewater flow rate affects the TOC removal of the restaurant wastewater. This is demonstrated in Figures 6 and 7. Because of fixed total volume of the continuous mixed flow reactor at 18 ml, an increase in the wastewater flow rate from 0.10 to 0.15 ml/min translates to a proportional decrease in the wastewater hydraulic residence time from 3 to 2 hr.

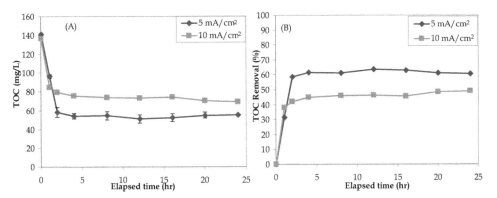

Fig. 4. Effect of current density on TOC removal (A) and TOC removal efficiency (B) in continuous restaurant wastewater treatment by using of $SnO_2/Ir/Ti$ electrode, SnO_2 thickness of 1.8 micron.

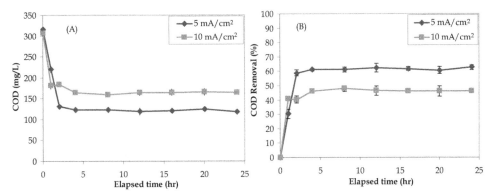

Fig. 5. Effect of current density on COD removal (A) and COD removal efficiency (B) in continuous restaurant wastewater treatment by using of $SnO_2/Ir/Ti$ electrode, SnO_2 thickness of 1.8 micron.

Normally, a reduction in residence time would expectedly lead to a decrease in the wastewater TOC removal. But, in this case, increasing of residence time does not proportionally increase TOC removal. As seen in Figure 6 ((A) and (B)), the TOC removal increases from around 55 to 62 % with the increase in the residence time from 2 to 3 hr. These results were also observed in the removal of COD and represented in Figure 7 (A) and (B). The COD removal increased from around 54 to 62 % with the increase in residence time from 2 to 3 hr.

It could be explained by the increasing of residence time from 2 to 3 hr has not strongly affected on the TOC and COD removal due to the fast reaction with zero-order reaction occurred in the first 2 hr. Then, the reaction was changed to the slower step with the first-order reaction as we found in the batch experiments.

Hence, it would be more economical to operate the electrochemical treatment at a lower residence time as long as the pollutant concentration of the treated wastewater meets the safe discharge requirement.

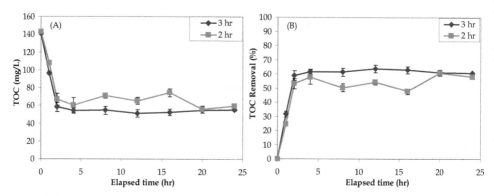

Fig. 6. Effect of residence time on TOC removal (A) and TOC removal efficiency (B) in continuous restaurant wastewater treatment by using of $SnO_2/Ir/Ti$ electrode, SnO_2 thickness of 1.8 micron and current density 5 mA/cm^2.

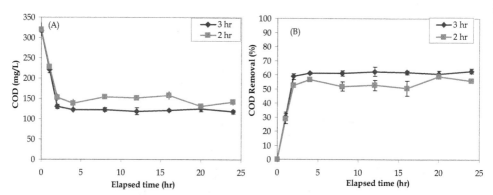

Fig. 7. Effect of residence time on COD removal (A) and on COD removal efficiency (B) in continuous restaurant wastewater treatment by using of $SnO_2/Ir/Ti$ electrode, SnO_2 thickness of 1.8 micron and current density of 5 mA/cm^2.

Influence of SnO_2 active layer thickness

Figures 8 (A) and (B) represent the effect of SnO_2 film thickness on the TOC degradation performance in continuous electrochemical oxidation. Similar to the pollutant degradation of organic pollutant in batch experiment, it shows that the SnO_2 active layer thickness was not a great influence on the TOC removal efficiency because the adsorbed hydroxyl radicals for organic pollutant degradation were produced only at the surface of electrode. However, the TOC removal efficiency was around 62% with the 1.8 micron of SnO_2 active layer while the efficiency was reduced to 51% with the SnO_2 active layer thickness of 3.6 micron. It agrees with the removal of COD from restaurant wastewater as presented in Figure 9. The COD removal efficiency was 62% when the thickness of SnO_2 active layer was 1.8 micron. However, the efficiency was decreased to 50% when the thickness of SnO_2 active layer was 3.6 micron. It should be explained that thickness of 3.6 micron has bigger grain size that leads to a less surface area; therefore, the reaction kinetic was decreased as found previously in the batch experiment.

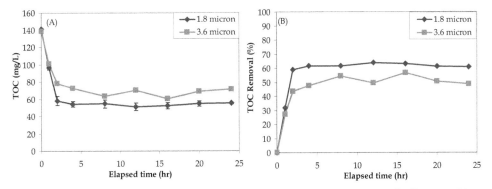

Fig. 8. Effect of SnO₂ layer thickness on TOC removal (A) and TOC removal efficiency (B) in continuous restaurant wastewater treatment by using of $SnO_2/Ir/Ti$ electrode and current density of 5 mA/cm².

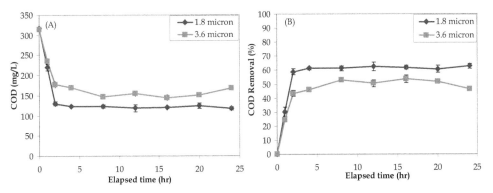

Fig. 9. Effect of SnO₂ layer thickness on COD removal (A) and COD removal efficiency (B) in continuous restaurant wastewater treatment by using of $SnO_2/Ir/Ti$ electrode and current density 5 mA/cm²

3.2 Color removal of pulp and paper mill wastewater by electrochemical incineration

Pulp and paper mill are considered as high water consumption and high wastewater discharge. The pollutants in pulp and paper mill industry are mainly presented in form of high-strength chemical oxygen demand (COD) and biochemical oxygen demand (BOD). Generally, the high-strength BOD and COD are totally removed in biological wastewater treatment unit. The high-strength color from lignin in pulp mill effluent is big problem for pulp and paper mill due to requirement of advanced treatment process with high investment and operating cost. In this chapter, the case study of high efficiency color removal by electrochemical incineration in pulp and paper mill wastewater treatment is explained. The electrochemical incineration is obtained with continuous electrochemical oxidation in a series of 4-simple 3-electrode electrochemical reactor with 5 liters of capacity. The Ti/IrO₂-RuO₂ mixed oxide electrodes were used as anode, the 316L stainless steel was used as cathode. The pulp and paper effluent is fed to the reactor by peristaltic pump and the effluent was collected at sample trap. The apparatus is presented in Figure 10.

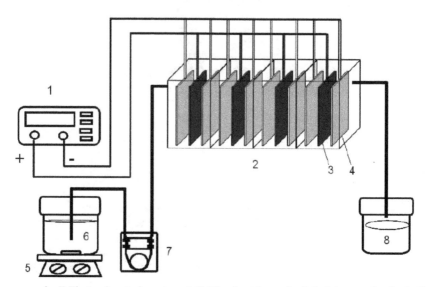

1) Power supply, 2) Electrochemical reactor set, 3) Mixed oxide anode, 4) Stainless steel cathode, 5) Magnetic stirrer, 6) Feed reservoir, 7) Peristaltic pump and 8) Sample trap

Fig. 10. Schematic diagrams of continuous electrochemical oxidation apparatus

Figure 11 show that electrochemical incineration is a powerful tool for color removal from pulp mill wastewater. The results presented that the color substrates in pulp mill effluent was immediately removed. The wastewater color was reduced from 938 Unit Pt/Co to 10-45 Unit Pt/Co or 95-99 % removal that has very good advantage over conventional biological wastewater treatment process that takes more that 24 hour for reducing pulp mill effluent color from 1,500 to 1,100 Unit Pt/Co. Figure 12 presents the pulp mill at various treatment time for different residence time after started electrochemical incineration process. It is clearly that electrochemical incineration is suitable for breaking color substrates in high color strength wastewater.

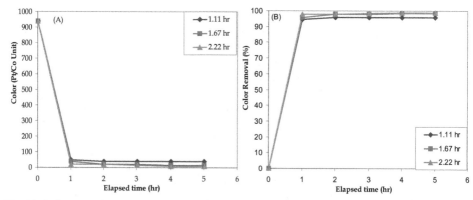

Fig. 11. Influence of treatment time on pulp mill effluent color removal by electrochemical incineration (A) Wastewater color and (B) Color removal efficiency.

Fig. 12. Electrochemical incineration of pulp mill effluent (A) Residence time of 1.1 hr, Anode-cathode distance of 1.2 cm, Treatment cost of 0.175 USD/m^3 and Color removal efficiency of 96%, (B) Residence time of 1.67 hr, Anode-cathode distance of 1.2 cm, Treatment cost of 0.255 USD/m^3 and Color removal efficiency of 98% and (C) Residence time of 2.22 hr, Anode-cathode distance of 1.2 cm, Treatment cost of 0.315 USD/m^3 and Color removal efficiency of 99%

4. Prospect & conclusion

Water treatment by electricity was used for several years, but electrochemical water or wastewater technologies are not yet a mature application in commercial scale by the limitation of relatively high capital investment and the expensive electricity supply. Nowadays, there are increasing on standard for water discharge and limitation of water resource that makes electrochemical treatment comparable with other water and wastewater treatment technologies. However, to mature the electrochemical water and wastewater treatment there are some requirements need to be fulfilled. SWOT (Strengths, Weaknesses, Opportunities and Threats) is commonly used for business analysis of internal and external factors that should support and sustain or collapse business project (Pinson, 2008). SWOT could help to pin point and predominate advantages of electrochemical technology over conventional technologies and drown out disadvantages. Side reactions are a killing trap of electrochemical technology that limits the efficiency of electrochemical systems. In electrochemical incineration, specific coating materials need to be developed to decrease side reaction and to promote desired reaction for attractive benefit and return of investment. During past two decades there are some development in coating technology such as chemical vapor deposition and thermal spray coating that should support the electrode fabrication with appreciate on both constant coating layer and long service life that will response to good return of investment. Electrochemical reactor design is big challenges in electrochemical technology due to it will response on mass transfer during operation. The lack of mass transfer and current distribution in electrochemical reactor will collapse all advantages of electrochemical technology in the battle of technologies.

5. References

Azzam, M.O., Tahboub, Y. and Al-Tarazi, M. (1999). Effect of counter electrode material on the anodic destruction of 4-Cl phenol solution, *Process Safety and Environmental Protection*, Vol 77, N° B4, pp 219-226, ISSN 0957-5820

Canizares, P.; Saez, C.; Lobato, J.; Rodrigo, M. A. (2004). Electrochemical treatment of 2,4-dinitrophenol aqueous wastes using boron-doped diamond anodes: Part II. Influence of waste characteristics and operating conditions. *Electrochimica Acta*, Vol 49, N° 26, pp 4641-4650,(2004), ISSN 0013-4686

Chen, G. (2004) Electrochemical technologies in wastewater treatment. *Separation and Purification Technology*, Vol 38, N° 1, pp 11-41, ISSN 1383-5866

Chen, X., Chen, G. and Yue, P.L. (2000). Separation of pollutants from restaurant wastewater by electrocoagulation. *Separation and Purification Technology*, Vol 19, N° 1-2, pp 5-76, (2000), ISSN 1383-5866

Comninellis, C. (1994). Electrocatalysis in the electrochemical conversion/ combustion of organic pollutants for waste water treatment. *Electrochimica Acta*. Vol. 39, No. 11/12, pp. 1857-1862

Comninellis, C. and Nerini, A. (1995). Anodic oxidation of phenol in the presence of NaCl for wastewater treatment. *Journal of Applied Electrochemistry* ,Vol. 25, N°1, pp 23-28, (1995), ISSN 0021-891

Comninellis, C. and Vercesi, G. P. (1991). Characterization of DSA®-type oxygen evolving electrodes. Choice of a Coating. *Journal of Applied Electrochemistry*, Vol 21, N° 4, pp 335-45, (1991), ISSN 0021-891X

Couper, A. Mottram; Pletcher, Derek; Walsh, Frank C.. (1990). Electrode materials for electrosynthesis. *Chemical Reviews (Washington, DC, United States)*, Vol 90, N° 5, pp 837-65, (1990), ISSN 0009-2665

Diniz, A. V.; Ferreira, N. G.; Corat, E. J.; Trava-Airoldi, V. J. (2003). Efficiency study of perforated diamond electrodes for organic compounds oxidation process. *Diamond and Related Materials*, Vol 12, N° 3-7, pp 577-582, (2003), ISSN 0925-9635

Fernandes, A.; Morao, A.; Magrinho, M.; Lopes, A.; Goncalves, I. (2004). Electrochemical degradation of C. I. Acid Orange 7. *Dyes and Pigments* ,Vol 61, N° 3, pp 287-296, (2004), ISSN 0143-7208

Grimm, J. et al. (1998). *Sol*-gel film preparation electrodes for the electrocatalytic oxidation of organic pollutants in water. *Desalination* Vol. 115, N° 3, pp. 295-302, ISSN 0011-9164.

Hanze, M., Harremës, P., Jansen, J. C. and Arvin, E. (1995). *Wastewater Treatment: Biological and Chemical Processes*. Springer, CAN 127:298091, Berlin, Germany http://www.environmental-center.com/articles/article1149/ article1149.htm

Kapałka, A., Fóti, G. and Comninellis, C. (2009). Basic Principles of the Electrochemical Mineralization of Organic Pollutants for Wastewater Treatment.. *Journal of Applied Electrochemistry*, Vol 40, N° 12, pp 2203, (2010), ISSN 0021-891X

Kesselman J. M., Weres O., Lewis N. S., Hoffmann M. R. (1997). Electrochemical Production of Hydroxyl Radical at Polycrystalline Nb-Doped TiO_2 Electrodes and Estimation of the Partitioning between Hydroxyl Radical and Direct Hole Oxidation Pathways. *Journal of Physical Chemistry B*, Vol. 101, N° 14, pp 2637-2643, ISSN 1089-5647

Klamklang, S. (2006). Restaurant wasterwater treatment by electrochemical oxidation in continuous process. *Chulalongkorn University Ph.D. Dissertation*, ISBN 974-14-3476-6, Bangkok, Thailand

Klamklang, S. (2007). Restaurant wasterwater treatment by electrochemical oxidation in continuous process. *Institut National Polytechnique de Toulouse Dissertation*,. ISBN 974-14-3476-6, Toulouse, France

Klamklang, S. Vergnes, H., Secocq, F., Pruksathorn, K., Duverneuil, P. and Damronglerd, S. (2010). Deposition of tin oxide, iridium and iridium oxide films by metal-organic chemical vapor deposition for electrochemical wastewater treatment. *Journal of Applied Electrochemistry*, Vol 40, N° 5, pp 997-1004, (2010), ISSN 0021-891X

Kuhn, A. T. (1971). Electrolytic decomposition of cyanides, phenols and thiocyanates in effluents streams-a literature review. *Journal of Applied Chemistry & Biotechnology*, Vol 21, N° 2, pp 29-34 Journal; General Review (1971), ISSN 0375-9210

Lissens, G.; Pieters, J.; Verhaege, M.; Pinoy, L.; Verstraete, W. (2003). Electrochemical degradation of surfactants by intermediates of water discharge at carbon-based electrodes. *Electrochimica Acta* ,Vol. 48, N° 12, pp 1655-1663, (2003), ISSN 0013-4686

Morão, A., Lopes, A. Pessoa de Amorimb, M. T. and Gonçalves, I. C. (2004). Degradation of mixtures of phenols using boron doped diamond electrodes for wastewater treatment. *Electrochimica Acta*, Vol 49, N° 9-10, pp 1587-1595, (2004), ISSN 0013-4686

Panizza, M. and Cerisola, G. (2005). Application of diamond electrodes to electrochemical processes. *Electrochimica Acta*, Vol 51, N°2, pp 191-199, (2005), ISSN: 0013-4686

Pinson, L. (2008). *Anatomy of a Business Plan*, 7th Edtion.: Out of Your Mind and Into Marketplace, ISBN 0-944205-35-6 California, USA

Reemtsma, T. and Jekel, M.(2006). *Organic Pollutants in the Water Cycle*. Wiley-VCH, ISBN 3-527-31297-8, Weinheim, Germany

Stucki, S., Kötz, R., Carcer, B. and Suter, W. (1991). Electrochemical waste water treatment using high overvoltage anodes Part II: Anode performance and applications. *Journal of Applied Electrochemistry* Vol. 21, N° 2, pp 99-104, (1991), ISSN 0021-891X

8

Organic Pollutants Treatment from Air Using Electron Beam Generated Nonthermal Plasma – Overview

Yongxia Sun[1] and A. G. Chmielewski[1,2]
[1]Institute of Nuclear Chemistry and Technology, Warsaw,
[2]University of Technology, Warsaw,
Poland

1. Introduction

The municipal and industrial activities of man lead to environment degradation. The pollutants are emitted to the atmosphere with off-gases from industry, power stations, residential heating systems and vehicles. Organic pollutants, mainly volatile organic compounds (VOCs), which are emitted into atmosphere cause stratospheric ozone layer depletion, ground level photochemical ozone formation, and toxic or carcinogenic human health effects, contribute to the global greenhouse effect, accumulate and persist in environment. Regulation on organic pollutants emission into atmosphere has been enforced in many countries. Electron beam (EB) generated nonthermal plasma technology is one of the most promising technologies which has been successfully demonstrated on industrial scale coal fired power plants to remove SO_2 and NOx from waste off-gases; Meanwhile EB technology has been tested in pilot scale to remove dioxins and Polycyclic aromatic hydrocarbons (PAHs) from off-gases generated from solid waste incinerators and coal fired power plants, good results were obtained. It is a very promising technology to treat multiple pollutants including SO_2, NOx and organic pollutants simultaneously from industrial off-gases.
The principle of EB process to decompose pollutants is following. When the energy of the fast electrons is absorbed in the carrier gas, it causes ionization and excitation processes of the nitrogen, oxygen or water molecules in the carrier gas. Primary species and secondary electrons are formed, and the latter are thermalized within 1 ns in air at 1 bar pressure. These primary species such as ions, radicals or other oxidizing species and the thermalized secondary electrons react with pollutants by a series of reactions to cause pollutants decomposition.
Organic pollutants treatment using EB technology has been studied intensively in recent 30 years mainly in laboratory scale. However less work has been done to review this technology development on organic pollutants treatment. This chapter aims a comprehensive description of organic pollutants treatment using EB generated nonthermal plasma technology. General description of EB generated nonthermal plasma technology will be given in section 2, organic pollutants treatment from air and its recent development will be overviewed in sections 3 & 4, and general mechanism of organic pollutants decomposition in air will be discussed in section 5.

2. Electron beam (EB) generated nonthermal plasma technology

2.1 History

Wet flue gas desulphurisation (FGD) and selective catalytic reduction (SCR) can be applied for flue gas treatment and SO_2 and NOx emission control. VOCs are usually adsorbed on active carbon, but this process is rarely used for lean hydrocarbon concentrations up to now. All these technologies are complex chemcial processes and waste, like wastewater, gypsum and used catalyses, are generated(Srivastava et al., 2001).

EB technology is among the most promising advanced technologies of new generation. This is a dry-scrubbing process of simultaneous SO_2 and NOx removal, where no waste except the by-product is generated. EB technology for air treatment was first used by Japanese scientists in 1970-1971 to study SO_2 removal using an electron from linear accelerator (2-12 MeV, 1.2kW). A dose of 50 kGy at 100°C led the conversion of SO_2 to an aerosl of sulphuric acid droplets, which were easily removed (Machi, 1983). In 1981, Slater (1981) used EB technology to study the decomposition of low concentrations of vinyl chloride(VC) in different base gas (air, nitrogen, and argon).

2.2 EB accelerator

In physics and chemistry, plasma is a state of matter similar to gas in which a certain portion of particles are ionised. Nonthermal plasma means only a small fraction (for example 1%) of the gas molecules are ionized. The most common method for plasma generation is by applying an electric field to a neutral gas. Electrons emitted from electron beam accelerators can ionize gas mixture and generate nonthermal plasma.

More than 1000 accelerators haven been used in the field of radiation chemistry and radiation processing (Zimek, 1995). The reduction of SO_2 and NOx pollutants from flue gases, emitted during fuel combustion in electrical power and heat production, is one of the radiation process which were successfully demostrated in industrail scale in electric power station (EPS) Pomorzany, Szczecin, Poland (Chmielewski et al, 2004a) . A basic principle of an accelerator is that the electric field acts on electrons as charged particles and give them energy equal to the voltage difference accross the acceleration gap. The accelerator types are mainly determined by the method by which electron field is generated. There are three type accelerators used in air pollutants treatment: transformer accelerator, UHF accelerator and linear microwave accelerator (Zimek, 2005). High power accelerators have been developed to meet specific demands of environmental application and high throughput processes to increses the capacity and reduce unit cost of operation. Table 1 lists accelerators for radiation processing (Zimek, 2005).

Accelerator type	Direct DC	UHF 100-200 MHz	Linear 1.3-5.8GHz
Beam current	< 1.5A	< 100mA	< 100 Ma
Energy range	0.1-5 MeV	0.3-10 MeV	2-10MeV
Beam power	400kW	700 kW	150kW
Efficiency	60-80%	25-50%	10-20%

Table 1. Accelerator for radiation processing (recent development)

2.3 Terminology

In radiation application in environmental protection, there are three terms to be mentioned, **dose, G-value** and **removal efficiency (Re)** or **decomposition efficiency(De).**
In radiation process, it is very important to consider energy consumption for decomposition of pollutants, how much energy (unit: kJ) is consumed/absorbed to decompose amount of pollutants in the base gas (unit: kg). Energy absorbed by per amount of gas is defined as a term of **dose**, unit is **kGy**. 1 kGy = 1 kJ/ kg
G-value is defined as the number of molecules of product formed, or of starting material changed, for every 100 eV of energy absorbed. The G value is related to the ionic yield (M / N) by the expression (Willis and Boyd, 1976):

$$G \text{ (molecules / 100 eV)} = (M / N) \times (100 / W)$$

Where W (measured in electronvolts) is the mean energy required to from an ion pair in the material being irradiated. G value of **1 molecule /100 eV** is equal to a radiation chemical yield of **0.1036 μmol. J^{-1}**.
Removal efficiency (Re) or decomposition efficiency (De) of organic pollutants is defined as below:

$$Re = (C_0 - C_i) / C_0$$

where C_0 is initial concentration of organic pollutants, unit: ppm (v/v);
C_i is concentration of organic pollutants at i kGy absorbed dose, unit: ppm (v/v).

3. EB treatment organic pollutants

There are two systems applied to study organic pollutants in laboratory scale by using EB generated nonthermal plasma - flow system and batch system. Flow system contains one step: preparation and irradiation of the gas mixture which contains organic pollutants are carried out in on-line system. Batch system contains two steps: first step is to prepare gas mixture which contains organic pollutants into a sealed container, the second step is to put this sealed container under electron beam accelerator for irradiation.

3.1 Aliphatic organic pollutants degradation in flow system under EB-irradiation
3.1.1 Concentration of aliphatic organic pollutants vs. dose
There have been some previous studies of chlorinate hydrocarbons' decomposition in plasma reactors. Slater(1981) studied the decomposition of low concentrations of vinyl chloride(VC) in air, nitrogen, and argon in an electron beam generated plasma reactor. It was found that VC can be effectively removed by electron-beam irradiation at concentrations 3-500 ppm from room-temperature host-gas streams of argon, nitrogen and air. And at low dose the specific energy required falled in the range 2.5< G < 10 molecules removed per 100 eV. HCl was one of main products.
Vitale et al.(1997a) studied decomposition low concentration of ethyl chloride (EC) and vinyl chloride(VC) in atmospheric air streams by an electron beam generated plasma reactor. The gas was prepared by mixing dry air with standard VC (3925ppm VC in air) or EC (3717ppm EC in air). The gas entered the reactor at atmospheric pressure and ambient temperature. The electrons entered the front of the reactor, and VOC contaminated gas entered the rear of the reactor. The VOC contaminated gas thus flowed counter-current to

the electron beam. A Hewlett-Packard 5890 gas chromatograph and a HP-5971-A mass spectrometer were used to analyze VOCs concentration. The energy requirements for 90% decomposition of VC and EC were reported as a function of inlet concentration. VC requires less energy for decomposition than EC.

Similar experiments were carried out to decompose 1,1-dichloroethane (1,1-DCA), 1,1-dichloroethylene (1,1-DCE),1,1,1-trichloroethane (1,1,1-TCA), trichloroethylene (TCE) using EB by the same research group (Vitale et al., 1996, 1997b-d). It was found that decomposition efficiency of chlorinated compounds was : TCE > 1,1-DCE > 1,1-DCA, 1,1,1-TCA.

Won et al. (2002) studied the decomposition of perchloroethylene (PCE), trichloroethylene (TCE), dichloroethylene (DCE) in dry air. An electron accelerator of ELV type, with electron energy 0.7 MeV, maximum beam current 35 mA, maximum power 25 kW was used for irradiation. Over 80% TCE was decomposed at 20 kGy dose at initial concentration below 2000 ppm. The order of decomposition efficiency of these compounds was: TCE > PCE > DCE. Hirota et al. (2004) studied dichloromethane decomposition under EB irradiation and found that it was very difficult to treat dichloromethane.

For non-chlorinated organic compounds, 20 VOCs divided into five groups were investigated by Hirota et al.(2004), among them, 13 VOCs were alipahtic organic compounds. The order of decomposition VOCs in air was: cyclohexadiene > cyclohexane > benzene (group I); trans-hexane > 1-hexane (group II); heptane > hexane > pentane (group IV); and trichloroethylene > methanol >> acetone > CH_2Cl_2 (group V). Organic substances with long carbon chains readily succumbed to electron-beam treatment.

3.1.2 Different base gas mixtures influence on the decomposition efficiency of aliphatic organic pollutants

Won et al. (2002) studied TCE decomposition in different gas mixtures and found that the order of decomposition efficiency of TCE in different gas mixtures was: oxygen >air > H_2 > He.

3.1.3 Water concentration

In order to clarify OH radical influence on the chlorinated hydrocarbons (Cl-HC) decomposition, Won et al. (2002) tested TCE and PCE decomposition of air mixtures with different water vapor concentrations, and found that the decomposition efficiency of TCE and PCE increased less than 10% in the presence of water vapor compared with that in the dry air.

3.1.4 Irradiation products

The irradiation products of DCE, TCE and PCE in dry air under EB-irradiation were investigated by Won et al. (2002) and it was found that CO and CO_2 were the irradiation products. For PCE, CO_2 formation was above 40% at 15 kGy absorbed dose. Vitale et al (1997a) also reported that CO, CO_2 and HCl as main irradiation products when they studied ethyl chloride and vinyl chloride decomposition in air. Prager et al. (1995) studied DCE, TCE and PCE degradation in dry or humidified synthetic air, they identified HCl, CO, chloromethanes, chloroacetyl chloride and phosgene as main products.

3.2 Aliphatic organic pollutants degradation in batch system under EB-irradiation
3.2.1 Concentration of aliphatic organic pollutants vs. dose

Chloroethylene can be effectively decomposed by EB irradiation in the order of PCE > TCE > trans-DCE > cis-DCE (Hakoda et al., 1998a, 1998b, 1999, 2000, 2001; Hashimoto et al.,

2000). Sun et al. (2001, 2003) and Sun and Chmielewski (2004) studied 1,1-DCE, cis-DCE, trans-DCE decomposition under EB irradiation and found that the order of decomposition DCEs in air was 1,1-DCE > trans-DCE > cis-DCE. Decomposition efficiency of chloroethylene increases with the absorbed dose increase. The initial concentration of chlorinated ethylene was in below 2000 ppm and the water concentration in the air mixture was 200–300 ppm. Son et al. (2010a) studied decomposition of butane in EB irradiation in batch system, it was found that removal efficiencies of butane were 40% at 2.5 kGy and 66% at 10 kGy, when the initial concentration of butane was 60 ppm.

3.2.2 Different base gas mixtures influence on the decomposition efficiency of aliphatic organic pollutants

Different base gases influencing on the decomposition efficiency of butane were studied (Son et al, 2010a), it was found that decomposition efficiency of butane was extremely low when the background gas was He, in contrast to the efficiencies with background gases of N_2 and air. Decomposition efficiencies of butane was 23% in He, 63% in N_2 and 70% in air at 10 kGy absorbed dose.

3.2.3 Water concentration

Water influence on the TCE decomposition and irradiation products of TCE under EB-irradiation were studied by Hakoda et al. (2000). It was found that when water concentration was below 1000 ppm, there was no big difference between process efficiency for dry and humid air for TCE decomposition and dichloroacetyl chloride, carbon monoxide, carbon dioxide, phosgene and small amount of chloroform irradiation products' formation.
This result agrees well with that water vapor effect on the TCE decomposition under EB-irradiation in a flow system (Won et al., 2002). Sun et al. (2001) made a computer simulation of 1,1-DCE decomposition in air in a batch system and found that a reaction pathway of OH radical contributes less than 10% for 1,1-DCE decomposition. When water vapor concentration increased to 2.5%, yield of gaseous products decreased, that means the aerosol products are possibly formed (Hakoda et al., 2000).

3.2.4 Ozone

Hakoda et al. (1999, 2000, 2001) investigated O_3 influence on the trans-DCE, cis-DCE and TCE decomposition by using EB-irradiation, it was found that O_3 enhanced decomposition of trans-DCE only, cis-DCE and TCE were not affected.

3.2.5 Irradiation products

From environmental protection point of view, it is very important to identify by-products formation from Cl–HC degradation. Radiolytic products of trans-DCE, cis-DCE, TCE, PCE under EB-irradiation were reported by Hakoda et al. (1999, 2000, 2001). Chloroacetyl chloride and dichloroacetyl chloride were the main organic products for DCE (trans and cis) and TCE degradation; CO and CO_2 were inorganic products and their formations were below 25% based on carbon balance. Chmielewski et al. (2004b), Sun et al. (2003) and Sun & Chmielewski (2004) studied cis-DCE and trans-DCE degradation under EB-irradiation. Chloroacetyl chloride was not observed as degradation products, but it was a degradation product for 1,1-DCE (Sun et al., 2001). Son et al. (2010a) studied butane decomposition under EB irradiation and identified CO_2, acetaldehyde, acetone, 2,3-butandione, 2-butanone, and 2-butanedinitrile as degradation products of butane.

3.3 Aromatic organic pollutants degradation in flow system under EB-irradiation

In this section, besides aromatic chlorinated hydrocarbons degradation, decomposition of some nonchlorinated aromatic organic compounds will be discussed, too.

3.3.1 Concentration of aromatic organic pollutants vs. dose

Xylene and chlorobenzene decompositions in a flow system under EB-irradiation (Hirota et al. 2000, 2002) were studied, it was found that decomposition efficiency of xylene was higher than that of chlorobenzene, and about 50% chlorobenzene was decomposed at an absorbed dose of about 10 kGy at the initial concentration of chlorobenzene being 10–40 ppm. Kim (2002) studied decomposition of benzene and toluene, it was found that the decomposition efficiency of toluene was higher than benzene, and about 80% benzene was decomposed at 16 kGy when the initial concentration of benzene was smaller than or equal to 130 ppm. Han et al. (2003) studied toluene, ethylbenzene, o-, m-, p-xylenes and chlorobenzene decomposition in air. The order of the decomposition efficiency of selected VOCs from high to low was : toluene > ethylbenzene > benzene; p-xylene> m-xylene > o-xylene. About 44.7 % toluene and 43.2% ethylbenzene was decomposed at 10 kGy, while 85% chlorobenzene was decomposed. The decomposition efficiency of ethylbenzene and toluene was significantly increased about 50% with the addition of chlorobenzene comparing without chlorobenzene addition.

3.3.2 NH₃ influence on the decomposition efficiency of aromatic organic pollutants

Effect of ammonia on the decomposition of PAHs was observed when an electron beam process was applied to treat multiple pollutants (SO_2, NOx , PAHs) with NH_3 addition from industrial off-gases emitted from EPS, Kawęczyn, Poland (Chmielewski et al., 2002). NH_3 addition enhanced PAHs removal efficiency. Hirota et al. (2000) studied chlorobenzene decomposition in air mixture, it was found that addition of NH_3 enhanced the dechlorination of chlorobenzene. About 65% of chlorine in reacted chlorobenzene was dissociated from carbon with electron beam at doses of 4 and 8 kGy. Ammonia addition enhanced the dechlorination to 80%.

3.3.3 Water concentration

Effect of water vapor on the decomposition of toluene was investigated by Kim (2002). It was found that the water vapor injection leads to 15–20% removal efficiency increase for toluene compared to the process without water injection. Water influences decomposition of toluene higher than TCE and PCE, OH radical plays an important role for aromatic hydrocarbon decomposition.

3.3.4 Irradiation products

Degradation products of chlorobenzene and xylene in an air mixture were studied by Hirota et al. (2000, 2002). The gaseous products of xylene degradation were identified to be formic, acetic, propionic, and butyric acids and/or the corresponding esters with CO and CO_2. Approximately 30% of the reacted xylene was the gaseous products at a dose of 8 kGy. Organic and inorganic chlorine presence in gaseous, aerosols and residues were investigated by same authors for chlorobenzene degradation. Inorganic chlorine was mainly presented in gaseous products, while organic chlorine was presented in aerosols and residues aerosols and residues. Some aerosol products were formed from chlorobenzene degradation, and 3% of the aerosol products were identified to be carboxylic acids or esters (Hirota et al., 2000). Aerosols, benzaldehyde, dipropyl 1,2-benzenedicarboxylic acid, nitromethane were reported as toluene main degradation products in dry air under EB irradiation; while methyl chloride, dipropyl 1,2-benzenedicarboxylic acid, toluene, nitromethane were reported as main degradation products of ethylbenzene/chlorobenzene mixture in dry air. Trace amount of acetone, hexane, benzene was also observed (Han et al., 2003).

3.4 Aromatic organic pollutants degradation in batch system under EB-irradiation
3.4.1 Concentration of aromatic organic pollutants vs. dose

Decomposition of aromatic organic compounds in a batch system vs. dose under EB-irradiation was studied by Hirota et al. (2002, 2004), Ostapczuk et al.(1999), Sun et al. (2008) and Hashimoto et al. (2000). Decomposition efficiency of these compounds increase with the absorbed dose increase, 4-chlorotoluene (4-CTO) decomposition as an example was presented in Fig. 1 . The order of decomposition efficiency of these compounds was: xylene >chlorobenzene > benzene > hexane > cyclohexane. For 4-chlorotolunene (4-CTO) and 1,4-dichlorobenzene (1,4-DCB), no apparent decomposition efficiency of these two compounds was observed (Fig.2).

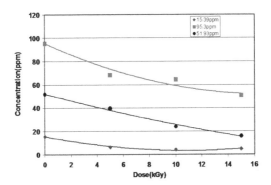

Fig. 1. 4-Chlorotoluene decomposition in air mixture in an electron beam generated non-thermal reactor.

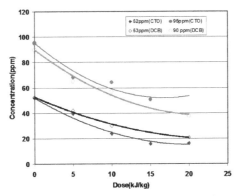

Fig. 2. Decomposition efficiency comparison between 4-chlorotoluene and 1,4-dichlorobenzene.

The decomposition efficiency of chlorinated aliphatic hydrocarbons using EB irradiation is more efficient than that of chlorinated aromatic hydrocarbons. 25.0 kGy is sufficient to remove over 97% 1,1-DCE and 98.0% trans-DCE at initial concentration of 1,1-dichloroethylene (DCE) and trans-dichloroethylene being 903.8 ppm and 342.0 ppm, respectively; while 60% 1,4-dichlorobenzene (DCB) at intial concentration being 90ppm was removed at 57.9 kGy. This result is comparable with decomposition of chlorobenzene. Hakoda et al. (1998b) and Hashimoto et al. (2000) studied degradation of chlorobenzene/air using EB irradiation in batch system, it was found that 40% chlorobenzene was removed under EB-irradiation at 37.7 kGy dose (calculated by N_2O gas dosimeter) for initial concentration of chlorobenzene being 102 ppm. Sun et al. (2007a) studied 1-chloronaphathalene and found that over 80% 1-chloronapthalene was removed at 57.9 kGy under EB-irradiation for low initial concentration of 1-chloronaphthalene (12~30 mg/m³) in air mixture. Energy consumption for decomposition 1,4-dichlorobenzene was lower than that of 1-chloronaphthalene. Therefore, the observed order in easily decomposition chlorinated hydrocarbons is: 1,1-DCE > trans-DCE > cis-DCE >1,4-DCB > 1-chloronaphthalene. Based on this work and other's work, we learn that: For chlorinated aliphatic hydrocarbons, the more chlorinated compounds is, the more it is easy to be decomposed.

Aliphatic hydrocarbons is more easily decomposed than aromatic hydrocarbons. For aromatic hydrocarbons, compounds with less benzene ring are easily to be decomposed.

3.4.2 Different base gas mixtures influence on the decomposition efficiency of aromatic organic pollutants

Toluene decomposition at different background gases in a batch system was studied by Kim (2002). The order of decomposition efficiency of toluene in different background gases is: N_2 > air >O_2 >He. This order is different from the order for TCE decomposition in a flow system (Won et al., 2002).

We studied 1,4-dichlorobenzene (1,4-DCB) decomposition in different base gas mixtures at the initial concentration of 1,4-DCB being 50 ppm, the similar phenomenon was observed (Sun et al, 2006). The decomposition efficiency of 1,4-DCB in nitrogen is higher than that in air and much more higher than in 1.027% NO-N_2 mixture(N_2 as balance gas) (Fig.3) , this phenomenon agrees well with toluene decomposition in different gases (Kim, 2002).

3.4.3 Water concentration

Effect of water vapor on the decomposition of aromatic compounds in a batch system under EB-irradiation was investigated by Kim (2002). Four percent water vapor injection leads to 5~10% increase of VOC removal efficiency for both toluene and benzene, and effect of water vapor influence on the decomposition of toluene under EB irradiation in a flow system is higher than that in a batch system.

3.4.4 Irradiation products

Benzaldehyde and phenol were reported as products when Ostapczuk et al. (1999) studied styrene decomposition in air under EB irradiation, the removal efficiency of styrene was ranged from 83-95%. The humidity in air mixture was ranged from 0.3% to 1.6%. In order to obtain information of by-products produced from toluene destruction, we carried out experiment at higher inlet concentration of toluene at higher absorbed dose. More than 97%

toluene was removed from gas phase at 53.6 kGy absorbed dose when inlet concentration of toluene was 151.9 ppm. A GC-MS spectrum of toluene/air mixture after EB-irradiation is presented in figure 4. A trace amount of benzaldehyde was eluted at retention time 7.735 min in figure 4 and was identified by our carefully comparing mass spectrum of this compound with a reference mass spectrum of benzaldehyde provided by Wiley library (figures 5a & 5b). Trace amount acetone was also found in our experimental condition (Sun et al., 2009a).

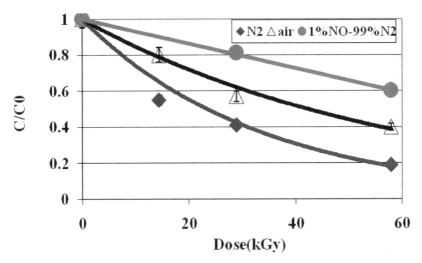

Fig. 3. 1,4-Dichlorobenzene decomposition in different gas mixture

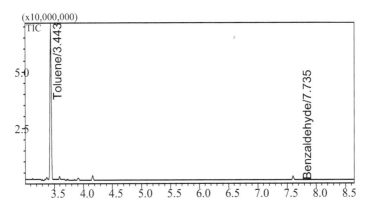

Fig. 4. A GC-MS spectrum of toluene/air mixture after EB-irradiation (inlet concentration of toluene was 151.9 ppm, dose was 53.6 kGy).

Benzaldehyde as by-product of degradation of toluene was also reported(Han et al., 2003; Kim et al., 2005). Trace amount of acetone was found based on Han et al.'s work (2003). Besides these, Aerosols and benzene were reported as by-products in both works (Han et al.,

2003; Kim et al., 2005). Han et al.(2003) also identified dipropyl 1,2-benzenedicarboxylic acid, nitromethane and trace amount of hexane as by-products of degradation of toluene. For degradition of 4-chlorotoluene in air mixture, chlorobenzene (C_6H_5Cl, retention time was 4.910 min) and 4-chlorobenzaldehyde (ClC_6H_4CHO, retention time was 12.502 min) were identified as by-products. A GC-MS spectrum of 4-chlorotoluene/air mixture after EB-irradiation was presented in figure 6, a compound eluted at retention time 7.590 min of the GC-MS spectrum was identified as 4-chlorotoluene (Sun et al., 2008).

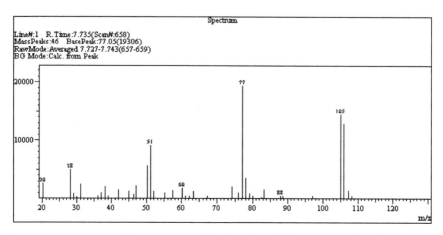

Fig. 5a. A mass spectrum of by-product which eluted at 7.735 min retention time

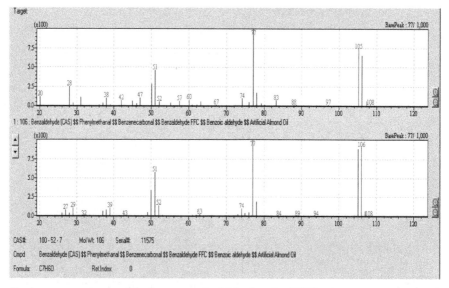

Fig. 5b. A mass spectrum of the compound which eluted at 7.735 min retention time and its reference mass spectrum of benzaldehyde

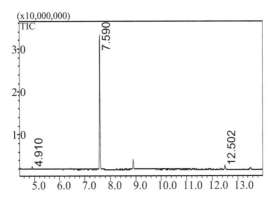

Fig. 6. A GC-MS spectrum of 4-chlorotoluene/air mixture after EB-irradiation

3.5 PAHs and Dioxin removal from waste off-gas under EB-irradiation

Dioxins reduction from waste incinerator was studied using EB technology in Japan (Hirota et al., 2003) and in Germany (Paur et al., 1998). Hirota et al. (2003) studied reduction the emission of polychlorinated dibenzo- p-dioxins (PCDD) and polychlorinated dibenzofurans (PCDF) in a flue gas of 1000 m^3N/h from the municipal solid waste incinerator (MSWI), located at Takohama Clean Center which treats 450 t (150 t * 3 furnaces) of solid waste in 1 day, at a temperature of 200 °C. they found that more than 90% PCDD/Fs was removed at 14 kGy when initial concentration of PCDD was in the range of 0.22-0.88 ng-TEQ/m^3N and PCDF in the range of 0.35-12.4 ng-TEQ/m^3N. Paul et al. (1998) also reported that over 90% PCDD was removed at 12 kGy dose for inital concentration of PCDD being 21-110 ng/m^3N (AGATE-M plant, Germany).

16 kinds of toxic PAHs were investigated under electron beam irradiation in the pilot plant in Electric Power Station Kawęczyn, Poland (Chmielewski, et al., 2003). The investigation was carried out under the following experimental conditions: flue gas flow rate 5000 Nm^3/h; humidity 4.5%; inlet concentrations of SO_2 and NOx that were emitted from the power station were 192 and 106 ppm, respectively; ammonia addition was 2.75 Nm^3/h; alcohol addition was 600 l/h, the absorbed dose was 8 kGy. The results was presented in Fig.7. It was found that under these experimental conditions the concentrations of naphthalene (NL, $C_{10}H_8$), acenaphthene (AC, $C_{12}H_{10}$), fluorene ($C_{13}H_{10}$), phenanthrene ($C_{14}H_{10}$), anthracene ($C_{14}H_{10}$) were decreased, while the concentrations of acenaphthylene ($C_{12}H_8$), fluoranthene ($C_{16}H_{10}$), pyrene ($C_{16}H_{10}$), benzo(a)anthracene ($C_{18}H_{12}$), chrysene ($C_{18}H_{12}$), benzo(b þ k)fluoranthene ($C_{20}H_{12}$), benzo(e)pyrene ($C_{20}H_{12}$), benzo(a)pyrene ($C_{20}H_{12}$), perylene ($C_{20}H_{12}$), dibenzo(a; h)anthracene+indeno(1,2,3-cd) pyrene ($C_{22}H_{14}$), benzo(g; h; l)perylene ($C_{22}H_{12}$) were increased. Removal efficiencies of SO_2 and NOx were 61.6% and 70.9%, respectively. The concentration of hydrocarbons of small aromatic ring (PAHs, like naphthalene ($C_{10}H_8$), acenaphthene ($C_{12}H_{10}$), fluorene ($C_{13}H_{10}$), anthracene ($C_{14}H_{10}$)) was reduced, while the concentration of fluoranthene was increased remarkably after irradiation.

Similar experiments were carried out in EPS Kawęczyn with ammonia presence but without alcohol addition (Chmielewski et al, 2002; Ostapczuk et al, 2008a). It was found that removal efficiency of PAHs ranges from 40% up to 98%.

Callén et al. (2007) studied PAH removal from lignite-combustion flue gas from Bulgarian Maritza-East thermal power plant (TPP) and obtained that PAHs concentration after EB irradiation resulted in ~ 10 fold decrease in studied PAHs emissions. The removal efficiency of PAH removal at the dose of 4 kGy was 85% (weight/weight). High PAH removal efficiency was obtained especially for 2 and 3 rings PAH, this result was similar to that obtained in our previous work (Chmielewski et al., 2003).

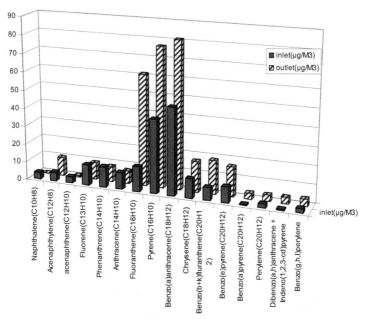

Fig. 7. EB irradiation influence on PAHs removal.

The concentration of PAHs in by-product was also examined. It was relatively low, varied from few up to 12 µg per kg of fertilizer for the experimental work carried out in EPS Kawęczyn, Poland. Less than 3% of PAH were removed in adsorption on the by-product surface (Ostapczuk, et al., 2008a). The study in Maritza-East TPP allowed PAH assessment in solid by-products obtained from EB lignite-combustion flue gas. The determined PAH content was reasonable, ~ 60 microg/kg and was lower than PAHs background in Bulgarian soils. These results demonstrated the insignificant role of adsorption for PAHs removal (Callén et al., 2007).

Naphthalene (NL) and acenaphthene (AC) decomposition in gas mixture was studied in lab scale experiment (Ostapczuk, et al., 2008b). It was found that NL was more easily decomposed than AC, G-values for these two compounds were 1.66 and 3.72 mol/100 eV for NL and AC at the dose of 1 kGy , respectively. Humidity influencing on NL and AC decomposition was studied. About 26% and 50% NL were decomposed at 1 kGy dose in dry air (90 vol% N_2; 10 vol% O_2 and 160 ppmv NO) and in humid air (84 vol% N_2; 10 vol% O_2 and 6 vol% H_2O), respectively. NL concentration in both mixtures was on the level of 10–11 ppmv. For AC, about 45% (in O_2) and 82% AC (in humid O_2, 94% O_2 + 6% H_2O) were decomposed at 1 kGy dose for the initial concentration of AC being 160 ppm. Two-ringed naphthol and nitronaphthalene; one-ringed 2,6-dietylbenzoquinone, indane, 1,2- and 1,4-dimetoxybenzenes and carbon oxides were identified as by-products of NL decomposition.

4. A novel hybrid EB-catalyst technology to treat organic pollutants

Electron beam (EB) irradiation is one of the most promising technologies for purification of dilute pollutants, mostly VOCs , with high flow-rate gas stream released from industrial off-gases. Under EB irradiation, VOCs are oxidized into irradiation by-products as well as CO_2 and CO. However, some of these irradiation by-products have adverse effect on environment and human beings.

A new technology which combines EB and catalyst together to treat aromatic VOCs, e.g., toluene (Kim et al., 2004, 2005; Jeon et al., 2008) styrene(Kim, 2005) , o-Xylene (Hakoda et al., 2008a, 2008b) and ethylbenzene (Son, et al, 2010b) was developed in the aim of enhancing higher oxidation efficiencies of VOCs into CO_2. Removal efficiency of toluene , styrene and ethylbenzene increased by 10%, 20%, and 20% in an EB-catalyst hybrid system in comparison with that achieved in catalyst-only method at approximately 10 kGy absorbed dose (Kim, et al, 2005). Removal efficiency of ethylbenzene in the EB-catalyst hybrid was 30% higher than that of EB-only treatment. Ethylbenzene was decomposed more easily than toluene by EB irradiation. The G-values for ethylbenzene increased with initial concentration and reactor type: the G-values vary in the range of 7.5-10.9 (EB-only) and 12.9-25.7 (EB-catalyst hybrid) by reactor type at the initial concentration of ethylbenzene being 2800 ppm. Son et al. (2008) and Jeon et al. (2008) also studied different catalysts (Pt, Pd , Cu and Mn) and humidity influence on removal efficiency of toluene using EB-catalyst hybrid system. It was found that removal efficiency of toluene was increased by 36.9% , 35.3% and 22% in the presence of Pt, Pd , Mn and Cu catalysts comparing with EB only for initial concentration of toluene being 1500 ppm, the selectivity to CO_2 with Pt and Pd coupling were relatively higher than those of Cu and Mn. Especially the CO_2 selectivity of EB–Pt coupling was significantly high at a relatively low absorbed dose. The catalytic activity for EB–catalyst coupling system was in the order of Pt, Pd, Mn and Cu. There was no significant difference of removal efficiency of toluene among 0.1, 0.5 and 1.0 wt% loading of catalyst. No significant water effect was observed in EB-catalyst hybrid system (Son et al, 2008).

Other type of catalysts such as TiO_2 (Hakoda, et al., 2008a) was used to study xylene decomposition under EB irradiation in lab scale experimental work. It was found that removal efficiency of xylene and CO_2 formation were increased with the presence of TiO_2 catalyst, the similar phenomenon was observed when Kim studied toluene decomposition using Pt as catalyst (Kim et al., 2005).

Hakoda, et al.(2008b) also studied xylene decomposition using MnO_2 (an O_3 decomposition catalyst), γ-Al_2O_3 was selected as a base material of the catalyst. The combination process at temperatures of about 100°C using MnO_2 placed downstream enhanced the oxidation of the by-products of xylene into CO_2 by active oxygen produced from the O_3 decomposition when the MnO_2 bed was placed downstream of an irradiation space. Furthermore, EB-irradiated γ-Al_2O_3 surface was found to be active, and the oxidation of organics was enhanced by primary electrons. The combination process using γ-Al_2O_3 reduced dose to 33% of a single EB process to obtain the same conversion of xylene to CO_2.

Ighigeanu (et al., 2008) studied VOCs (Toluene, hexane + toluene mixture diluted in air) decomposition by using combination of three different technologies (EB, microwave (MW) and catalysts): (EB + MW+ catalyst); (MW + catalyst) and (EB+catalyst). They found that decomposition efficiency (De) and oxidation efficiency (Eo) of toluene increased significantly for the (EB+MW+catalyst) treatment as compared with (MW + catalyst) and (EB + catalyst) treatments, at initial concentration of toluene being in the range of 180 ppm –

523 ppm; and CO_2 and CO concentrations after treatment were higher for the (EB+MW+catalyst) treatment than for (MW+catalyst) and (EB+catalyst) treatments. De and Eo of toluene were, respectively, as follows: 59.5% and 82.2% for the (MW + catalysis), 77.2% and 87.1% for the (EB + catalyst) and 92.8% and 90.5% for the(EB + MW + catalyst). For air mixture contained toluene and hexane, removal efficiency of toluene and hexane in (EB + MW + catalyst) system was higher than that in (MW + catalyst) system or in (EB + catalyst) system, about 88.5% toluene and 87.8% hexane were decomposed for initial toluene and hexane concentration being 250 ppm, respectively.

5. Mechanism of organic pollutants degradation by using EB technology

In order to obtain high decomposition efficiency of organic pollutants and less toxic by-products, it is very important to understand mechanism of organic pollutants degradation under EB irradiation. In this section, we will discuss mechanism of two groups (chlorinated and nonchlorinated) organic pollutants. General mechanism of organic pollutants decomposition in gas phase under EB irradiation is illustrated in Fig. 8.

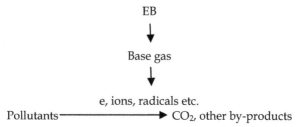

Fig. 8. General mechanism of organic pollutants decomposition under EB irradiation

5.1 Chlorinated organic compounds (Cl-HC)
5.1.1 General mechanism of chlorinated aliphatic hydrocarbon decomposition in air mixture

Computer simulations of chlorinated aliphatic hydrocabons'decomposition in air mixture were carried out and discussed in details (Nichipor et al., 2000, 2002, 2003, 2008; Sun et al., 2001, 2007b, 2009b) . The general mechanism of aliphatic hydrocarbon decomposition in an air mixture is described below:

When fast electrons from electron beams are absorbed in the carrier gas, they cause ionization and excitation processes of the nitrogen, H_2O and oxygen molecules in the carrier gas. Primary species and secondary electrons are formed. The secondary electrons are thermalized fast within 1 ns in air at 1 atmosphere .

The G-values (molecules/100 eV) of main primary species are simplified as follows (Mätzing, 1989):

$$4.43N_2 \rightarrow 0.29N_2^* + 0.885N(^2D) + 0.295N(^2P) + 1.87N + 2.27N_2^+ + 0.69N^+ + 2.96e \quad (1)$$

$$5.377O_2 \rightarrow 0.077O_2^* + 2.25O(^1D) + 2.8O + 0.18O^* + 2.07O_2^+ + 1.23O^+ + 3.3e \quad (2)$$

$$7.33H_2O \rightarrow 0.51H_2 + 0.46O(^3P) + 4.25OH + 4.15H + 1.99H_2O^+ + 0.01H_2^+ + 0.57OH^+$$
$$+ 0.67H^+ + 0.06O^+ + 3.3e \quad (3)$$

Where G-values of molecules decomposed are listed in the left side of the arrows, and G-values of species formed from the pure k type molecules that absorb an energy of 100 eV are listed in the right side of the arrows. These primary species and thermalized secondary electrons cause Cl-HC decomposition. Based on our and others published work , we know that several type reactions cause Cl-HC degradation.

Positive ions charge transfer and particle dissociation reactions

It is well known that : when air component and molecule are ionized and excited, a large amount of N_2^+, O_2^+ , N^+, O^+, H_3O^+(if water concentration is high) are formed, and their ionization potential energy (IE) is higher than that of Cl-HC (see Table 2). The positive charge transfer reaction, positive ion cluster reaction, or particle dissociation reaction occur (Spanel, et al., 1999a, 1999b).

Molecule	IE(eV)	EA(eV)	PA(kJ.mol^{-1})
N_2	15.58		493.8
O_2	12.07	0.45	421
H_2O	12.62		691
NO	9.26	0.03	531.8
O_3	12.53	2.10	625.5
CCl_4	11.47	0.80	
$CHCl_3$	11.37	0.62	650.6
CH_2ClCH_2Cl	11.07		
$CHCl_2CH_3$	11.04		
C_2H_3Cl	9.99		
$1,1$-$C_2H_2Cl_2$	9.81	0.1	
cis-$C_2H_2Cl_2$	9.65		
trans- $C_2H_2Cl_2$	9.64		
C_2HCl_3	9.46	0.40	
C_2Cl_4	9.32	0.64	
1,4-dichlorobenzene	8.92		

Table 2. Ionization energy (IE,eV) , electron affinity(EA, eV) and proton affinity(PA, kJ.mol^{-1}) data.

In general, the H_3O^+ reactions with the aliphatic chloride more varies in their rate constants and products, and in some reactions $H_3O^+.M$ ions (M=Cl-HC) are formed. The NO^+ reaction with the aliphatic compounds is generally slow association reactions and form $NO^+.M$ ions (for e.g., $NO^+ + CHClCCl_2 = NO^+.CHClCCl_2$). The O_2^+ reactions are fast mainly proceeding via nondissociative charge transfer reactions to produce the cations M^+ only (for e.g., $C_2HCl_3 + O_2^+ = C_2HCl_3^+ + O_2$), but in some of these reactions minority dissociative charge transfer reactions take place to eliminate Cl/HCl and leave hydrocarbon ion (for e.g., $CH_2ClCH_2Cl + O_2^+ = C_2H_3Cl^+(95\%)+HCl+O_2$; and $CH_2ClCH_2Cl + O_2^+ = C_2H_4Cl^+(5\%)+Cl+O_2$).

From the calculation results of dichloroethylene (Sun et al., 2001, 2007b), trichloroethylene (Nichipor et al, 2008) and tetrachloroethylene (Sun et al, 2009b), we learn that positive charge transfer reactions contributing to chlorinated ethylenes decomposition less than 10%

Secondary electron attachment, Cl dissociated reaction and negative ions charge transfer reactions

Reaction pathway of secondary electron attachment, Cl dissociative reactions followed by peroxyl radical reactions is the main reaction pathway which causes Cl-HC decomposition. The rate constants of electron and Cl with Cl-HC , and the products of these reactions are listed in table 3, respectively (Atkinson, R., 1987a).

For chlorinated methane, the products formed by Cl with CH_3Cl, CH_2Cl_2, $CHCl_3$, reactions are: CH_2Cl, $CHCl_2$, CCl_3 , and HCl; for chlorinated methane CCl_4, CCl_3 and Cl_2 are formed.

If we assume CH_2Cl, $CHCl_2$, CCl_3 as radical R, generalized mechanism of peroxyl radical reactions could be written as follows:

$$R+ O_2 = RO_2 \tag{4}$$

$$RO_2+ RO_2 = 2 RO + O_2 \tag{5}$$

$$RO = products \tag{6}$$

This is a main reaction pathway for Cl-HC decomposition (Bryukov et al., 2002;).

For chlorinated aliphatic ethylene, the mechanism of its degradation in air mixture can be generalized as follows:

$$e + C_2X_3Cl = Cl^- + \bullet C_2X_3 \text{ (Szamrej, et al., 1996;)} \tag{7}$$

$$X_2CCX_2 \text{ (X=H, Cl, at least 1 Cl inside)} \tag{8}$$

$$X_2CCX_2 + Cl \rightarrow (Cl)X_2CCX_2 \text{ (Cl adds to heavily chlorinated carbon side)} \tag{9}$$

$$(Cl)X_2CCX_2 + O_2 \rightarrow (Cl)X_2CCX_2(O_2) \tag{10}$$

$$2(Cl)X2CCX2(O) + O2 \tag{11}$$

$$(Cl)X_2CCX_2(O) \rightarrow (Cl)X_2CCXO + Cl \text{ (if one X = Cl)} \tag{12}$$

$$\text{or } (Cl)X_2CCX_2(O) \rightarrow CX_2Cl + COX_2 \tag{13}$$

Molecule	Electron [cm³.s⁻¹]		Cl [cm³.s⁻¹]	
CH_3Cl	6.1×10^{-11},	$Cl^- + CH_3$	4.78×10^{-13},	$CH_2Cl + HCl$
CH_2Cl_2	1.6×10^{-10},	$Cl^- + CH_2Cl$	3.5×10^{-13},	$CHCl_2 + HCl$
$CHCl_3$	4.9×10^{-9},	$Cl^- + CHCl_2$	1.2×10^{-13},	$CCl_3 + HCl$
CCl_4	1×10^{-7},	$Cl^- + CCl_3$	1.4×10^{-10},	$CCl_3 + Cl_2$
C_2H_3Cl	$(1{\sim}8) \times 10^{-10}$,	$Cl^- + C_2H_3$	1.27×10^{-10},	$CH_2ClCHCl$
$C_2H_2Cl_2$	1×10^{-9},	$Cl^- + C_2H_2Cl$	1.4×10^{-10},	CH_2ClCCl_2
C_2HCl_3	$(0.29{\sim}1) \times 10^{-8}$,	$Cl^- + C_2HCl_2$	9.3×10^{-12},	CCl_3CHCl
C_2Cl_4	1×10^{-7},	$Cl^- + C_2Cl_3$	$(4{\sim}6) \times 10^{-11}$,	C_2Cl_5
C_2H_5Cl	$(2{\sim}7) \times 10^{-13}$,	$Cl^- + C_2H_5$	6.8×10^{-12},	$CH_3CHCl + HCl$

Table 3. Rate constants and products for the reactions of electron, Cl with chlorinated aliphatic compounds.

The decomposition efficiency of Cl-HC mainly depends on the rate constants of secondary electron attachment, and Cl addition reaction followed by peroxyl radical reactions (Knox et al., 1966, 1969; Thűner et al., 1999). This decomposition pathway has been confirmed experimentally (Hirota et al., 2002).

O_2^- cause Cl-HC decomposition, $O_2^- + M = O_2 + M^-$ (M= Cl-HC) (14)

O atom, OH radical , and other radical reactions with Cl-HC

Other decomposition pathways for chlorinated aliphatic hydrocarbons are: O atom decomposition pathway (Sanhueza et al., 1974a, 1974b; Teruel, et al., 2001), OH radical decomposition pathway (Atkinson, R., 1987b; Howard, et al., 1976; Liu et al., 1989; Chandra et al., 1999; Chang et al., 1977) and other species decomposition pathway, such as O_3 and NO_3. The rate constants of O and OH with Cl-HC , and the products of these reactions are listed in table 4, respectively (http://kinetics.nist.gov/kinetics/index.jsp). By-products of irradiation vary with reactants.

Molecule	O [cm³.s⁻¹]		OH [cm³.s⁻¹]	
CH_3Cl	1.18×10^{-16},	$OH + CH_2Cl$	4.2×10^{-14},	$CH_2Cl + H_2O$
CH_2Cl_2	6.48×10^{-16},	$OH + CHCl_2$	1.4×10^{-13},	$CHCl_2 + H_2O$
$CHCl_3$	1.02×10^{-15},	$OH + CCl_3$	1.0×10^{-13},	$CCl_3 + H_2O$
CCl_4	1.89×10^{-16},	$ClO + CCl_3$	$< 4 \times 10^{-16}$,	$CCl_3 + HOCl$
C_2H_3Cl	5.96×10^{-13},	products	8.06×10^{-12},	$CHClCH_2OH$
$1,1-C_2H_2Cl_2$	9.8×10^{-13},	products	8.10×10^{-12},	CH_2OHCCl_2
C_2HCl_3	1.4×10^{-13},	products	2.2×10^{-12},	products
C_2Cl_4	1.9×10^{-13},	products	1.7×10^{-13},	products
C_2H_5Cl	1.12×10^{-15},	OH+ other products	6.42×10^{-13},	H_2O + other products

Table 4. Rate constants and products for the reactions of O, OH with chlorinated aliphatic compounds

The mechanism of decomposition of chlorinated aliphatic hydrocarbons under EB irradiation could be described as follows: Cl⁻ dissociative secondary electron attachment followed by peroxyl radicals reaction is a main path for Cl-HC decomposition, positive and negative charge transfer reactions with Cl-HC, O atoms and other radicals reactions with Cl-HC cause Cl-HC degradation too.

5.1.2 General mechanism of chlorinated aromatic hydrocarbons (Cl-AH) decomposition under EB-irradiation

Similar to the mechanism of chlorinated aliphatic hydrocarbons under EB-irradiation, the mechanism of chlorinated aromatic hydrocarbons go through secondary electron attachment and positive charge transfer reactions at the beginning stage of irradiation. At the late stage of irradiation, radical reactions play very important role for chlorinated aromatic hydrocarbon decomposition. Because rate constants of Cl radicals with chlorinated aromatic hydrocarbons (usually $1.0 \times 10^{-15} \sim 1.0 \times 10^{-16}$) (Shi & Bernhard, 1997) are much smaller than those of OH radicals ($1.0 \times 10^{-12} \sim 1.0 \times 10^{-13}$), Cl radical addition reaction followed by peroxyl radical reaction pathway is not so important for chlorinated aromatic hydrocarbon decomposition in air mixture; on the contrary, OH radical reaction pathway is more important for chlorinated aromatic hydrocarbon decomposition in low or high humidity air mixture (Sun et al., 2007c).

Some positive charge transfer reactions, such as N_2^+, cause benzene ring cleavage of chlorinated aromatic hydrocarbons. Aliphatic byproducts are formed. Hirota et al.(2000) and Han et al. (2003) observed some aliphatic organic compounds formed from chlorobenzene decomposition in air mixture under EB-irradiation.

The generalized chemical reactions could be written as follows:

$$Cl\text{-}AH + M^+ = M + (Cl\text{-}AH)^+ \quad (M^+ = N_2^+, O_2^+, N^+, O^+, NO^+, H_3O^+ \text{ ect.}) \tag{15}$$

$$(Cl\text{-}AH)^+ = \text{products (including ring cleavege reactions)} \tag{16}$$

$$(Cl\text{-}AH)^+ + (\text{radicals / neutral}) = \text{products} \tag{17}$$

$$e + (Cl\text{-}AH) = Cl^- + (AH) \tag{18}$$

$$OH + (Cl\text{-}AH) = \text{products} \tag{19}$$

$$Cl + (Cl\text{-}AH) = \text{products} \tag{20}$$

$$(AH). + (\text{Radicals , neutral or } M^+) = \text{products} \tag{21}$$

5.2 Nonchlorinated organic compounds

For nonchlorinated aromatic organic compounds, VOCs decomposition mainly go through:

• Positive ions' charge transfer reactions:

$$M^+ + RH \ (RH=VOC) = M + RH^+ \tag{22}$$

Because RH has lower ionisation energy (IE) (for eg., IE $_\text{benzene}$ = 9.24 eV; IE$_\text{PAHs}$ < 10 eV) than most primary positive ions (IE > 11 eV), such as N_2^+, O_2^+ formed from radiolysis of base gas, part of VOC will be decomposed by rapid charge transfer reactions.

• Radical-neutral particles reactions

OH radicals play very important role for VOC decomposition, especially when water concentration is above 1%. OH radicals react with VOC in two ways:

OH radicals addition to the aromatic ring or H atom abstraction (e.g. toluene)

$$OH + C_6H_5CH_3 = R1 \ (\text{OH radical addition}) \tag{23}$$

$$C_6H_5CH_3 + OH = R2 + H_2O \ (\text{H atom abstraction}) \tag{24}$$

Radicals (R1, R2) formed above go through very complex reactions: O_2 addition, O atom release, aromatic-CHO (-dehydes), -OH compounds formed or ring cleavage products:

$$R + O_2 = RO_2 \tag{25}$$

$$2RO_2 = 2RO + O_2 \tag{26}$$

$$RO_2 + NO = RO + NO_2 \tag{27}$$

$$RO + O_2 = HO_2 + \text{products (aromatic –CHO, -OH)} \tag{28}$$

$$RO \rightarrow \text{aliphatic products} \tag{29}$$

6. Conclusion

Electron beam technology to treat organic compounds has been studied for many years. Based on experiments of lab scale in batch system and flow system and experiments of pilot scale, it was shown that aliphatic organic compounds (C≤4) are easily to be decomposed by electron beam technology, the enegy necessary to decompose aliphatic hydrocarbons in the order of lower to higher: chlorinated unsaturated hydrocarbons, chlorinated saturated hydrocarbons, hydrocarbons. For aromatic hydrocarbons decomposition in gas phase, energy is much higher than that used to decompose aliphatic hydrocarbons. About 70% aromatic VOCs are decomposed at 20 kGy absorbed dose for most single ring aromatic hydrocabons.

Organic compounds in gas mixture can be decomposed by EB-irradiation, and the decomposition efficiency of organic pollutants increases with the absorbed dose. For chlorinated aliphatic hydrocarbons, the decomposition efficiency of unsaturated (with double C=C bond) hydrocarbons is higher than that of saturated hydrocarbons, and the decomposition efficiency of chlorinated compounds with higher numbers of chlorine groups is higher than observed for the compounds with lower number of chlorine groups. Decomposition efficiency of chlorinated aromatic hydrocarbons is lower than chlorinated unsaturated (with double C = C bonds) aliphatic hydrocarbons.

Different matrix gas and some additives influence the organic pollutants decomposition. For chlorinated aliphatic hydrocarbons, the decomposition efficiency of Cl–HC in oxygen or air is higher thanthat observed in nitrogen; and for chlorinated aromatic hydrocarbons (such as 1,4-DCB) the decomposition efficiency of Cl–HC in nitrogen is higher than that in air. The reason for this can be explained by their different decomposition mechanisms. Water vapor injection and NH_3 addition increase decomposition efficiency of organic pollutants.

Removal efficiency of organic pollutants in hybrid system (EB + catalyst) is higher than that in EB or catalyst system only.

Mechanism of organic pollutants decomposition is composed of following steps. At the early stage of EB irradiation, secondary electrons interact with the base gas mixture components and positive and negative charge transfer reactions play important roles for organic pollutants decomposition. At the latter stage of EB-irradiation, radical reactions play important roles for organic pollutants decomposition.

7. Acknowledgment

This contribution is financed by "PlasTEP: Dissemination and fostering of plasma based technological innovation for environment protection in BSR" (Project No #033 of the Baltic Sea Region Program 2007-2013)", and this financial support is greatly acknowledged.

8. References

Atkinson, R.(1987a). Kinetics of the Gas-Phase Reactions of Cl Atoms with Chloroethenes at 298.72K and Atmosphere Pressure. *International Journal of Chemical Kinetics*, Vol.19, No.12, pp.1097–1105, ISSN 0538-8066

Atkinson, R. (1987b). A Structure–Activity Relationship for the Estimation of Rate Constants for the Gas-Phase Reactions of OH Radicals with Organic Compounds. *International Journal of Chemical Kinetics*, Vol.19, No.9, pp.799–828, ISSN 0538-8066

Bryukov, M.G.; Slagle, I.R.& Knyazev, V.D.(2002). Kinetics of Reactions of Cl Atoms with Methane and Chlorinated Methanes. *The journal of physical chemistry A* , Vol. 106, No.44, pp.10532 –10542, ISSN 1089-5639

Callén, M.S.; de la Cruz, M.T.; Marinov, S.; Stefanova, M.; Murillo, R. & Mastral, A.M. (2007). Flue Gas Cleaning in Power Stations by Using Electron Beam Technology. Influence on PAH Emissions. *Fuel Processing Technology*, Vol. 88, No.3, pp. 251-258, ISSN 0378-3820

Chandra, A.K. & Uchimaru, T.(1999). An Ab Initio Investigation of the Reactions of 1,1- and 1,2- Dichloroethane with Hydroxyl Radical. *The journal of physical chemistry A*, Vol.103, No.50, pp. 10847-10883, ISSN 1089-5639.

Chang, J.S. & Kaufman, F.(1977). Kinetics of the Reactions of Hydroxyl Radicals with Some Halocarbons: $CHFCl_2$, CHF_2Cl, CH_3CCl_3, C_2HCl_3, and C_2Cl_4. *Journal of Chemical Physics*, Vol.66, No.11, pp.4989-4994, ISSN 0021-9606

Chmielewski, A.G.; Ostapczuk, A.; Zimek, Z.; Licki, J. & Kubica, K. (2002). Reduction of VOCs in Flue Gas from Coal Combustion by Electron Beam Treatment. *Radiation Physics and Chemistry*, Vol. 63, No.3-6 , pp. 653-655, ISSN 0969-806X

Chmielewski, A.G.; Sun, Y.; Licki, J.; Bulka, S.; Kubica, K. & Zimek, Z. (2003). NOx and PAHs Removal from Industrial Flue Gas by using Electron Beam Technology with Alcohol Addition, *Radiation Physics and Chemistry*, Vol. 67,No.3-4, pp. 555-560, ISSN 0969-806X

Chmielewski, A.G.; Licki, J.; Pawelec, A.; Tymiński, B. & Zimek, Z. (2004a). Operational Experience of the Industrial Plant for Electron Beam Flue Gas Treatment. *Radiation Physics and Chemistry*, Vol.71, No.1-2, pp.439-442, ISSN 0969-806X.

Chmielewski, A.G.; Sun, Y.-X.; Bulka, S. & Zimek, Z. (2004b). Chlorinated Aliphatic and Aromatic VOC Decomposition in Air Mixture by Using Electron Beam Irradiation. *Radiation Physics and Chemistry*, Vol. 71,No.1-2, pp. 435-438, ISSN 0969-806X

Hakoda, T.; Hirota, K. & Hashimoto, S. (1998a). Decomposition of Tetrachloroethylene by Ionizing Radiation (IAEA-SM-350/4). *Radiation Technology for Conservation of the Environment, Proceeding of a symposium held in Zakopane, Poland, 8–12 September 1997*, IAEA-TECDOC-1023, pp. 55-66, ISSN 1011-4289

Hakoda, T.; Yang, M.; Hirota, K. & Hashimoto, S. (1998b). Decomposition of Volatile Organic Compounds in Air by Electron Beam and Gamma Ray Irradiation. *Journal of Advanced Oxidation Technologies*, Vol. 3, pp.79–86, ISSN 1203-8407

Hakoda, T.; Zhang, G. & Hashimoto, S. (1999). Decomposition of Chloroethenes in Electron Beam Irradiation. *Radiation Physics and Chemistry*, Vol. 55, No.5-6, pp. 541-546, ISSN 0969-806X

Hakoda, T.; Hashimoto, S.; Fujiyama, Y. & Mizuno, A. (2000). Decomposition Mechanism for Electron Beam Irradiation of Vaporized Trichloroethylene–Air Mixtures. *The journal of physical chemistry. A*, Vol. 104, No.1, pp.59–66, ISSN 1089-5639

Hakoda, T.; Zhang, G. & Hashimoto, S. (2001). Chain Oxidation Initiated by OH, O(^{3}P) Radicals, Thermal Electrons, and O_3 in Electron Beam Irradiation of 1,2-Dichloroethylenes and Air Mixtures. *Radiation Physics and Chemistry*, Vol. 62, No.2-3, pp.243–252, ISSN 0969-806X

Hakoda, T.; Matsumoto, K.; Mizuno, A.; Kojima, T. & Hirota, K. (2008a). Catalytic Oxidation of Xylene in Air using TiO_2 under Electron Beam Irradiation. *Radiation Physics and Chemistry*, Vol. 28, No.1, pp. 25-37, ISSN 0969-806X

Hakoda, T.; Shimada, A. & Hirota, K. (2008b). Development of Removal Technology for Volatile Organic Compounds (VOCs) using Electron Beams, *International Conference on Recent Developments and Applications of Nuclear Technologies*, pp. 200, ISBN 978-83-909690-8-4, Białowieża, Poland, September 15-17, 2008.

Han, D.H.; Stuchinskaya, T.; Won, Y.S.; Park, W.S. & Lim, J.K. (2003). Oxidative Decomposition of Aromatic Hydrocarbons by Electron Beam Irradiation. *Radiation Physics and Chemistry*, Vol. 67,No.1, pp. 51-60, ISSN 0969-806X

Hashimoto, S.; Hakoda, T.; Hitora, K. & Arai, H. (2000). Low Energy Electron Beam Treatment of VOCs. *Radiation Physics and Chemistry*, Vol. 57, No.3-6, pp. 485-488, ISSN 0969-806X

Hirota, K.; Hakoda, T.; Arai, H.& Hashimoto, S.(2000). Dechlorination of Chlorobenzene in Air with Electron Beam. *Radiation Physics and Chemistry*, Vol. 57, No.1, pp. 63-73, ISSN 0969-806X

Hirota, K.; Hakoda, T.; Arai, H.& Hashimoto, S.(2002). Electron-Beam Decomposition of Vaporized VOCs in Air. *Radiation Physics and Chemistry*, Vol. 65, No. 4-5, pp. 415-427, ISSN 0969-806X

Hirota, K.; Hakoda, T.; Taguchi, M.; Takigami, M.; Kim, H. & Kojima, T. (2003). Application of Electron Beam for the Reduction of PCDD/F Emission from Municipal Solid Waste Incinerators. *Environmental Science & Technology*, Vol.37, No.14, pp. 3164-3170, ISSN 0013-936X

Hirota, H.; Sakai, H.; Washio, M. & Takuji, K. (2004). Application of Electron Beams for the Treatment of VOC Streams. *Industrial & Engineering Chemistry Research*, Vol.43, No.5, pp. 1185–1191, ISSN 0888-5885

Howard, C.J. (1976). Rate Constants for the Gas-Phase Reactions of OH Radicals with Ethylene and Halogenated Ethylene Compounds. *Journal of Chemical Physics*, Vol.65, No.11, pp. 4771-4777, ISSN 0021-9606

Ighigeanu, D.; Calinescu , I.; Martin, D. & Matei, C. (2008). A New Hybrid Technique for the Volatile Organic Compounds Removal by Combined Use of Electron Beams, Microwaves and Catalysts. *Nuclear Instruments and Methods in Physics Research B*, Vol. 266, No.10, pp. 2524–2528, ISSN 0168-583X

Jeon, E.C.; Kim, K.J.; Kim, J.C.; Kim, K.H.; Chung, S.G.; Sunwoo, Y. & Park, Y.K. (2008). Novel Hybrid Technology for VOC Control using an Electron Beam and Catalyst. *Research on Chemical Intermediates* , Vol.34, No.8–9, pp. 863–870, ISSN 0922-6168

Kim, J.C. (2002). Factors Affecting Aromatic VOC Removal by Electron Beam Treatment. *Radiation Physics and Chemistry*, Vol. 65, No. 4-5 , pp. 429-435, ISSN 0969-806X

Kim, J.; Han, B.; Kim, Y.; Lee, J.H.; Park, C.R.; Kim, J.C. & Kim, K.J. (2004). Removal of VOCs by Hybrid Electron Beam Reactor with Catalyst Bed. *Radiation Physics and Chemistry*, Vol. 71, No.1-2 , pp. 427-430, ISSN 0969-806X

Kim, K.J.; Kim, J.C.; Kim, J. & Sunwoo, Y. (2005). Development of Hybrid Technology using E-Beam and Catalyst for Aromatic VOCs Control. *Radiation Physics and Chemistry*, Vol. 73, No.2, pp. 85-90, ISSN 0969-806X

Knox, J. H. & Riddick, J. (1966). Activated Chloroethyl Radicals in the Chlorination of 1,2-Dichloroethylenes. *Transactions of the Faraday Society*, Vol. 62, pp. 1190-1205, ISSN 0956-5000

Knox, J. & Waugh, K.C. (1969). Activated Chloroalkyl Radicals in the Chlorination of Trichloroethylene and Other Olefins. *Transactions of the Faraday Society* ,Vol.65, pp.1585-1594, ISSN 0956-5000

Liu, A.; Mulac, W.A. & Jonah, C. D. (1989). Pulse Radiolysis Study of the Gas-Phase Reaction of OH Radicals with Vinyl Chloride at 1 atm and Over the Temperature Range 313-1173 K. *The Journal of Physical Chemistry*, Vol.93, No.10, pp. 4092-4094, ISSN 1089-5639

Machi, S. (1983). Radiation Technology for Environmental Conservation. *Radiation Physics and Chemistry*, Vol. 22, No. 1-2, pp. 91-97, ISSN 0969-806X

Mätzing, H. (1989). Chemical Kinetics of Flue Gas Cleaning by Irradiation with Electrons. In: *Advances in Chemical Physics Volume LXXX* , I. Prigogine, & S.A.Rice, (Ed.), 315–402, John Wiley & Sons. Inc., ISBN 0-471-53281-9

Nichipor, H.; Dashouk, E.; Chmielewski, A.G.; Zimek, Z. & Bułka, S. (2000). A Theoretical Study on Decomposition of Carbon Tetrachloride, Trichloroethylene and Ethyl Chloride in Dry Air under the Influence of an Electron Beam. *Radiation Physics and Chemistry*, Vol. 57, No.3-6, pp. 519-525, ISSN 0969-806X

Nichipor, H.; Dashouk, E.; Yacko, S.; Chmielewski, A.G.; Zimek, Z. & Sun, Y. (2002). Chlorinated Hydrocarbons and PAHs Decomposition in Dry and Humid Air by Electron Beam Irradiation. *Radiation Physics and Chemistry*, Vol. 65, No.4-5, pp. 423-427, ISSN 0969-806X

Nichipor,H.; Dashouk, E.; Yacko, S.; Chmielewski, A.G.; Zimek, Z.; Sun, Y. & Vitale, S.A. (2003). The Kinetics of 1,1-Dichloroethene($CCl_2=CH_2$) and Trichloroethene ($HClC=CCl_2$) Decomposition in Dry and Humid Air under the Influence of Electron Beam. *Nukleonik*, Vol. 48, No.1, pp.45-50, ISSN 0029-5922

Nichipor, H.; Yacko,S.; Sun, Y.; Chmielewski, A.G. & Zimek, Z.(2008). Theoretical Study of Dose and Dose Rate Effect on Trichloroethylene ($HClC=CCl_2$) Decomposition in Dry and Humid Air under Electron Beam Irradiation. *Nukleonik*, Vol. 53, No.1, pp.11-16, ISSN 0029-5922

Ostapczuk, A.; Chmielewski, A.G.; Honkonen, V.; Ruuskanen,J.; Tarhanen, J. & Svarfvar, B.(1999). Preliminary Test in Decomposition of Styrene by Electron Beam Treatment. *Radiation Physics and Chemistry*, vol 56, No.4, pp. 369-371, ISSN 0969-806X

Ostapczuk, A.; Licki, J. & Chmielewski, A. (2008a). Polycyclic Aromatic Hydrocarbons in Coal Combustion Flue Gas under Electron Beam Irradiation. *Radiation Physics and Chemistry*, vol 77, No.4, pp. 490-496, ISSN 0969-806X

Ostapczuk, A.; Hakoda, T.; Shimada, A. & Kojima, T. (2008b). Naphthalene and Acenaphthene Decomposition by Electron Beam Generated Plasma Application. *Plasma Chemistry and Plasma Processing*, Vol. 28, No. 4, pp. 483-494, ISSN 0272-4324

Paur, H.-R. (1998). Decomposition of Volatile Organic Compounds and Polycyclic Aromatic hydrocarbons in industrial off gas by electron beam—a review (IAEA-SM-350/52). *Radiation Technology for Conservation of the Environment, Proceeding of a symposium held in Zakopane, Poland, 8–12 September 1997*, IAEA-TECDOC-1023, pp. 67-85, ISSN 1011-4289

Prager, L.; Langguth, H.; Rummel, S. & Mehnert, R. (1995). Electron Beam Degradation of Chlorinated Hydrocarbons in Air. *Radiation Physics and Chemistry*, vol 46, No.4-6, pp. 1137-1142, ISSN 0969-806X

Sanhueza, E. & Heickien, J. (1974a). The Reaction of $O(^3P)$ with C_2Cl_4. *Canadian Journal of Chemistry*, Vol.52, No.23, pp.3870-3878, ISSN 0008-4042

Sanhueza, E. & Heickien, J. (1974b). The Reaction of $O(^3P)$ with C_2HCl_3. *International Journal of Chemical Kinetics*, Vol.6, No.4, pp.553-565, ISSN (printed) 0538-8066, ISSN (electronic) 1097-4601

Shi, J. & Bernhard, M.J. (1997). Kinetic Studies of Cl-atom Reactions with Selected Aromatic Compounds using the Photochemical Reactor-FTIR Spectroscopy Technique. *International Journal of Chemical Kinetics*, Vol. 29, No.5, pp. 349–358, ISSN 0538-8066

Slater, R.C. & Douglas-Hamilton, D.H. (1981). Electron-Beam-Initiated Destruction of Low Concentrations of Vinyl Chloride in Carrier Gases. *Journal of Applied Physics*, Vol. 52, No.9, pp. 5820–5828. ISSN 0021-8979

Son, Y.S.; Kim, J.; Kim, K. & Son, Y.S. (2008). VOC Removal Characteristics for E-Beam-Catalyst Coupling with Respect to Catalysts and Humidity, *Recent developments and Applications of Nuclear Technologies*, pp.201-201, ISBN 978-83-909690-8-4, Białowieża, Poland, Sept 15-17, 2008.

Son, Y.S.; Park, K.N. & Kim, J.C. (2010a). Control Factors and By-Products During Decomposition of Butane in Electron Beam Irradiation. *Radiation Physics and Chemistry*, Vol.79, No.12, pp. 1255-1258, ISSN 0969-806X

Son, Y.S.; Kim, K.J.; Kim, J.Y. & Kim, J.C. (2010b). Comparison of the Decomposition Characteristics of Aromatic VOCs Using an Electron Beam Hybrid System. *Radiation Physics and Chemistry*, Vol.79, No.12, pp. 1270-1274, ISSN 0969-806X

Spanel, P. & Smith, D. (1999a). Seleted Ion Flow Tube Studies of the Reactions of H_3O^+, NO^+, and O_2^+ with Some Chloroalkanes and Chloroalkenes. *International Journal of Mass Spectrometry*. Vol. 184, No. 2-3, pp.175-181, ISSN 1387-3806

Spanel, P. & Smith, D. (1999b). Seleted Ion Flow Tube Studies of the Reactions of H_3O^+, NO^+, and O_2^+ with Several Aromatic and Aliphatic Monosubstituted Halocarbons. *International Journal of Mass Spectrometry*. Vol.189, No. 2-3, pp. 213-223, ISSN 1387-3806

Srivastava, R.K.; Jozewicz, W. & Singer, C.(2001). SO_2 Scrubbing Technologies: a Review. *Environmental Progress*, Vol. 20, No.4, pp. 219-228, ISSN 0278-4491

Sun, Y.; Hakoda, T.; Chmielewski, A.G.; Hashimoto, S.; Zimek, Z.; Bułka, S.; Ostapczuk, A. & Nichipor, H. (2001). Mechanism of Decomposition of 1,1-Dichloroethylene in Humid Air under Electron beam Irradiation. *Radiation Physics and Chemistry*, Vol. 62, No.4, pp. 353-360, ISSN 0969-806X

Sun, Y.; Hakoda, T.; Chmielewski, A.G. & Hashimoto, S. (2003). Trans-1,2- Dichloroethylene Decomposition in Low-Humidity Air under Electron Beam Irradiation. *Radiation Physics and Chemistry*, Vol. 68, No.5, pp. 843-850, ISSN 0969-806X

Sun, Y. & Chmielewski, A.G. (2004). 1,2-Dichlroethylene Decomposition in Air Mixture Using Ionization Technology. *Radiation Physics and Chemistry*, Vol. 71, No.1-2, pp. 433-436, ISSN 0969-806X

Sun, Y.; Chmielewski,A.G.; Bułka, S.& Zimek, Z.(2006). Influence of Base Gas Mixture on Decomposition of 1,4-Dichlorobenzene in an Electron Beam Generated Plasma Reactor. *Plasma Chemistry and Plasma Processing*, Vol. 26, No.4, pp. 347 – 359, ISSN: 0272-4324.

Sun, Y.; Chmielewski, A.G.; Bułka, S. & Zimek, Z.(2007a). 1-Chloronaphthalene Decomposition in Different Gas Mixtures under Electron Beam Irradiation. *Radiation Physics and Chemistry*, Vol. 76, No.11-12, pp. 1802-1805, ISSN 0969-806X

Sun, Y.; Chmielewski, A.G.; Bułka, S.; Zimek, Z. & Nichipor,H. (2007b). Simulation of Decomposition of Dichloroethylenes (trans-DCE, cis-DCE, 1,1-DCE)/Air under Electron Beam Irradiation. *Nukleonik*, Vol. 52, No.2, pp.59-67, ISSN 0029-5922

Sun, Y.; Chmielewski, A.G.; Bułka, S.; Zimek, Z. & Nichipor,H. (2007c). Mechanism of Decomposition of 1,4-Dichlorobenzene/Air in an Electron Beam Generated Plasma Reactor. *Radiation Physics and Chemistry*, Vol. 76, No.7, pp. 1132-1139, ISSN 0969-806X

Sun, Y.; Chmielewski, A.G.; Bułka, S.& Zimek, Z.(2008). Organic Pollutants Treatment in Gas Phase by Using Electron Beam Generated Non-Thermal Plasma Reactor. *Chemické Listy* 102, s1524-s1528, ISSN 1803-2389

Sun, Y.; Chmielewski, A.G.; Bułka, S.& Zimek, Z. (2009a). Decomposition of Toluene in Air Mixtures Under Electron Beam Irradiation. *Nukleonika*, Vol. 54, No.2, pp.65-70, ISSN 0029-5922

Sun, Y.; Chmielewski, A.G.; Bułka, S.; Zimek, Z. & Nichipor,H.(2009b). Simulation Calculations of Tetrachloroethylene Decomposition in Air Mixtures Under Electron Beam Irradiation. *Radiation Physics and Chemistry*, Vol. 78, No.7-8, pp. 715-719, ISSN 0969-806X

Szamrej, I.; Tchórzewska, W.; Kość, H. & Foryś, M. (1996). Thermal Electron Attachment Processes in Halomethanes - Part I. CH_2Cl_2, $CHFCl_2$ and CF_2Cl_2. *Radiation Physics and Chemistry*, Vol. 47, No.2, pp. 269-273, ISSN 0969-806X

Teruel, M.A.; Taccone, R.A. & Lane, S.I. (2001). Gas-Phase Reactivity Study of $O(^3P)$ atoms with Trans-CHCl = CHCl and CHCl = CCl_2 at 298 K: Comparison to Reactions with Some Other Substituted Ethenes. *International Journal of Chemical Kinetics* , Vol. 33, No.7, pp. 415-421, ISSN 0538-8066

Thüner, L.P.; Barnes, I.; Becker, K.H.; Wallington, T.J.; Chrisensen, L.K.; Orlando, J.J. & Ramacher, B. (1999). Atmospheric Chemistry of Tetrachloroethylene ($Cl_2C=CCl_2$): Products of Chlorine Atom Initiated Oxidation. *The journal of physical chemistry. A*, Vol. 103, No.43, pp. 8657-8663, ISSN 1089-5639

Vitale, S.A.; Hadidi, K.; Cohn, D.R.; Falkos, P. & Bromberg, L. (1996). Electron Beam Generated Plasma Decompositio of 1,1,1-Trichloroethane. *Plasma Chemistry and Plasma Processing*, Vol.16, No.4, pp. 651–668, ISSN 0272-4324

Vitale, S.A.; Hadidi, K.; Cohn, D.R. & Bromberg, L. (1997a). Decomposition of Ethyl Chloride and Vinyl Chloride in an Electron Beam Generated Plasma Reactor. *Radiation Physics and Chemistry*, Vol. 49, No.4 , pp. 421-428, ISSN 0969-806X

Vitale, S.A.; Hadidi, K.; Cohn, D.R. &, Bromberg, L. (1997b). Evaluation of the Reaction Rate Constants for Chlorinated Ethylene and Ethane Decomposition in Attachment-Dominated Atmospheric Pressure Dry Air Plasmas. *Physics Letter A*, Vol. 232, No.6, pp. 447–455, ISSN 0375-9601

Vitale, S.A.; Hadidi, K.; Cohn, D.R. & Bromberg, L. (1997c). Decomposition of 1,1-Dichloroethane and 1,1-Dichloroethene in an Electron Beam Generated Plasma Reactor. *Journal of Applied Physics*, Vol.81, No.1, pp. 2863–2868, ISSN 0021-8979

Vitale, S.A.; Hadidi, K.; Cohn, D.R. & Falkos, P. (1997d). The Effect of a Carbon–Carbon Double Bond on Electron Beam-Generated Plasma Decomposition of Trichloroethylene and 1,1,1-Trichloroethane. *Plasma Chemistry and Plasma Processing*, Vol.17, No.1, pp. 59–78, ISSN 0272-4324

Willis C. & Boyd.A.W.(1976). Excitation in the Radiation Chemistry of Inorganic Gases. *International Journal for Radiation Physics and Chemistry*, Vol. 8, No.1-2, pp. 71-112, ISSN 0020-7055

Won, Y.-S.; Han, D.-H.; Stuchinskaya, T.; Park, W.-S. & Lee, H.-S. (2002). Electron Beam Treatment of Chloroethylenes/Air Mixture in a Flow Reactor. *Radiation Physics and Chemistry*, Vol. 63, No.2 , pp. 165-175, ISSN 0969-806X

Zimek, Z. (1995). High Power Electron Accelerators for Flue Gas Treatment. *Radiation Physics and Chemistry*, Vol. 45, No.6, pp. 1013-1015, ISSN 0969-806X

Zimek, Z. (2005). High Power Accelerators and Processing Systems for Environmental Application. *Radiaiton Treatment of Gaseous and Liquid Effluents for Contaminant Removal*, IAEA-TECDOC-1473, pp. 125-137, ISBN 92-0-110405-7, ISSN 1011-4289

9

Vapor Phase Hydrogen Peroxide – Method for Decontamination of Surfaces and Working Areas from Organic Pollutants

Petr Kačer[1], Jiří Švrček[1], Kamila Syslová[1], Jiří Václavík[1],
Dušan Pavlík[2], Jaroslav Červený[2] and Marek Kuzma[2]
[1]*Institute of Chemical Technology*,
[2]*Institute of Microbiology*,
Prague
Czech Republic

1. Introduction

Decontamination, i.e. cleaning by removal of chemicals or germs, is a term commonly used for the process of treating devices, instruments and surfaces in order to ensure their safe operation. It is of exceptional importance in healthcare, food and pharmaceutical industry as well as in the areas of army and public defense. The decontamination process covers several steps like simple washing with water and soap and final disinfection or sterilization. Sterilization is a process which reduces the microbial contamination by 6 logs and can thus prevent the spread of infectious diseases in medical centers, where it forms a part of daily cleaning routines. Disinfection is a very similar process to sterilization, but the reduction rate of microbial contamination is only 5 logs. (Favero & Bond, 1991; Sagripanti & Bonifacino, 1996)
Decontamination is not only elimination of biological pollution but it also comprises detoxification and removal of dangerous chemical compounds. It should be applied whenever a real threat of microbial or chemical contamination exists.

2. History

While the sophisticated ("scientific") ways of decontamination only started to emerge about 150 years ago, essentially similar process are already mentioned in the Bible, in the works of the poet Homer and in the files of Aristotle. An important milestone concerning decontamination was the year 1438 when Sanitary council in Venice was found to provide fumigation of cargo delivered to the port. This institution represented the fundamental prevention and active defense against infectious diseases and parasites. Maturity of Italian health service was demonstrated in the works of poet, philosopher and physician Girolam Fracastor (1478-1553). He first promoted an idea that epidemics had been caused by very small particles which can be transferred among people by three different ways: by contact with infected patients, by contact with the contaminated staff (medicinal staff) and by air transport. Although this great man's work can be considered a huge milestone in the fight with infectious diseases, in his time it was forgotten and only rediscovered in the 20th century (1930 - W. C. Wright, 1960 - W. Bulloch). In 1676, a chemical (vinegar) was used to kill germs for the first time by Antonie van Leeuwenhoek who observed them with his microscope, calling them „animalcules". However, the breakpoint in this topic came in the second half of the 18th century with the discovery of chlorine (1774, C. W. Scheele) and hypochlorites (1789, C. L. Berthollet). These compounds quickly found their application in deadhouses, sewers, hospitals areas, ships, prisons and mainly in drinking water treatment. In 1810 Nicolas Appert discovered the modern food sterilization method by temperature conservation. Shortly after this, the founder of microbiology Louis Pasteur discovered the sterilization effect of overheated steam. This further inspired Charles Chamberland to construct the first steam autoclave (1879). Parallel with the development of this excellent technique, in 1877 A. Downes and T. P. Blunt discovered the antimicrobial effects of ultraviolet light and M. Wald (1892) continued in this work describing the relation of light wavelength and its germicidal effect (blue light is more effective than red one). Another famous man connected with the decontamination process was Robert Koch who in his book "On Disinfection" (1881) described the potential of 70 chemical compounds at different concentrations, temperatures and various mixtures to eliminate the spores of anthrax. In 1897 B. Krönig and T. Paul developed the grounds of chemical disinfection and these principles were applied in the famous "phenol coefficient method" to test the effectiveness of disinfection compounds. The following 20th century meant a great improvement in chemistry – mainly organic – which lead to the discovery of many disinfection compounds (Block, 1991; Fraise, 2004).

2.1 Reasons for decontamination
The need of decontamination already appeared during the army operations in the Antiquity. Aristotle (384 – 332 B.C.) revealed the danger of infectious diseases and recommended preventive measures for the army troops of Alexander the Great. The progress of medicine and the study of infectious diseases lead to formulation of strict rules to prevent infections. During epidemic periods, the application of protective means was becoming common, e.g. application of antiseptic compounds against gangrene (1750 J. Pringle), using of hypochlorite solutions before surgeries (O. W. Holmese - 1843 a I. P. Semmelweise - 1861), treatment of surgery instruments with flame, sterilization of bandage by heat (L. Pasteur) and many others. Despite all these advances, more soldiers died during the Second World War due to infections and diseases than as a consequence of fight injuries (Block, 1991).

Decontamination is an important part of the whole modern medicine system and is based on strict rules and application of several procedures like cleaning, disinfection, sterilization etc. Nonetheless, nowadays a real danger of pandemic (epidemic spreading in parallel in several states or continents) also exists, which was evident in the case of the two recent pandemics of flu (bird flu, pig flu) that proceeded very fast. Therefore, hospital decontamination is based on proper decontamination equipment (built-in or mobile), protective items and educated staff.

Another very important reason for the progress of decontamination is chemical war as a new military strategy. In the 19th and 20th centuries, the great progress of chemistry led to the development of several poisonous substances (weapons of mass destruction) (Duffy, 2009). A warfare agent of such kind was used for the first time during the WWI when the

Germans used chlorine (April 22, 1915). Later, phosgene, benzyl bromide and others were used (Duffy, 2009). The destructive force of these weapons was improved by new stable, more potent and easily spreading compounds. From the beginning of the WWI to the end of WWII, blistering agents were developed such as yperite and even more dangerous nerve agents like somane, sarine or tabune. Fortunately, they have never been used in war. After the WWII, research was focused on the development of nerve agents and effective defense against them. The new types of V-agents were developed in 1995 and represent the most toxic compounds ever synthesized.

Although the application of chemical weapons is currently considered a war crime (1993 Paris convention) (International Committee of the Red Cross, 2005), they are used by terrorists against civilian population. For instance, one can mention the sarine terroristic attacks in Japan (1995 Tokyo subway, 1994 Matsumoto town) (Okumura et al., 1998, 2003) or the bio-terrorist attacks by Anthrax spores delivered by mail service in the USA in 2001 (24 buildings were contaminated and the remedies cost as much as 200 million USD) (Jernigan et al., 2002). These incidents showed global unpreparedness for large contamination and improper decontamination methods for such spaces.

Decontamination is therefore highly important, both in the defense of an individual person and the defense of a country mainly against the pandemics or terroristic attacks. Beside this, the huge amounts of toxic compounds which are daily manufactured, modified and transported need to be considered. In case of an accident or improper manipulation, these substances can endanger the safety of a particular person or the whole environment. As a few examples of the 20th century, we would like to mention the outflow of toxic dioxin in Italian town Seveso in 1976 (initiated two prevention guidelines SEVESO I and SEVESO II), the nuclear accident in Chernobyl (1986) and the biggest industrial accident in the town Bhopal, India, (1984) where approximately 20,000 people died (Sharma, 2005).

For the sake of global prevention, it is extremely desirable to develop novel effective decontamination methods appropriate for application in large areas like rooms, buildings, airplanes, subways or airports. Despite the fact that decontamination of such premises represents an issue of foremost importance, it is not solved satisfactorily at present. An example of such solution is the patent of the United Technologies Corporation (Watkins, 2006) dealing with easy distribution of hydrogen peroxide aerosol or gas to large areas. A list of important methods usable for large and closed areas can be found in the Compilation of Available Data on Building Decontamination Alternatives issued by the U.S. Environmental Protection Agency (EPA) in 2005 (U.S. Environmental Protection Agency, 2005). In the last three decades, chemical contamination has most frequently been caused by pollution with toxic and usually carcinogenic pesticides or industrial intermediates while accidents or terrorist attacks have only formed a minority of cases. These compounds are mainly characterized by greater stability and usually persist in the environment for a long time. A second important group of polluting compounds are pharmaceuticals, diagnostics, flavoring substances and other bioactive compounds. Presently, the production, distribution and application of bioactive substances indeed represent a fast-growing branch of industry. These compounds and their bioactive metabolites are mainly concentrated in wastewaters, which are subsequently drained into the environment. Although their concentration is rather limited (ng l^{-1} – ug l^{-1}, corresponds to ppt – ppb) and they do not seem to be an actual danger, their final influence can be very dangerous. The active pharmaceutical substances are designed for a very specific effect, but their final side effects can often be unpredictable.

Lists of pharmaceuticals which contaminate the environment can be found in the literature. These compounds come either in the unchanged form or in the form of their biotransformation products (metabolites, which can be more active than the original substances (Daughton & Ternes, 1999)). The lists include for example analgesics, antidepressants, antiepileptics, antihypertensives, antiseptics, cytostatics, hormones, cholesterol reducing substances, radiocontrast substances, steroids, tranquilizers and others. (Daughton & Ternes, 1999; Lopez et al., 2003; Pereira et al., 2007). Big sources of these substances are of course pharmaceutical companies, research laboratories, hospitals, pharmacies and households in which these compounds are used.

National Institute for Occupational Safety and Health, NIOSH, estimates that 5.5 million of workers can be indirectly exposed to dangerous pharmaceutical substances which are commonly labeled "cytotoxic". These data are based on evaluation between 1996-1998 in the USA (National Institute for Occupational Safety and Health, 2004) among research staff, pharmacists, physicians, nurses and other supporting staff. Staff from pharmaceutical factories were not considered because of the strict rules for clean premises which reduced the risk of contamination to minimum. On the contrary, in hospitals the staff and the patient relatives can be in touch with hazardous pharmaceutical substances like cytostatics, antivirotics, hormones etc. (National Institute for Occupational Safety and Health, 2004). These substances can cause undesirable effects (Castegnaro et al., 1997) by transport into their bodies through skin absorption, inhalation of aerosols, syringe needles or open strokes. The danger of indirect contamination with pharmaceutical substances on the inner surfaces of 14 hospitals in Germany is described in the work of Schmaus (Schmaus et al., 2002). Better results were naturally achieved in those hospitals which strictly followed the safety measures and where their staff worked properly with cytotoxic substances.

Methods of protection against dangerous cytotoxic substances are nowadays undergoing a fast development. The most important measure is to strictly follow proper working rules to avoid contact of staff with these compounds, which means e.g. working in isolators with intrinsic decontamination system, proper waste management, safe storage or periodic evaluation of the level of contamination. (Fisher & Caputo, 2004; National Institute for Occupational Safety and Health, 2004) Currently, the development is focused on new decontamination methods to provide perfect cleanup and inactivation of dangerous substances in various kinds of waste material packaging and biological liquids before their disposal (Cazin & Gosselin, 1999; Hansel et al., 1997).

2.2 State of the art of decontamination techniques

Up to now, a lot of decontamination methods to inactivate biological pollution on different surfaces have been developed. Surface decontamination of chemical (mainly cytotoxic) substances was studied less intensively but nowadays it is also becoming a priority. Decontamination can be carried out by several ways depending on the contaminant, environment, area size or target (people, tools, indoor or outdoor space). Among the main factors affecting the right selection of suitable decontamination techniques are: the method of distribution of the decontamination substance (washing, wiping, spraying, foaming, using aerosol, fumigation etc.); the operation range (selectivity to the microorganism or chemical pollutants); the influence of working conditions (temperature, humidity, presence of other compounds); the operation time (time necessary for proper reaction – minutes or hours) and the influence of the decontaminant upon target materials (possibility of damage).

It is possible to categorize the decontamination methods according to their principle of action to mechanical, chemical, physical and physicochemical. The most widespread are chemical decontamination procedures (an action of a chemical agent to a decontaminated item) that can be applied in two different ways – the wet approach, which uses water or other solution of active agent and the dry method which uses gas or vapor phase of active substance.

Application of the aforementioned methods is connected with serious danger because of toxicity, carcinogenicity, flammability or explosiveness of agents, irradiation or burn caused by rays, and potential toxic residues which are harmful to the environment. A big disadvantage of liquid agents' application is a non-uniform distribution on the surfaces in all target areas. This disadvantage was solved by spraying of the agents or using fumigation (vapor or gas of the active agent). It is evident (see Table 1), that chemical agents are a very heterogeneous group with different mechanisms of action. It is also important that several chemical agents can be mixed with detergents, which can support the deactivation process, or other compounds that improve their properties (anticorrosives, aromatic additives and others). These additives can e.g. substantially reduce the surveillance of germs (bacteria, viruses or fungi). According to their influence, these agents can be divided into two types:

- (Bacteria) -cide- meaning permanent dispatch
- (Bacteria) -static- meaning temporary loss of any ability, i.e. multiplication or growing.

Type of decontamination		Active agent
Mechanical ways		Sucking, washing, wiping etc.
Chemical ways	Wet	Water solution of ClO_2, CH_3COOOH, H_2O_2, NaOCl, liquid detergents (presence of quaternary ammonia salts), alcohols, aldehydes, phenol derivatives, iodoform, Fenton agent and others.
	Vapor or gas phase	Ethylene oxide, formaldehyde, ClO_2, O_3, CH_3COOOH, H_2O_2, propylene oxide, β-propiolactone, methylene bromide and others.
Physical ways		X-ray, gamma ray, microwave and UV radiation, heat (dry, wet – water steam), freezing, plasma, photochemical reaction, hydrostatic pressure and others.
Physicochemical ways		Heat or radiation combined with chemical agents

Table 1. A list of basic decontamination methods (Kuzma et al., 2008; McDonnell, 2004; McDonnell & Russell, 1999; Rogers et al., 2005; Russell, 1990, 1991).

It is necessary to regularly change the decontamination agent with respect to a different active substance in order to avoid the potential resistance of the microorganisms to the agent used. An ideal procedure for the cleanup of chemical contaminants using chemical agents should consist of their physical removal from the surface followed by effective degradation to nontoxic or at least less toxic compounds. Not a single decontamination agent with such a wide spectrum of reactivity exists, which could be used for the decontamination of all biologically active substances. Their degradation can even lead to formation of products which are more toxic than the original substance. Identification of these products is very difficult. Therefore, it is preferred to apply "one use" surfaces during the operation with biologically active substances. These surfaces are sequentially washed but their decontamination is quite risky due to the removal of highly toxic contaminants like cytostatics or immunosuppresives. Moreover, they can be drained to the environment which is dangerous not only for the staff but, more importantly, to the whole population. It was already mentioned that these substances are strictly designed for a specific application but they can have numerous side effects on people. Thus, the contact with these compounds has to be avoided as much as possible (Roberts et al., 2006).

NIOSH recommends to decontaminate all surfaces which have been in touch with cytotoxics according a protocol that includes appropriate agent able to deactivate or remove chemical or biological contaminants (National Institute for Occupational Safety and Health, 2004). The basic question to be solved in the field of chemical decontamination is the criterion determining the level at which the contaminant can be considered deactivated. In the case of biologically active substances like warfare agents or pharmaceuticals, this criterion represents the loss of their biological activity, i.e. changes in the chemical structure of the contaminant leading to biological inactivity. An ideal degradation leads to gaseous, nontoxic products which can be easily exhausted – oxides of elements commonly contained in organic molecules (CO_2, H_2O, NO_x). Since they are products of oxidative reactions, it is appropriate that strong oxidative agents be used. Even though the application of $KMnO_4$ seems to be very effective, due to safety reasons it is not acceptable in common places like hospitals (Barek et al., 1998). Other well-known oxidative substances like $Ca(OCl)_2$, $NaOCl$ or H_2O_2 are widely applicable mainly in the liquid form and thus they can only be used in local areas. Unfortunately, the methods described above are inapplicable for treatment of large areas in routine application. In the development of novel decontamination technologies, it is necessary to approach the "ideal decontamination agent" (Rutala & Weber, 1999).

3. Ideal decontamination agents

Based on the above-mentioned facts, the "ideal decontamination agent" can be defined. It should possess high activity against a wide spectrum of biological and chemical contaminants, quick start of action and long-lasting effect. It should be nontoxic to humans and the environment, compatible with a variety of materials, resistant to organic materials, have a non-limited disposal and long term stability during storage. It should also be easily detectable, have a pleasant or no smell and a reasonable cost. Its handling (application and storage) should be easy and safe. Another useful property of the ideal agent can be e.g. applicability to large areas or whole buildings mainly in case of pandemics, chemical accidents or terrorist attacks. The contaminants can be found in places difficult to clean like cracks in walls, carpets, woods, ventilation pipelines and air conditioning units. The common methods are inapplicable in large areas where the only solution is application of a gaseous decontamination agent because it provides easy distribution and penetration in broken surfaces. However, the most commonly known gas phase decontamination agents are connected with a number of disadvantages such as toxicity, material incompatibility, concentration requirements, time of exposition and aeration time (Rogers et al., 2005). From the aforementioned chemical agents, hydrogen peroxide looks like an ideal agent due to the non-toxic products of its decomposition – water and oxygen. The well-described antimicrobial activity and strong oxidative potential are also a good precondition for its wide application mainly against biological and chemical contaminants.

4. Vapor phase hydrogen peroxide

Very close to "the ideal decontamination agent" seems to be Vapor Phase Hydrogen Peroxide (VPHP). It is a relatively new but very progressive method with many advantages:
-   approved sterilization of a wide range of microorganisms
-   higher germicidal activity than what can be achieved with a liquid solution of hydrogen peroxide
-   environmental friendliness – the decomposition products are water and oxygen and leave no toxic residues on the surfaces
-   possibility of its usage under common conditions - atmospheric pressure, laboratory temperature
-   applicability in larger areas and rugged surfaces.

For its excellent antimicrobial activity and nontoxic decomposition residues, the VPHP process tends to replace especially toxic, carcinogenic and potentially explosive formaldehyde and ethylene oxide used for sterilization of heat-sensitive materials (Block, 1991; Heckert et al., 1997).

VPHP is typically generated from a water solution of hydrogen peroxide (35% w/w). A common way of vapor generation is controlled heating of the solution under proper conditions avoiding decomposition of the VPHP. Like all other decontamination agents in vapor phase, also VPHP is decomposed during the operation (in fact, the rate of decomposition of hydrogen peroxide is higher than the decomposition rate of ClO_2) and it is thus necessary to refill "fresh" VPHP to the target area. The dose of new VPHP keeps the required concentration during the whole process. At the end of decontamination, the generator is switched off and the rest of hydrogen peroxide vapor is ventilated out by aseptic air. This exhaust goes through the catalyst to decompose hydrogen peroxide (U.S. Environmental Protection Agency, 2005).

The sterilization properties of VPHP were for the first time mentioned in the 70's of the 20th century, but the modern concept is dated 1989 when this method was used for quick sterilization of rugged dental instruments (Block, 1991). In the same year, EPA approved the usage of VPHP in closed premises like isolators, closed rooms or operation boxes (U.S. Environmental Protection Agency, 2005; McDonnell et al., 2007). Since then, fast-growing application of this agent has started, focusing mainly on bio-decontamination in pharmaceutical industry, health service and food industry (Block, 1991; Kahnert et al., 2005; Klapes & Vesley, 1990). In 2001, VPHP was used for the first time for decontamination of two post office buildings (the General Services Administration's Buildings 410 in Washington, D.C. and the U.S. State Department Mail Facility in Sterling, Virginia; contaminated space 30 000 – 60 000 m³). These buildings were contaminated by Anthrax spores released from "Anthrax letters" by terroristic attacks in the USA (U.S. Environmental Protection Agency, 2005).

In the literature (Block, 1991; Heckert et al., 1997; Johnston et al., 2005; Klapes & Vesley, 1990; Roberts et al., 2006), there are numerous applications of VPHP in the decontamination of fermenters, dialysers, incubators, isolators (Fisher & Caputo, 2004; Lysfjord & Porter,

1998), glove boxes, hazard boxes (Hall et al., 2007), animal houses (Kahnert et al., 2005; Krause et al., 2001), hospital wards (French et al., 2004; Hardy et al., 2007), inner space of airplanes (Krieger & Mielnik, 2005; Shaffstall et al., 2011), ambulances, various large spaces (Krause et al., 2001), lyophylisators (Johnson et al., 1992), ultra centrifuges, sterilization tests (Kokubo et al., 1998), product and pipe lines (Hatanaka & Schibauchi, 1989), dental and surgery instruments (catheters, endoscopes, etc.) (Bathina et al., 1998), contact lenses, hardware space systems (Chung et al., 2008), and food commodities (Forney et al., 1991; Gruhn et al., 1995; Sapers et al., 2003; Simmons et al., 1997). It is a method by which a high selectivity of the process can be achieved due to very precise control of the sterilization conditions (concentration of VPHP, temperature, time of sterilization) and thus only pathogenic microorganisms are destroyed leaving normal living cells unharmed. Therefore, this method is feasible for the decontamination of the surface of living cell cultures. The world-known companies like Tetra Pak International, PepsiCo Inc. or Tetra Laval Holding & Finance are dealing with research and application of in-line sterilization of food packaging by VPHP, which clearly demonstrates how important this technology is in this branch. An interesting and effective application of this method is bleaching of textile materials, which allows to decrease the operating temperature and thus to increase the economics of the whole process. The main leaders in bio-decontamination of surfaces by VPHP are at present companies Steris (Mentor, USA) and Bioquell (Andover, UK). Beside them, several other companies work on VPHP technology, e.g. Pharmaceutical Systems (Franklin Lakes, USA), American Sterilizer Company (Mentor, USA), Johnson & Johnson and division Ethicon (New Brunswick, New Jersey, USA) or Surgikos (New Brunswick, USA).

5. VPHP – Mechanism of action

Information about the mechanism of VPHP action is actually very limited as the process is still in the focus of basic research (Klapes & Vesley, 1990). Current literature data (Chung et al., 2008; Unger-Bimczok et al., 2008; U.S. Environmental Protection Agency, 2005) indicate that the VPHP process is a multi-parameter problem, the effectiveness of which is mainly influenced by the concentration of gaseous hydrogen peroxide, temperature, relative humidity, and condensation of hydrogen peroxide on the decontaminated surfaces. Similar behavior was found for gaseous formaldehyde as a decontamination agent which has been described in detail (Hoffman & Spiner, 1970). Unger et al. (Unger et al., 2007) for first time studied the influence of all these conditions on the sporicidal effect of VPHP. The results suggest that the main parameter for microbial deactivation is the molecular distribution of water and hydrogen peroxide on the surface while the concentration of hydrogen peroxide only plays a secondary role. It was also found that the decontamination cycle using a relatively lower concentration of hydrogen peroxide and higher relative humidity gave very similar results as an experiment with higher concentration of hydrogen peroxide and lower relative humidity. This confirms the possibility of conducting the VPHP process in two ways: "wet" or "dry".

6. VPHP – Operational conditions

Decontamination of closed areas by VPHP is carried out in 4 consecutive steps (Fisher & Caputo, 2004; Heckert et al., 1997; Roberts et al., 2006; Watling et al., 2002). The first phase is dehumidification, i.e. reduction of humidity to an acceptable level, and also temperature stabilization of the VPHP generator. The second phase is conditioning which includes the transport of evaporated hydrogen peroxide by the carrying medium (air) to the decontaminated area and achievement of the required hydrogen peroxide concentration. The third phase is the decontamination itself – that means steady evaporation of hydrogen peroxide and its transport to the decontaminated area to maintain a constant concentration during the whole process. Aeration represents the final phase which consists in feeding of aseptic air in the decontaminated area to exhaust hydrogen peroxide vapor and to keep its concentration at a safe level. The process is illustrated in Figure 1.

Duration of the whole decontamination cycle depends on many parameters. The main ones are the size of area, the profile of surfaces, the way of VPHP generation, endurance of the contaminant against VPHP and the method of space aeration. The whole decontamination cycle should not exceed 10 hours as this is a limit of application of this chemical compound as a sterilization agent (Gurevich, 1991). The decontamination cycle based on biocidal properties of VPHP meets this requirement. There are two different ideas about how to carry out the surface decontamination with VPHP (Fisher & Caputo, 2004; Unger-Bimczok et al., 2008; Watling et al., 2002). The traditional one prefers to perform the decontamination under "dry" conditions without condensation of hydrogen peroxide or water (preferred by Steris (Fisher & Caputo, 2004)). Condensation is unwanted because of corrosion of many materials and prolonged aeration time. The process is not under control and thus in case of condensation the decontamination is not homogeneous (Unger-Bimczok et al., 2008; Watling et al., 2002). In case of the "dry" way of the VPHP process, decontamination of closed indoor spaces and surfaces or quick inactivation of contaminants due to high concentration of hydrogen peroxide vapors is quickly achieved. However, the atmosphere in a defined space can absorb only a limited amount of water and hydrogen peroxide, so it is necessary to remove humidity out of this space by desiccators to avoid condensation (Fisher & Caputo, 2004). On the contrary, the second popular opinion says that the hydrogen peroxide vapors are stable and condensation is necessary. In this view, also condensation is the primary reason of the VPHP decontamination and thus condensation is necessary in order to carry out surface decontamination by hydrogen peroxide vapors (Watling et al., 2002). This theory is supported by theoretical and experimental analysis which clearly show that condensation, and mainly "microcondensation" (i.e. non-visible condensation in small amounts), are key and critical parameters for quick and reproducible inactivation of microorganisms by VPHP (Unger-Bimczok et al., 2008). Condensation (preferred by Bioquel (Fisher & Caputo, 2004)) in case of the "wet" VPHP process is the basic requirement of the technology. A thin condensation film is formed on the decontaminated surfaces. It is necessary to control the condensation level of decontamination agents during the whole process because it inhibits the process (Sheth & Upchurch, 1996). Determination of the vapor mixture dew point (i.e. hydrogen peroxide and water vapours) is very difficult. Its value depends on the temperature and pressure and cannot be predicted easily. This dependence is well described by the Raoult's law which requires the input knowledge of activity coefficients that are also dependent on temperature and concentration. For better understanding of the difference between "wet" and "dry" processes, it is good to study the thermodynamics of the hydrogen peroxide-water solution and its behavior during evaporation and condensation (Manatt & Manatt, 2004; Scatchard et al., 1952). The pressure of saturated vapor of water and hydrogen peroxide is below atmospheric so both compounds start evaporating under atmospheric conditions. Lower pressure of saturated vapor of hydrogen peroxide causes

Concentration H_2O_2 at 25 °C [mass. %]	
Vapor	Liquid
1,9	32,1
8,0	55,7
24,1	73,8
35,0	77,8
58,4	88,3

Table 2. The equilibrium concentration of vapor and liquid phase of H_2O_2 reached by evaporation (Hultman et al., 2007)

The hydrogen peroxide vapor can be generated in two ways – by controlled or flash evaporation. If the liquid solution of hydrogen peroxide evaporates in a dry closed space at 25°C (normal conditions), the concentration of hydrogen peroxide in the gas phase is much lower than in the liquid phase because of faster water evaporation from the hydrogen peroxide solution. For example, by evaporation of 35% (w/w) hydrogen peroxide solution, the final gas phase contains 2.15 % (w/w) H_2O_2 and 65% of H_2O (w/w) (Hultman et al., 2007). Saturation is a state when no more hydrogen peroxide and water vapor can be absorbed and thus condensation occurs.

A different situation occurs when water and hydrogen peroxide vapor condense at 25 °C. In Table 2, it is shown that the equilibrium condensate concentration formed from the vapor above 35% (w/w) hydrogen peroxide solution is 77.8 %, which is about two times higher than in the parent solution. Hydrogen peroxide in the vapor phase condenses preferentially over water. If the system contains vapors of hydrogen peroxide, its condensation occurs and thus its concentration in the vapor phase decreases.

The rate of evaporation can be increased by supplying heat. However, this must be done with a particular care because of safety reasons – hydrogen peroxide is extremely unstable at higher temperatures. Heat supply to the concentrated evaporated hydrogen peroxide slightly above normal conditions is safe and can be used for proper evaporation of concentrated hydrogen peroxide. The condensate is a highly concentrated solution that is not compatible with a wide range of materials and can cause corrosion.

Flash evaporation is another kind of a process related to hydrogen peroxide evaporation. The solution of hydrogen peroxide can be directly applied on a heated surface and thus evaporated. During the flash evaporation, hydrogen peroxide and water are evaporated from the solution simultaneously, so the concentration in the vapor phase is approximately the same as the concentration of the starting solution (decomposition of hydrogen peroxide is not considered). Thus, the concentration in the condensate is the same as in the parent solution.

In case of the "wet" VPHP process, the high concentration of hydrogen peroxide in the condensate can have a positive effect with respect to faster microbial decontamination, but only if the condensate covers the entire surface homogeneously. This is, however, nearly impossible to ensure, owing to different surface profiles of materials. In large rooms, temperature differences and different circulation of the atmosphere also play an important role. These factors, together with surface properties like wettability, sorption and catalytic activity, lead to formation of heterogeneous condensate in the form of drops or a thin film (depends on the wettability).

The application of VPHP in the solely "dry" process is advantageous because gas has a uniform contact with all exposed surfaces. All types of surfaces can thus be decontaminated to the same degree, including those with complex geometry – horizontal, vertical, cracks and curved surfaces. Moreover, it is possible to quickly remove the gaseous hydrogen peroxide from the area at the end of decontamination and thus save time of the whole cycle. A theoretical model of decontamination by VPHP was presented (Watling et al., 2002) and it described the concentration profile of hydrogen peroxide vapor in a closed space during all four phases of the decontamination cycle and compared it with experimental results. The goal of this work was to create the model that could predict the main parameters of the decontamination process (concentration of VPHP, dew point, etc.) on the basis of operation conditions and other parameters (space dimensions), and thus control and conduct this process under optimal conditions with the highest efficiency.

7. VPHP – An excellent biocidal agent

Similarly as in its liquid solution, hydrogen peroxide also has sterilization properties in the vapor phase against vegetative bacteria and highly resistant bacteria endospores (Block, 1991; French et al., 2004; Hall et al., 2007; Johnston et al., 2005; Kahnert et al., 2005; Klapes & Vesley, 1990; Kokubo et al., 1998; Rogers et al., 2005; Sapers et al., 2003; Unger-Bimczok et al., 2008), viruses (Heckert et al., 1997), fungi (Forney et al., 1991), yeast, amoebae, infective proteins and other microorganisms (Fichet et al., 2004; Klapes & Vesley, 1990; Vassal et al., 1998). As a model organism for the validation, *Bacillus stearothermophilus* (*Geobacillus stearothermophilus*) which is mainly deposited on stainless steel is commonly used (Block, 1991; Bounoure et al., 2006; Chung et al., 2008, Fisher & Caputo, 2004; Johnston et al., 2005; Klapes & Vesley, 1990; Unger et al., 2007). This microorganism is very resistive against VPHP and serves as a surrogate of anthrax (*Bacillus anthracis*) because of their very similar behavior. The VPHP process is considered successful when all these microorganisms are deactivated. A wide range of commercial biological indicators designed for VPHP (spores of *Bacillus subtilis*) exist that are often used for VPHP validation (Klapes & Vesley, 1990; Kokubo et al., 1998).

Sporicidal efficiency of chemical decontamination agents is often expressed as the D-value, which represents the time (minutes) necessary to kill 90 % of the starting amount of microorganisms (or logarithms of the amount) at a constant temperature (Gould, 2004; Unger-Bimczok et al., 2008). In the next table, D-values of selected bacterial spores are compared and evaluated by liquid hydrogen peroxide and VPHP decontamination (U.S. Environmental Protection Agency, 2005). These results (Table 3) show that, in order to kill selected microorganisms, a 200-fold concentrated solution of hydrogen peroxide is necessary to get comparable results as in vapor phase.

Although it is well known that the bactericidal activity of liquid hydrogen peroxide solution grows with its increasing concentration, the linear dependence of its vapor-phase concentration on the killing activity of selected microorganisms is still a widely discussed topic. Some authors say that the antimicrobial activity grows with higher concentration of gaseous hydrogen peroxide, others express a completely opposite opinion and prefer the microbial deactivation at lower concentrations (Unger-Bimczok et al., 2008). In Table 4, all microorganisms tested for VPHP inactivation are summarized (Forney et al., 1991; Hall et al., 2007; Heckert et al., 1997; Johnston et al., 2005; Klapes & Vesley, 1990; Kokubo et al.,

1998; Reich & Caputo, 2004; Simmons et al., 1997). The table clearly shows that the application of VPHP as a biocidal agent is a well mapped topic.

Tested microorganism (Spores)	D-value [min]	
	Liquid solution of H_2O_2 $c(H_2O_2)$ = 370 mg l^{-1} T = 24 - 25 °C	VPHP $c(H_2O_2)$ = 1 - 2 mg l^{-1} T = 24 - 25 °C
Bacillus	1.5	1 - 2
Bacillus subtilis	2.0 - 7.3	0.5 - 1
Clostridium sporogenes	0.8	0.5 - 1

Table 3. Comparison of sporicidal effect of liquid and gaseous hydrogen peroxide (VPHP)

Bacteria + spores
Aeromonas sp.; Acholeplasma laidlawii; Acinetobacter baumannii; Acinetobacter calcoaceticus; Anaerobic cocci; Asperigillus spores; Bacillus anthracis (antrax illness); Bacillus alvei; Bacillus cereus; Bacillus circulans; Bacillus firmus; Bacillus licheniformis; Bacillus megaterium; Bacillus pumilus; Bacillus sphaericus; Bacillus (resp. Geobacillus) stearothermophilus; Bacillus subtilis; Bacillus thuringiensis; Bacteroides fragilis; Campylobacter sp.; Clostridium botulinum; Clostridium difficile; Clostridium perfringens; Clostridium piliforme; Clostridium sporogenes; Clostridium tetani; Deinococcus radiodurans; Enterobacter cloacae; Enterococcus faecium/faecalis; Escherichia coli; Fusobacterium sp.; Lactobacillus caesei; Legionella pneumoniae; Listeria monocytogenes; Klebsiella pneumoniae; Methicillin-resistant Staphylococcus aureus (MRSA); Micrococcus sp.; Moroxelia osloensis; Mycobacterium bovis; Mycobacterium chelonei; Mycobacterium smegmatis; Mycobacterium tuberculosis; Pseudomonas aeruginosa; Pseudomonas cepacia; Salmonella choleraesuis; Salmonella typhimurium; Shigella sp.; Staphylococcus
Viruses (family: type of virus)
Adenoviridae: Adenovirus, Canine adenovirus; Caliciviridae: Feline calicivirus, Vesicular exanthema virus; Coronaviridae: Infectious bronchitis virus; Flaviviridae: Dengue virus; Hog cholera virus; Herpesviridae: Herpes simplex Type 1, Pseudorabies virus; Iridoviridae: African swine fever virus; Orthomyxoviridae: Influenza A2, Avian infuenza virus; Paramyxoviridae: Newcastle disease virus; Parvoviridae: Parvovirus, Canine parvovirus, Feline parvovirus; Picornaviridae: Rhinovirus 14, Polio type 1, Swine vesicular disease; Poxviridae: Vaccinia; Reoviridae: Bluetongue virus; Rhabdoviridae: Vesicular stomatitis virus
Fungi
Alternaria; Aspergillus niger; Aspergillus sp., Blastomyces dermatitidis; Botrytis cinerea; Candida albicans; Candida parapsilosis; Coccidioides immitis; Histoplasma capsulatum; Penicillium sp.; Trichophyton mentagrophytes
Other microorganisms
Caenrohabditis elegans; Cryptosporidium parvum, Lactococcal bacteriophage; Syphacia muris

Table 4. List of microorganisms which have been inactivated by VPHP

8. Synergism of VPHP and related chemicophysical factors

Most of the works concerning the synergism of hydrogen peroxide were focused on liquid-phase reactions (water disposal treatment) and the synergism of VPHP is a considerably less studied topic.

In order to boost up the effects of hydrogen peroxide vapor, its ionization by plasma can be performed (Bathina et al., 1998; U.S. Environmental Protection Agency, 2005; Vassal et al., 1998). Commonly, plasma is generated from gases (argon, helium, nitrogen, etc.) by an electric pulse, radiofrequency or microwave irradiation. It can be formed at atmospheric pressure and higher temperature (105 °C) or at reduced pressure (~ 40 Pa) and substantially lower temperature (55 – 60 °C) which is called low-temperature plasma (Crow & Smith, 1995). It consists of free radicals (mainly hydroxyl or hydroperoxyl radicals), ions, neutral particles and excited atoms or molecules which show high activity in deactivation of contaminants (U.S. Environmental Protection Agency, 2005). The Johnson & Johnson Medical company and its division Surgikos Inc. patented for the first time a decontamination device in the form of a vacuum chamber where the contaminated instruments were treated with hydrogen peroxide plasma (Parisi & Young, 1991). This kind of decontamination, marked as „STERAD sterilization system", is recommended by the Food and Drug Administration (FDA) as an advanced sterilization technique for enclosed spaces (Crow & Smith, 1995). Although the systems using VPHP together with plasma are very useful for temperature- or water-sensitive materials, their big disadvantage is that they can only be applied in small closed spaces because of the high vacuum or temperature required in the chamber (Adams et al., 1998).

Other synergistic effect can be observed by combination of VPHP with UV irradiation (Klapes & Vesley, 1990). Application of UV irradiation alone requires a relatively long time and thus its combination with VPHP can significantly shorten the operations. The UV/H_2O_2 combination (photo-oxidation) is very effective in destroying microorganisms and heavy decomposable organic pollutants, i.e. volatile organic compounds (VOC) like benzene, toluene, phenol, *tert*-butyl methyl ether, halogenated compounds, pharmaceutical substances and pesticides, mainly in water (Esplugas et al., 2002; Kang & Lee, 1997; Lopez et al., 2003; Pereira et al., 2007; Prousek, 1996). It is known that hydrogen peroxide absorbs UV light within the range of 185 - 400 nm (Esplugas et al., 2002). The radiation energy at these wavelengths is sufficient to provoke photo excitation of the hydrogen peroxide molecule and a subsequent cleavage (photolysis) of the –O–O– peroxide bond, forming hydroxyl radicals which can initiate radical chain reactions of contaminants. The homolytic splitting of the hydrogen peroxide molecule leads to generation of hydroxyl radicals due to the absorption of a photon (Dionysiou et al., 2004). Hydroxyl radicals can be formed by the wide range of low wavelength ultraviolet radiation within 200-280 nm (Lopez et al., 2003). This interval of wavelengths is called UV-C and because of its germicidal properties it is commonly used for water and air sterilization and also for the degradation of photo unstable organic pollutants (Pereira et al., 2007). The UV radiation not only consists of UV-C (100-280 nm) but also contains the UV-B (280-315 nm) and UV-A (315-400 nm) ranges. These two last radiation areas are not used for the activation of hydrogen peroxide. The most common source (Lopez et al., 2003) of the UV-C radiation are low-pressure mercury lamps with the emission maximum at 254 nm since at this wavelength, the quantum yield of hydroxyl radicals equals to 1. The synergistic combination of liquid hydrogen peroxide and UV-C at the 254 nm emission maximum (indirect photolysis) used for the degradation of organic pollutants shows that it is highly effective and that the rate of reaction is higher compared to individual application of hydrogen peroxide or UV-C irradiation (Esplugas et al., 2002; Lopez et al., 2003; Pereira et al., 2007). The total rate of chemical degradation of contaminants by UV-C and hydrogen peroxide is dependent on the mechanism of the reaction of OH radicals with the contaminants, on the reaction rate of direct photolysis of contaminant (absorptivity of substrate), on the absorptivity of by-products and other absorbents of UV-C radiation at 254 nm (competitive absorption), on the intensity of the UV-C source and on the concentration of hydrogen peroxide (Kang & Lee, 1997). Degradation of organic substances by this effective combination proceeds by a radical oxidative reaction (Ray, 2000).

There are several practical applications of the VPHP decontamination process employing the synergistic effect of UV-C (VPHP/UV-C) for the sterilization of continual filing processes of food in liquid form (milk, water, juice). VPHP is applied in the first step of decontamination to treat surface and then the UV-C radiation treatment follows. It is also possible to combine the above-mentioned VPHP/UV-C system with other oxidative agents such as ozone, or the catalytic properties of TiO_2 can be used. Decontamination processes combining O_3/UV-C/VPHP show the most effective contaminant degradation and allow complete mineralization of pollutants (Esplugas et al., 2002). In the USA, the Department of Energy patented a portable device for surface chemical and biological decontamination which uses ozone as the main agent, reinforced by UV radiation and hydrogen peroxide added to reach higher reactivity (O'Neill & Brubaker, 2003).

In 1972, Fujishima and Honda (Fujishima & Honda, 1972) discovered the photocatalytic properties of nanocrystalline TiO_2 and predicted its possible application in chemical decontamination (Linsebigler et al., 1995; Wold, 1993). The TiO_2/UV system was described in several studies (Mills & Hunte, 1997; Peral et al., 1997; Rauf & Ashraf, 2009; Tschirch et al., 2008; Zhao et al., 2005), mainly with regard to a higher efficiency of degradation of chemical pollutants in water and air. Total mineralization of pollutants was observed, producing only CO_2, H_2O and inorganic ions (Zhao et al., 2005). Thus, the process is highly environmentally friendly. Its reasonable price, compared to other Advanced Oxidative Processes (Rauf & Ashraf, 2009), is also highly important.

The decontamination by TiO_2/UV/H_2O_2 is studied mainly in the liquid phase, preferentially in water (Domínguez et al., 2005), and thus there is no information on a VPHP-based modification of this system. The addition of a small amount of hydrogen peroxide shows a synergistic effect, i.e. a notable speedup of the photocatalytic TiO_2/UV/H_2O_2 degradation of pollutants (Tschirch et al., 2008). This positive effect is due to increased formation of hydroxyl radicals as a result of direct photolysis of hydrogen peroxide and its interaction with the active TiO_2 surface. However, higher concentrations of hydrogen peroxide inhibit the degradation because under these conditions, recombination of hydroxyl radicals is preferred (Dionysiou et al., 2004; Esplugas et al., 2002). Thus, it is necessary to work with optimal concentration of hydrogen peroxide to achieve the highest efficiency of contaminant degradation in the TiO_2/UV/H_2O_2 system (Elmolla & Chaudhuri, 2010). Although similar properties as TiO_2 are also provided by other semiconductor materials as ZnO, ZnS, CdS, Fe_2O_3, WO_3, these materials do not show the efficiency of degradation as high as TiO_2 does (Ray, 2000).

There are also other possibilities to promote VPHP process, e.g. by the addition of other oxidative agents like ozone, peracetic acid, or concentrated solution of hydrogen peroxide before evaporation. The main problem of hydrogen peroxide evaporation is water which can condensate on rugged surfaces like medical instruments and it impedes the surface penetration by the hydrogen peroxide. Hydrogen peroxide is able to form anhydrous complexes with a wide range of organic compounds: polyvinylpyrrolidone, urea, glycine anhydride-peroxide complex, and inorganic compounds: $Na_4P_2O_7 \cdot 3H_2O_2$, $KH_2PO_4 \cdot H_2O_2$, which can be prepared easily using known procedures based on their crystallization from solutions. In these complexes, the hydrogen peroxide moiety only binds to the electronegative atom of the other molecule via two H-bonds, which greatly facilitates its release from these substances. For example, thermal decomposition or vacuum can be used to perform this. The vapors of hydrogen peroxide formed this way can be generated directly in the decontaminated area or in another place, in which case they can be transported by a pipeline to the decontaminated space. Another interesting kind of hydrogen peroxide potentiation is the addition of metals (Ag, Al, Ca, Ce, Cu Mg, Sr, Sn, Ti, Zn) (Carnes et al., 2004), oxides or hydroxides, the particles of which (must) have a relatively high surface (at least 15 m^2 g^{-1}). The application of transition metals for hydrogen peroxide activation (i.e. for the formation of OH or other radicals) is mainly used in the liquid phase, like Fenton oxidation (Fenton, 1894) or photo-Fenton oxidation (Prousek, 1996), where Fe^{3+} ions (Xu, 2001) or Cu complexes are used (Martínez et al., 2008). These systems are mainly used for the degradation of resistant organic contaminants like dyes. Another (and also very interesting) combination is mixing of VPHP with volatile basic compounds like ammonia. This method seems to be very promising in the area of deactivation of warfare agents (Wagner et al., 2007).

9. VPHP decontamination of chemicals – Molecular structure effects of decontaminants

There is only limited information concerning the effects of the contaminant chemical structure upon decontamination by VPHP (McVey et al., 2006; Roberts et al., 2006). Several functional groups which are sensitive to VPHP have been found (Švrček, 2010), namely the aldehyde group, aliphatic tertiary nitrogen and the sulfide group (thioethers). Since all of the tested substances contain one of these groups, they were successfully degraded by VPHP. Other VPHP-sensitive compounds seem to be phenols, out of which mainly their hydroxy- and amino-derivatives. It can be expected that the decontamination of more complicated structures (pharmaceutical substances) will be sufficient if these compounds contain one or more such reactive groups in their structure.

9.1 Aldehyde group

Substances containing the aldehyde group undergo preferentially an oxidation process leading to carboxylic acid but other reactions can also take place, for example Dakin reaction (Pan et al., 1999), decarbonylation, substitution, or cracking.

The mechanism of Dakin reaction is illustrated in Figure 3 on an example of vanillin degradation, which in principle proceeds by insertion of an oxygen atom as a result of ketone oxidation (Baeyer-Villiger oxidation). In the first step, an OH radical attacks the carbonyl group (1) and an alkoxy radical (2) is generated, which is then transformed to an unstable alkylhydroperoxide (3) that is rearranged into a more stable product. A subsequent hydrogen shift (a) leads to the final product – carboxylic acid (4). In some cases, migration of a different group can occur, e.g. the aryl group that forms a formate (5) as a key intermediate in the above-mentioned Dakin reaction (b). These compounds are very often unstable and are subject to hydrolysis (cleavage of formic acid) or decarboxylation (giving CO_2) and formation of a phenol derivative – in case of vanillin it is 2-methoxyhydroquinone (6). The ability of different groups to migrate can be sorted in the following order: tertiary alkyl > secondary alkyl, aryl > primary alkyl > methyl. Electron-donating substituents on the benzene ring of vanillin increase migration ability of this aryl group by the hydroperoxide rearrangement. Electron-withdrawing substituents have opposite effects and impede migration. Therefore, vanillin can easily undergo Dakin reaction that leads to 4-hydroxy-3-methoxyfenyl formate (5). The key factor of Dakin reaction is the presence of strong electron-donating substituents (-OH or -NH_2) in the structure of aromatic aldehydes in the *ortho*- or *para*- position to the –CHO group (Švrček, 2010).

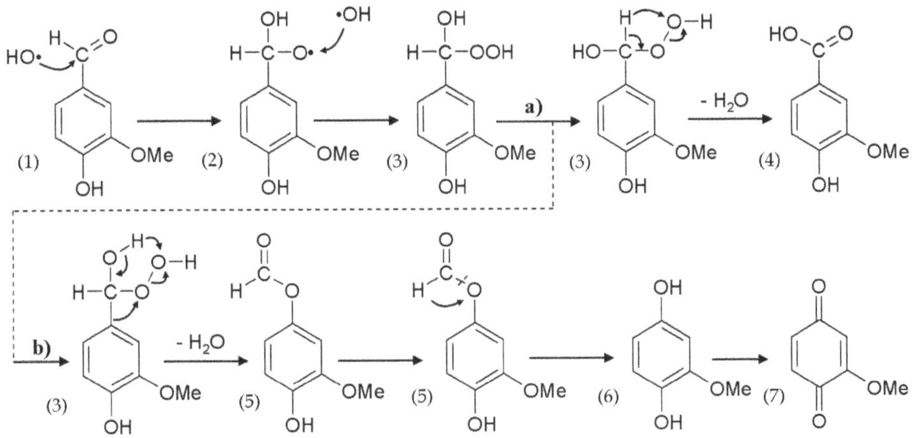

Fig. 3. Expected mechanism of the VPHP-induced degradation of vanillin

9.2 Aliphatic tertiary nitrogen

Other compounds which are highly sensitive to VPHP seem to be those containing an aliphatic tertiary nitrogen atom. All tested substances (Švrček, 2010) were decomposed by VPHP, which could be deduced from notable color changes, decrease in the sample weight, and detected products of degradation. However, the high volatility of starting compounds or degradation products prevented proper analysis by GC-MS and NMR. It is known that in the presence of H_2O_2, tertiary amines are converted to N-oxides that can be decomposed by the Cope elimination reaction leading to an alkene and N-hydroxylamine. This elimination is mainly carried out at high temperature but we can assume that it can proceed under laboratory conditions (Cope et al., 1949).

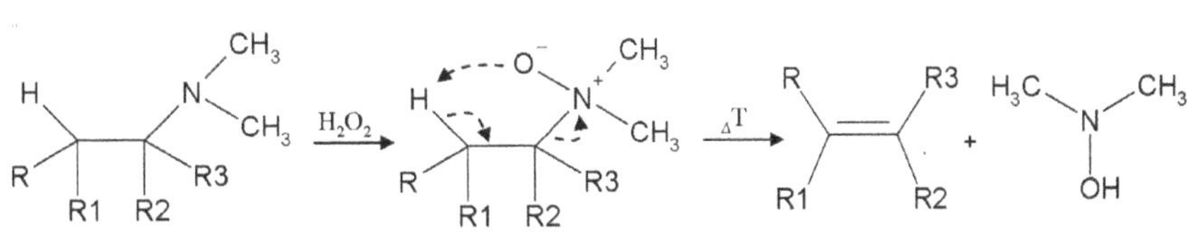

Fig. 4. Mechanism of Cope elimination

9.3 Sulfidic sulphur

In the presence of hydrogen peroxide, oxidation of sulfidic compounds (thioethers) proceeds and more stable higher oxidative compounds are formed (sulfoxides and sulfones) in a rate depending on the concentration of the oxidative agent.

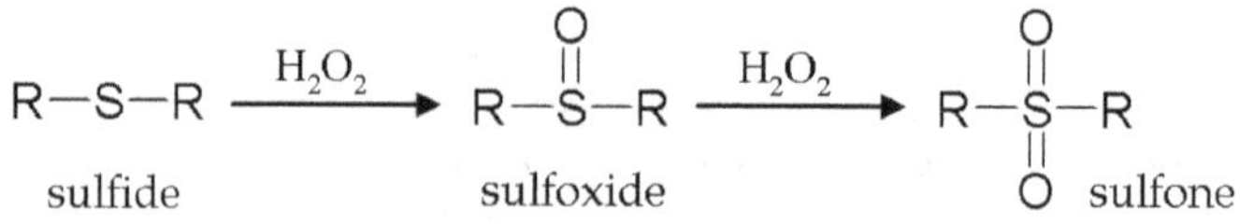

Fig. 5. Mechanism of oxidation of sulfidic compounds (thioethers)

The molecules containing sulfidic sulphur are oxidized by VPHP to both products. In case of thioanisol (phenyl methyl sulfide), after a VPHP degradation process only phenyl methyl sulfoxide was detected by MS. The final oxidative product (phenyl methyl sulfone) was not detected because of its high volatility and thus quick evaporation during the decontamination test, but we can assume its formation. Oxidation of dimethyl sulfoxide to dimethyl sulfone is driven by VPHP to a total conversion (Švrček, 2010).

9.4 Phenol derivatives

VPHP can also decompose phenol derivatives substituted in the *para-* position with electron-donating groups – mainly hydroxyl and amino. On the contrary, electron-withdrawing substituents (-CN, -NO$_2$, -COOH) make the contaminants intact to the VPHP atmosphere (Švrček, 2010).

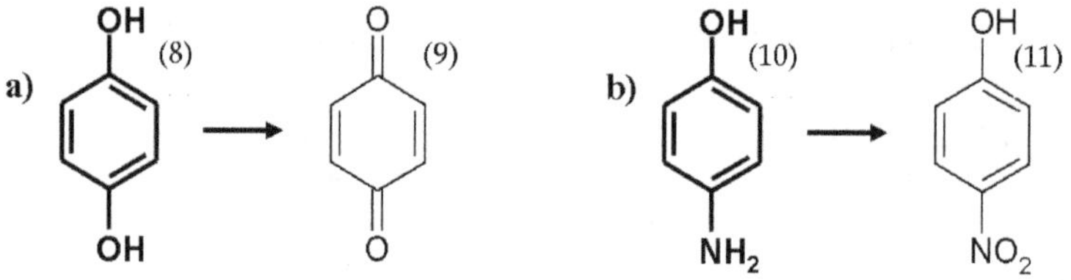

Fig. 6. Oxidation of hydroquinone and *p*-aminophenol by VPHP vapors

Hydroquinone (8) is oxidized by VPHP to 1,4-benzoquinone (9) which was also true for vanillin degradation. In case of *p*-aminophenol (10), a solid, weak and soluble mixture of degradation products was formed in the VPHP atmosphere. One of the products was *p*-nitrophenol (11) and the other compounds were products of oligo- and polymerization reactions between the starting compound and the products of degradation.

9.5 Molecular structure effects promoting the VPHP-induced degradation

A group of model chemical substances with similar structures (approximately 30 benzaldehyde and phenol derivatives bearing different types, numbers and position of substituents on the aromatic ring) was tested to study the influence of structure on the sensitivity to the VPHP process and reaction schemes were evaluated (Švrček, 2010). It is clear that the presence of electron-donating group on the benzaldehyde skeleton in many cases leads to an increase in its sensitivity to degradation by VPHP. The type, number and position of substituents on the aromatic ring with respect to the –CHO group play an important role. The highest sensitivity is shown with substances bearing high electro-donating substituents (like -NR$_2$) located in the *para*- position to –CHO. With the decreasing number of substituents, the time necessary for the VPHP degradation also decreased. The lowest sensitivity was found with benzaldehyde derivatives containing a halogen atom, an electron-withdrawing substituent or sterically demanding substituents.

10. VPHP as an agent for decontamination of biologically active compounds

In 1985, the International Agency for Research on Cancer (IARC) issued recommendations on the application of oxidative processes for the decontamination of cytotoxic compounds in waste waters (Roberts et al., 2006). Following studies (Barek et al., 1998; Hansel et al., 1997) tried to find the ideal oxidative agent that could be used in hospitals for the purposes described above. Three systems were tested: water solution of hydrogen peroxide (< 30 %, w/w), NaOCl and Fenton reagent for the degradation of selected cytostatics. Hydrogen peroxide showed the lowest efficiency of degradation compares to other oxidative systems, but it still appeared to be a promising agent.

The efficiency of 35% (w/w) liquid solution of hydrogen peroxide, VPHP and eight liquid detergents containing NaOCl for the deactivation of cytotoxic compounds present on the inner surface of pharmaceutical isolators was compared (McVey et al., 2006; Roberts et al., 2006). The contaminants tested were cyclophosphamide, doxorubicin and 5-fluorouracil. The liquid detergents removed all substances from the surfaces except for doxorubicin that showed high resistance to alkaline detergents. However, the contaminants were only washed away from the surface and their bioactivity was not eliminated. Cyclophosphamide and 5-fluorouracil were completely intact after a one-hour exposure to concentrated solutions of detergents which means that they preserved their full biological activity. Degradation of doxorubicin only occurred in a strong alkaline detergent (pH = 13.2, 80% conversion over 1 hour) and it remained unchanged in the water solution of hydrogen peroxide over 1 hour. The results from similar studies (Castegnaro et al., 1997; Hansel et al., 1997) showed that the application of 30% (w/w) hydrogen peroxide solution led to slow degradation of doxorubicin (60% conversion after 24 hours) and total degradation of cyclophosphamide after 1 hour. The VPHP "dry" process at temperature below 30 °C and reaction time of 1 hour exhibited slight degradation of 5-fluorouracil and significant degradation of doxorubicin (44 – 92 %). Unfortunately, cyclophosphamide remained intact. The mechanism was not studied (Roberts et al., 2006).

On the contrary, the application of VPHP as a deactivator of warfare agents is much more studied. A main leader on this topic is the STERIS company that found a quick and safe means of decontamination of large polluted areas (McVey et al., 2006). Their method seems to be very promising mainly because of a danger of terrorist attacks. The application of hydrogen peroxide to deactivate chemical weapons of mass destruction was studied (Wagner & Yang, 2001) which resulted in creation of a universal liquid deactivation agent called "Decon Green". This agent is composed of a hydrogen peroxide water solution, alcohol (ethanol, isopropyl alcohol, *tert*-butanol, polypropylene glycol) and carbonates, bicarbonates or monoperoxocarbonate - Na$_2$CO$_3$, K$_2$CO$_3$, Li$_2$CO$_3$, NH$_4$HCO$_3$). „Decon Green" provides high efficiency for the degradation of nerve agents and blistering agent like sarine (GB), soman (GD), sulfide yperite (HD) and VX compounds. The addition of carbonate to the liquid solution of peroxide leads to the formation of an OOH- anion that is highly reactive, mainly with nerve agents. It speeds up the per-hydrolysis of VX and G compound families. The addition of peroxocarbonate, which is highly active for the oxidation of HD substances, leads to the formation of nontoxic products (Wagner & Yang, 2001). The addition of alcohol serves as the nonfreezing part of this mixture.

Carbonates are very good activators of hydrogen peroxide but they cannot be used in the VPHP process as they are not volatile. The special arrangement can distribute carbonates in the gaseous phase which supports vapor-phase decontamination. A pilot work (McVey et al., 2006) studied the application of VPHP to the degradation of warfare agents. It was shown that the VPHP process was highly potent in the degradation of HD and VX compounds even without any activator present. The G type agents are very stable under the VPHP conditions. It is known that they can be decomposed easily under alkaline conditions (Wagner et al., 2007). Hydrogen peroxide is relatively unstable in alkaline solutions but it has been shown (McVey et al., 2006; Wagner et al., 2007) that under these conditions the GD compounds can be decomposed to nontoxic products. Furthermore, the combination of VPHP and a base also accelerated the degradation of VX and HD compounds compared with VPHP alone (McVey et al., 2006; Wagner et al., 2007). The degradation started at an ammonia level that was below the safety limit of 25 ppm. It was observed that the rate of degradation was increasing with the increasing ammonia concentration (Wagner et al., 2007). A detailed study on the degradation of GD compounds using a mixture of VPHP/NH_3 shows a different mechanism compared with liquid "Decon Green". In case of application of the VPHP/NH_3 mixture to HD and VX, higher reaction rate was observed together with some differences in the decomposition mechanism (Wagner et al., 2007). It is interesting that the commercial decontamination agents containing ammonia are only active for the degradation of G but not for VX and HD.

The higher efficiency of VPHP for the degradation of warfare agents compared to water solution of hydrogen peroxide is explained as follows. The acidic products formed by per-hydrolysis are volatile and are ventilated out. On the contrary, in the water solution the acidic products are concentrated and thus the pH value decreases under a level which is necessary for the formation of peroxide anions and the reaction is thus stopped. Higher efficiency of VPHP is also explained by a model simulating the distribution of contaminants on contaminated surfaces.

Hydrogen peroxide is due to its properties closer to organic solvents than to water and thus it is selectively adsorbed and concentrated inside drops or film of the contaminant. Molecular distribution results in a higher local concentration of H_2O_2 inside the contaminant and exclusion of water vapor out of the contaminant organic layer (Wagner et al., 2007). When using the liquid phase, competitive adsorption of water and hydrogen peroxide to contaminant molecules occurs which slows down its degradation noticeably.

It was showed (Wagner et al., 2007) that the exposition time necessary to decontaminate inner premises of buildings contaminated by the warfare agents using VPHP/NH_3 is 24 hours. These results clearly show that such modified VPHP process (VPHP/ NH_3) is the right choice for quick and effective decontamination of large areas and objects (military vehicles, airplanes) contaminated by dangerous compounds.

11. VPHP – Material compatibility

The extent of contaminant elimination by hydrogen peroxide in vapor phase is mainly influenced by the concentration of VPHP, exposure time, temperature, humidity, condensation and also the physical and chemical properties of materials (Chung et al., 2008). The influence of the contaminated surface upon the efficiency of VPHP-induced inactivation of the spores of *Bacillus stearothermophilus*, which serves as a biological indicator, was studied (Unger et al., 2007). Among the notable influences are the chemical composition of the selected material, its potential catalytic activity, absorption of gaseous hydrogen peroxide in the material and, last but not least, the differences in the surface production or its treatment. The relationships between the deactivation rate and porosity and wettability are documented (Unger et al., 2007) – VPHP is mainly effective for the sterilization of smooth surfaces and therefore stainless steel and glass coupons are used for the evaluation of the process efficiency (Bounoure et al., 2006; Unger et al., 2007). Spores on a porous surface can be hidden in cavities and in such a case the penetration of hydrogen peroxide to the material plays an important role in decontamination (Bounoure et al., 2006). Some materials have demonstrated an inhibition effect to the microorganisms without any decontamination agent, such as ethylene-propylene-diene rubbers. All knowledge mentioned above was implemented by FDA in the Guidelines for industry. It is recommended to choose suitable materials for the design of any aseptic process. They have to provide proper compatibility with the chemicals used and thus can be easily cleaned and decontaminated. It is highly important to choose construction materials with appropriate texture and porosity by the build-up of the aseptic process mainly when the validation of decontamination process is required (Unger et al., 2007). The VPHP decontamination was tested on various surfaces but the toxicological evaluation reports on surfaces treated in such a way are very limited. Hydrogen peroxide is characterized by high cytotoxicity (i.e. harmfulness to cells) so its residual amounts in the materials can cause irritation of eyes, skin, mucous membrane and also acute lung dropsy in cases of long-time inhalation. Therefore, the construction materials have to be selected not only due to their compatibility with the VPHP process but also with respect to the efficiency of residual hydrogen peroxide aeration out of these materials. Up to now, the studies have mainly focused on polymeric materials because of their popularity in the pharmaceutical industry. The permeation of gases through plastic materials occurs in two steps: the first one consists in dissolution of gas in the thin surface layer and the second is its diffusion to the material. It was found that polyethylene and polypropylene can easily release hydrogen peroxide because of its limited migration in these materials. In contrast, polystyrene, polyurethane, poly(methyl methacrylate) (PMMA), poly(2-hydroxyethylmethacrylate) (HEMA), fluorosilicone acrylate and mixture of polyurethane and silicone demonstrated strong cytotoxicity after standard aeration (Ikarashi et al., 1995). The PMMA and HEMA materials are used for the production of contact lenses and thus, after their sterilization by VPHP, the attention needs to be paid to residual hydrogen peroxide. Hydrogen peroxide can easily penetrate through polyolefins in general, out of which PVC is used for packaging of infusion solutions. Therefore, the VPHP process has to be applied very carefully in order to avoid the contact of residual hydrogen peroxide with the infusion solution due to resulting oxidative reactions (Ikarashi et al., 1995). Nonetheless, information on the resistance of other construction materials is very limited. Although there have been reported interactions of several materials with VPHP which can lead to their damaging (Hultman et al., 2007), the process is still considered non-corrosive for surface decontamination (U.S. Environmental Protection Agency, 2005). Several works tested the resistivity of commonly-used devices to VPHP but the results were evaluated only visually and by testing of functionality of the devices. Generally, it was concluded that VPHP was an acceptable sterilization method for the devices (Hall et al., 2007; Heckert et al., 1997).

In global, materials can be divided into four groups according to their tolerance to VPHP. Group 1 represents such materials that can be in the contact with hydrogen peroxide for a long time, like pure aluminum, tin, borosilicate glass or Teflon. In contrast to this group stands Group 4 which comprises materials that cannot be in any contact with hydrogen peroxide because of their fast decomposition or formation of explosive mixtures (copper, iron, carbon steel, magnesium alloys). The concentration of 45% (w/w) hydrogen peroxide is considered critical (Hultman et al., 2007) because, when exceeded, undesirable interaction or damage of different materials can occur. For chemical decontamination, higher-concentrated peroxide is necessary and in a proper arrangement it can be a safe system. It can be expected that during the "wet" VPHP process, the concentration of the condensate is above this critical concentration and thus the materials can be damaged. However, this concentration can also be exceeded during the "dry" VPHP process in the aeration phase when the absorbed hydrogen peroxide is concentrated on the surfaces. If the damage is not visible, it does not mean that the material has not been damaged. Microscopic lesions can appear and the macroscopic ones can become evident only after a long-time exposure to VPHP. The VPHP process was also tested for the interiors of airplanes and ambulances (Krieger & Mielnik, 2005; Shaffstall et al., 2006). As these vehicles are frequently in the contact with infection, quick and effective decontamination is important to avoid the propagation of infection (SARS, bird flu). Application of VPHP in these facilities is risky because several sensitive devices which are vital for their proper function can be damaged. The Federal Aviation Administration (FAA) studied the influence of the VPHP process on textile materials of airplane interiors and found significant changes. The leader in the VPHP technology, Bioquell, tried to map the influence of long-time application of VPHP to the hospital equipment. The company tried to fill in the information gap of the VPHP material compatibility. Some materials were shown to be highly resistant to VPHP (called „VPHP-resistant"), some were degradable by VPHP and some absorbed hydrogen peroxide substantially. There are also materials unsuitable for VPHP like untreated aluminum, copper, soft steels, coated steels and generally materials that have similar behavior in the liquid state. Materials containing cellulose tend to absorb hydrogen peroxide and are further degraded; therefore, they are also unsuitable for VPHP (von Woedtke et al., 2004). Furthermore, EPA studied the VPHP compatibility of common construction or decorative materials (U.S. Environmental Protection Agency, 2008). The visual inspection did not reveal any significant changes of the tested materials but the tensile strength was reduced, which was caused by changes in their inner structure.

Low-temperature H_2O_2 plasma material compatibility is actually a much more studied topic. The Johnson & Johnson company tested a wide range of materials and devices which were in a periodic contact with hydrogen peroxide plasma. It was found that the materials containing amines in their structure were unsuitable because of their oxidation that damaged the structure. Also the S–S bond in the materials was proven to be incompatible with this process.

It is necessary to select such materials that are fully compatible with VPHP by the design of the device or the process where it is expected to use VPHP for decontamination. In such case, it needs to be known how the materials interact with hydrogen peroxide, namely with respect to absorption and the rate of hydrogen peroxide decomposition. It is also important to know the aeration time required to vent out hydrogen peroxide in order to avoid undesirable cytotoxic effects (Ikarashi et al., 1995). It is necessary to evaluate the resistivity to hydrogen peroxide and the amount of residual hydrogen peroxide left in the material. There are several analytical methods for the determination of hydrogen peroxide such as reduction by SnCl2 (Egerton et al., 1954), thiocyanate method (Egerton et al., 1954; Ikarashi et al., 1995), enzymatic method of dimerization of p-hydroxyphenolic acid (Christensen, 2000), color changes of $Ti_2(SO_4)_3$ or $TiCl_4$ by spectrophotometric detection (Egerton et al., 1954), UV spectroscopy, titration of $KMnO_4$, I_3^-, or $Ce(SO_4)_2$ (Klassen et al., 1994), electrochemical, polarographic and other methods (Higashi et al., 2005).

12. Glory and pitfalls of VPHP

The VPHP technology is a very progressive method of sterilization and chemical decontamination. Its popularity was mainly achieved due to low toxicity of hydrogen peroxide and its non-toxic decomposition products, environmentally friendly behavior, relative flexibility and a wide spectrum of applications. Its high efficiency against a wide range of microorganisms makes this method universal in terms of bio-decontamination and thus it is very popular in medical and pharmaceutical industry. This application is widely studied and described. The absence of theoretical knowledge is the only drawback of this method. Proper understanding of the sterilization principle would allow to optimize the method and to make it exceptionally powerful. On the other hand, the chemical decontamination by VPHP is still a growing area with a wide potential of application. Mainly the combination and potentiation of VPHP by other chemicals or physical phenomena should improve it and make it a very powerful tool for the decontamination of dangerous chemicals. Also in this case, the absence of proper theoretical knowledge limits its application.

13. Future perspective of VPHP

As already mentioned above, the highly efficient VPHP process has found a great deal of new applications in bio and chemical decontamination. The first challenge seems to be theoretical understanding of its mechanism and thus obtaining the basics for finding new applications or improving existing processes. The next challenge is optimization of existing processes by application of new approaches and knowledge as a product of undergoing research and development. Another very important challenge is to discover novel possibilities of VPHP potentiation, mainly with respect to chemical decontamination. There is a great need for the decontamination of a wide range of chemical pollutants. The VPHP process can play a very important role in the case of homeland security as it can be easily implemented to the defense system, which provides greater security for the state and citizens against terrorists, spreading of infections, chemical and biological accidents, etc.

14. Conclusion

At the present time, a broad spectrum of decontamination techniques is acknowledged to be utilized to remove biological and chemical contaminants from different surfaces. Nevertheless, new physical and chemical processes are continuously being developed. The main reasons for the development of new decontamination methods are negative properties of many decontamination agents, especially their toxicity or toxicity of residues formed after their application. This review summarizes the recent findings concerning a new promising decontamination agent Vapor Phase Hydrogen Peroxide (VPHP) whose properties were observed to be very close to an ideal decontamination agent. VPHP has become the method of choice by meeting many bio-decontamination requirements in the pharmaceutical, biomedical and healthcare sectors for it is reliability, rapidness, the fact that it leaves no residues (breaks down into water and oxygen) and the advantage that it can be validated. It is also a decontamination method of choice for chemically and biologically active compounds. The application of VPHP as a potential decontamination agent is apparently still in its infancy. Therefore, it comes as no surprise that the knowledge of the cidal action mechanism(s) and the influential factors is yet to be completed.

15. Acknowledgement

The project is supported by the grant of Ministry of Defense of the Czech Republic (OVVSCHT200901).

16. References

Adams, D.; Brown, G. P.; Fritz, C. & Todd, T. R. (1998). Calibration of a Near-Infrared. (NIR) H_2O_2 Vapor Monitor. *Pharmaceutical Engineering*, Vol. 18, No. 3, (May/June 1998), pp.1-11, ISSN 0273-8139

Barek, J.; Cvačka, J.; Zima, J.; De Méo, M.; Laget, M.; Michelon, J. & Castegnaro, M. (1998). Chemical Degradation of Wastes of Antineoplastic Agents Amsacrine, Azathioprine, Asparaginase and Thiotepa. *The Annals of Occupational Hygiene*, Vol. 42, No. 4, (May 1998), pp. 259-266, ISSN 1475-3162

Bathina, M.N.; Mickelsen, S.; Brooks, C.; Jaramillo, J.; Hepton, T. & Kusumoto F.M. (1998). Safety and efficacy of hydrogen peroxide plasma sterilization for repeated use of electrophysiology catheters. *Journal of the American College of Cardiology*, Vol. 32, No. 5, (November 1998), pp. 1384-1388, ISSN 0735-1097

Block, S.S. (1991). Historical Review, In: *Disinfection, Sterilization, and Preservation* (fourth edition), S.S. Block, (Ed.), 3-17, Lea & Febiger, ISBN 978-081-2113-64-8 London, Philadelphia

Block S.S. (1991). Peroxygen Compounds, In: *Disinfection, Sterilization, and Preservation* (fourth edition), S.S. Block, (Ed.), 167-181, Lea & Febiger, ISBN 978-081-2113-64-8 London, Philadelphia

Bounoure, F.; Fiquet, H. & Arnaud, P. (2006). Comparison of hydrogen peroxide and peracetic acid as isolator sterilization agents in a hospital pharmacy. *American Journal of Health-System Pharmacy*, Vol. 63, No. 5, (March 2006), pp. 451-455, ISSN 1535-2900

Carnes, C.L.; Klabunde, K.J.; Koper, O.; Martin, L.S.; Knappenberger, K.; Malchesky, P.S. & Sanford, B.R. (2004). Decontaminating Systems Containing Reactive Nanoparticles and Biocides. *United States Patent*, US 2004/0067159 A1, filed (October 2002), issued (April 2004)

Castegnaro, M.; De Méo, M.; Laget, M.; Michelon, J.; Garren, L.; Sportouch, M.H. & Hansel, S. (1997). Chemical degradation of wastes of antineoplastic agents 2: Six anthracyclines: idarubicin, doxorubicin, epirubicin, pirarubicin, aclarubicin, and daunorubicin. *International Archives of Occupational and Environmental Health*, Vol. 70, No. 6, (December 1997), pp. 378-384, ISSN 1432-1246

Cazin, J.L. & Gosselin, P. (1999). Implementing a multiple-isolator unit for centralized preparation of cytotoxic drugs in a cancer center pharmacy. *Pharmacy World & Science*, Vol. 21, No. 4, (April 1999), pp. 177-183, ISSN 1573-739X

Christensen, C.S.; Brødsgaard, S.; Mortensen, P.; Egmose, K. & Linde, S.A. (2000). Determination of hydrogen peroxide in workplace air: interferences and method validation. *Journal of Environmental Monitoring*, Vol. 2, No. 4, (August 2000), pp. 339-343, ISSN 1464-0333

Chung, S.; Kern, R.; Koukol, R.; Barengoltz, J. & Cash, H. (2008). Vapor hydrogen peroxide as alternative to dry heat microbial reduction. *Advances in Space Research* Vol. 42, No. 6, (September 2008), pp. 1150-1160, ISSN 0273-1177

Cope, A.C.; Foster, T.T. & Towle, P.H. (1949). Thermal Decomposition of Amine Oxides to Olefins and Dialkylhydroxylamines, *Journal of the American Chemical Society*, Vol. 71, No. 12, (December 1949), pp. 3929–3934, ISSN 0002-7863

Crow, S. & Smith, J. H. (1995). Gas Plasma Sterilization: Application of Space-Age Technology. *Infection Control and Hospital Epidemiology*, Vol. 16, No. 8 (August 1995), pp. 483-487, ISSN 1559-6834

Daughton, C.G. & Ternes, T.A. (1999). Pharmaceuticals and Personal Care Products in the Environment: Agents of Subtle Change? *Environmental Health Perspectives*, Vol. 107, No. 6, (December 1999), 907-938, ISSN 0091-6765

Dionysiou, D. D.; Suidan, M. T.; Baudin, I. & Laîné, J. M. (2004). Effect of hydrogen peroxide on the destruction of organic contaminants-synergism and inhibition in a continuous-mode photocatalytic reaktor. *Applied Catalysis B: Environmental*, Vol. 50, No. 4, (July 2004), pp. 259-269, ISSN 0926-3373

Domínguez, J. R.; Beltrán, J. & Rodríguez, O. (2005). Vis and UV photocatalytic detoxification methods (using TiO_2, TiO_2/H_2O_2, TiO_2/O_3, $TiO_2/S_2O_8^{2-}$, O_3, H_2O_2, $S_2O_8^{2-}$, Fe^{3+}/H_2O_2 and $Fe^{3+}/H_2O_2/C_2O_4^{2-}$) for dyes treatment. *Catalysis Today*, Vol. 101, No. 3-4, (April 2005), pp. 389-395, ISSN 0920-5861

Duffy, M. (August 2009). Weapons of War - Poison Gas, In: *firstworldwar.com*, 20. 7. 2011, Available from: http://www.firstworldwar.com/weaponry/gas.htm

Egerton, A.C.; Everett, A.J.; Minkoff, G.J.; Rudrakanchana, S. & Salooja, K.C. (1954). The analysis of combustion products: Some improvements in the methods of analysis of peroxides. *Analytica Chimica Acta*, Vol. 10, No. 5, (1954), pp. 422-428, ISSN 0003-2670

Elmolla, E. S. & Chaudhuri, M. (2010). Photocatalytic degradation of amoxicillin, ampicillin and cloxacillin antibiotics in aqueous solution using UV/TiO_2 and $UV/H_2O_2/TiO_2$ photocatalysis. *Desalination*, Vol. 252, No. 1-3, (March 2010), pp. 46-52, ISSN: 00119164

Esplugas, S.; Giménez, J.; Contreras, S.; Pascual, E. & Rodríguez, M. (2002). Comparison of different advanced oxidation processes for phenol degradation. *Water Research*, Vol. 36, No. 4, (February 2002), pp.1034-1042. ISSN 0043-1354

Favero, M.S. & Bond, W.W. (1991). Chemical Disinfection of Medical and Surgical Materials, In: *Disinfection, Sterilization, and Preservation* (fourth edition), S.S. Block, (Ed.), 617-641, Lea & Febiger, ISBN 978-081-2113-64-8 London, Philadelphia

Fenton, H.J.H. (1894). Oxidation of Tartaric Acid in presence of Iron. *Journal of the Chemical Society, Transactions*, Vol. 65, No. 0, (1894), pp. 899-910, ISSN 0368-1645

Fichet, G.; Comoy, E.; Duval, Ch.; Antloga, K.; Dehen, C.; Charbonnier, A.; McDonnell, G.; Brown, P.; Lasmézas, C. I. & Deslys, J. P. (2004). Novel methods for disinfection of prion-contaminated medical devices. *The Lancet*, Vol. 364; No. 9433, (August 2004), pp. 521-526, ISSN 140-6736

Fisher, J. & Caputo, R.A. (2004). Comparing and Contrasting Barrier Isolator Decontamination Systems. *Pharmaceutical Technology*, Vol. 28, No. 11, (November 2004), pp. 68-82, ISSN 1543-2521

Forney, Ch. F.; Rij, R. E.; Denis-Arrue, R. & Smilanick, J. L. (1991). Vapor phase hydrogen peroxide inhibits postharvest decay of table grapes. *HortScience*, Vol. 26, No. 12, (December 1991), pp. 1512-1514, ISSN 0018-5345

French, G.L.; Otter, J.A.; Shannon, K.P.; Adams, N.M.T.; Watling, D. & Parks, M.J. (2004). Tackling contamination of the hospital environment by methicillin-resistant Staphylococcus aureus (MRSA): a comparison between conventional terminal cleaning and hydrogen peroxide vapour decontamination. *Journal of Hospital Infection*, Vol. 57, No. 1, (May 2004), pp. 31-37, ISSN 0195-6701

Fraise, A.P. (2004). Historical introduction, In: *Principles and Practice of Disinfection, Preservation & Sterilization* (fourth edition), A.P. Fraise, P.A. Lambert & J.Y. Maillard, (Ed.), 3-7, Russell, Hugo & Ayliffe´s. Blackwell Publishing, ISBN 1-4051-0199-7, Oxford

Fujishima, A. & Honda K. (1972). Electrochemical Photolysis of Water at a Semiconductor Electrode. *Nature*, Vol. 238, No. 5358, (July 1972), pp. 37 – 38, ISSN 0028-0836

Gould, G.W. (2004). Heat sterilization, In: *Principles and Practice of Disinfection, Preservation & Sterilization* (fourth edition), A.P. Fraise, P.A. Lambert & J.Y. Maillard, (Ed.), 361-383, Russell, Hugo & Ayliffe´s. Blackwell Publishing, ISBN 1-4051-0199-7, Oxford

Gruhn, R.; Baessler, H. J. & Werner, U. J (1995). Sterilization of chicken eggs for vaccine production with vaporized hydrogen peroxide. *Pharmazeutische Industrie*, Vol. 57, No.10, (October 1995), pp. 873-877, ISSN 0031-711X

Gurevich, I. (1991). Infection Control: Applying Theory to Clinical Practice, In: *Disinfection, Sterilization, and Preservation* (fourth edition), S.S. Block, (Ed.), 655-662, Lea & Febiger, ISBN 978-081-2113-64-8 London, Philadelphia

Hall, L.; Otter, J.A.; Chewins, J. & Wengenack, N.L. (2007). Use of Hydrogen Peroxide Vapor for Deactivation of Mycobacterium tuberculosis in a Biological Safety Cabinet and a Room. *Journal of Clinical Microbiology*, Vol. 45, No. 3, (March 2007), pp. 810–815, ISSN 1098-660X

Hansel, S.; Castegnaro, M.; Sportouch, M.H.; De Méo, M.; Milhavet, J.C.; Laget, M. & Duménil, G. (1997). Chemical degradation of wastes of antineoplastic agents: cyclophosphamide, ifosfamide and melphalan. *International Archives of Occupational and Environmental Health*, Vol. 69, No. 2, (January 1997), pp. 109-114, ISSN 1432-1246

Hardy, K.J.; Gossain, S.; Henderson, N.; Drugan, C.; Oppenheim, B.A.; Gao, F. & Hawkey, P.M. (2007). Rapid recontamination with MRSA of the environment of an intensive care unit after decontamination with hydrogen peroxide vapour. *Journal of Hospital Infection*, Vol. 66, No. 4, (August 2007), pp. 360-368, ISSN 0195-6701

Hatanaka, K. & Schibauchi, Y. (1989). Sterilization Method and Apparatus Therefor. *United States Patent*, US 4,797,255, filed (March 1987), issued (January 1989)

Heckert, R.A.; Best, M.; Jordan, L.T.; Dulac, G.C.; Eddington, D.L. & Sterritt, W.G. (1997). Efficacy of Vaporized Hydrogen Peroxide against Exotic Animal Viruses. *Applied and Environmental Microbiology*, Vol. 63, No. 10, (October 1997), pp. 3916–3918, ISSN 1098-5336

Higashi, N.; Yokota, H.; Hiraki S. & Ozaki, Y. (2005). Direct Determination of Peracetic Acid, Hydrogen Peroxide, and Acetic Acid in Disinfectant Solutions by Far-Ultraviolet Absorption Spectroscopy. *Analytical Chemistry*, Vol. 77, No. 7, (February 2005), pp. 2272–2277, ISSN 0003-2700

Hoffman, R.K. & Spiner, D.R. (1970). Effect of Relative Humidity on Penetrability and Sporicidal Activity of Formaldehyde. *Applied Microbiology*, Vol. 20, No.4, (October 1970), pp. 616-619, ISSN: 0003-6919

Hultman, C.; Hill, A. & McDonnell, G. (2007). The Physical Chemistry of Decontamination with Gaseous Hydrogen Peroxide. *Pharmaceutical Engineering*, Vol. 27, No. 1, (January/February2007), pp. 22, ISSN: 0273-8139

Ikarashi, Y.; Tsuchiya, T. & Nakamura, A. (1995). Cytotoxicity of medical materials sterilized with vapour-phase hydrogen peroxide. *Biomaterials*, Vol. 16, No. 3, (February 1995), pp. 177-183, ISSN 0142-9612

International Committee of the Red Cross (n.d. 2005). Convention on the prohibition of the development, production, stockpiling and use of chemical weapons and on their destruction, Paris 13 January 1993, In: *International Humanitarian Law - Treaties & Documents*, 20. 7. 2011, Available from: http://www.icrc.org/ihl.nsf/FULL/553?OpenDocument

Jernigan, D.B.; Raghunathan, P.L.; Bell, B.P.; Brechner, R.; Bresnitz, E.A.; Butler, J.C.; Cetron, M.; Cohen, M.; Doyle, T.; Fischer, M.; Greene, C.; Griffith, K.S.; Guarner, J.; Hadler, J.L.; Hayslett, J.A.; Meyer, R.; Petersen, L.R.; Phillips, M.; Pinner, R.; Popovic, T.; Quinn, C.P.; Reefhuis, J.; Reissman, D.; Rosenstein, N.; Schuchat, A.; Shieh, W.J.; Siegal, L.; Swerdlow, D.L.; Tenover, F.C.; Traeger, M.; Ward, J.W.; Weisfuse, I.; Wiersma, S.; Yeskey, K.; Zaki, S.; Ashford, D.A.; Perkins, B.A.; Ostroff, S.; Hughes, J.; Fleming, D.; Koplan, J.P.; Gerberding, J.L. & the National Anthrax Epidemiologic Investigation Team. (2002). Investigation of bioterrorism-related anthrax, United States, 2001: Epidemiologic findings. *Emerging Infectious Diseases*, Vol. 8, No. 10, (October 2002), pp. 1019-1028, ISSN 1080-6059

Johnson, J.W.; Arnold, J.F.; Nail, S.L. & Renzi, E. (1992). Vaporized hydrogen peroxide sterilization of freeze dryers. *Journal of Parenteral Sciencece and Technology*, Vol. 46, No. 6, (November-December 1992), pp. 215-225, ISSN 0279-7976

Johnston, M.D.; Lawson,S. & Otter, J.A. (2005). Evaluation of hydrogen peroxide vapour as a method for the decontamination of surfaces contaminated with Clostridium botulinum spores. *Journal of Microbiological Methods*, Vol. 60, No. 3, (March 2005), pp. 403–411, ISSN 0167-7012

Kahnert, A.; Seiler, P.; Stein, M.; Aze, B.; McDonnell, G. & Kaufmann, S.H.E. (2005). Decontamination with vaporized hydrogen peroxide is effective against Mycobacterium tuberculosis. *Letters in Applied Microbiology*, Vol. 40, No. 6, (June 2005) pp. 448-452, ISSN 1472-765X

Kang, J. W. & Lee, K. H. (1997). A Kinetic Model of the Hydrogen Peroxide/UV Process for the Treatment of Hazardous Waste Chemicals. *Environmental Engineering Science*, Vol. 14, No. 2, (January 1997), pp. 183-192, ISSN 1092-8758

Klapes, N.A. & Vesley, D. (1990). Vapor-Phase Hydrogen Peroxide as a Surface Decontaminant and Sterilant. *Applied and Environmental Microbiology*, Vol. 56, No. 2, (February 1990), pp. 503-506, ISSN: 1098-5336

Klassen, N.V.; Marchington, D. & McGowant, H.C.E. (1994). H_2O_2 Determination by the I_3^- Method and by $KMnO_4$ Titration. *Analytical Chemistry*, Vol. 66, No. 18, (September 1994), pp. 2921-2925, ISSN 0003-2700

Kokubo, M.; Inoue, T. & Akers, J. (1998). Resistance of Common Environmental Spores of the Genus Bacillus to Vapor Hydrogen Peroxide. *PDA Journal of Pharmaceutical Science and Technology*, Vol. 52, No. 5, (September/October 1998), pp. 228-231, ISSN 1948-2124

Krause, J.; McDonnell, G. & Riedesel, H. (2001). Biodecontamination of animal rooms and heat-sensitive equipment with vaporized hydrogen peroxide. *Contemporary Topics in Laboratory Animal Science*, Vol. 40, No. 6, (November 2001), pp. 18-21, ISSN 1060-0558

Krieger, E.W. & Mielnik, T.J. (2005). Aircraft and Passenger Decontamination System. *United States Patent*, US 2005/0074359 A1, filed (October 2003), issued (April 2005)

Kuzma, M.; Kačer, P.; Pánek, L. (2008). Decontamination with hydrogen peroxide vapors as technology of the future. *CHEMagazín*, Vol. 18, No. 6, (November/December 2008), pp. 12-13, ISSN 1210-7409

Linsebigler, A. L.; Lu, G. & Yates, J. T. (1995). Photocatalysis on TiOn Surfaces: Principles, Mechanisms, and Selected Results. *Chemical Reviews*, Vol. 95, No. 3, (May 1995), pp. 735-758, ISSN 00092665

Lopez, A.; Bozzi, A.; Mascolo, G. & Kiwi, J. (2003). Kinetic investigation on UV and UV/H_2O_2 degradations of pharmaceutical intermediates in aqueous solution. *Journal of Photochemistry and Photobiology A: Chemistry*, Vol. 156, No. 1-3, (March 2003), pp. 121-126, ISSN 1010--6030

Lysfjord, J. & Porter, M. (1998). Barrier Isolation History and Trends. *Pharmaceutical Engineering*, Vol. 18, No. 5, (September/October 1998), ISSN 0273-8139

Manatt, S.L. & Manatt, M.R.R. (2004). On the Analyses of Mixture Vapor Pressure Data: The Hydrogen Peroxide/Water System and Its Excess Thermodynamic Functions. *Chemistry - A European Journal*, Vol. 10, No. 24, (December 2004), pp. 6540–6557, ISSN 1521-3765

Martínez, J.M.L.; Denis, M.F.L.; Piehl, L.L.; de Celis, E.R.; Buldain, G.Y. & Orto, V.C.D. (2008). Studies on the activation of hydrogen peroxide for color removal in the presence of a new Cu(II)-polyampholyte heterogeneous catalyst. *Applied Catalysis B: Environmental*, Vol.82, No.3-4, (August 2008), pp. 273-283, ISSN 0926-3373

McDonnell, G.; Bonfield, P. & Hernandez, V.D. (2007). The Safe and Effective Fumigation of Hospital Areas with a New Fumigation Method Based on Vaporized Hydrogen Peroxide. *American Journal of Infection Control*, Vol. 35, No. 5, (June 2007), pp. E33-E34, ISSN 0196-6553

McDonnell, G. (June 2004). Large area decontamination, In: *Cleanroom Technology*, 22. 7. 2011, Available from:
http://www.cleanroom-technology.co.uk/technical/article_page/
Large_area_decontamination/52681

McDonnell, G. & Russell, A.D. (1999). Antiseptics and Disinfectants: Activity, Action, and Resistance. *Clinical Microbiology Reviews*, Vol. 12, No. 1, (January 1999), pp. 147–179, ISSN 1098-6618

McVey, I.F.; Schwartz, L.I.; Centanni, M.A. & Wagner, G.W. (2006). Activated Vapor Treatment for Neutralizing Warfare Agents. *United States Patent*, US 7,102,052 B2, filed (April 2003), issued (September 2006)

Mills, A. & Hunte, S. L. (1997). An overview of semiconductor photocatalysis. *Journal of Photochemistry and Photobiology A: Chemistry*, Vol. 108, No. 1, (1997), pp. 1-35, ISSN 1010-6030

National Institute for Occupational Safety and Health. (September 2004). Preventing Occupational Exposures to Antineoplastic and Other Hazardous Drugs in Health Care Settings, In: *DHHS (NIOSH), 2004–165*, 20. 7. 2011, Available from: http://www.cdc.gov/niosh/docs/2004-165/

Okumura, T.; Ninomiya, N. & Ohta, M. (2003). The Chemical Disaster Response System in Japan. *Prehospital and Disaster Medicine*, Vol. 18, No. 3, (July – September 2003), pp. 189–192, ISSN 1049-023X

Okumura, T.; Suzuki, K.; Fukuda, A.; Kohama, A.; Takasu, N.; Ishimatsu, S. & Hinohara, S. (1998). The Tokyo Subway Sarin Attack: Disaster Management, Part 1: Community Emergency Response. *Academic Emergency. Medicine,* Vol. 5, No. 6, (June 1998), pp. 613-617, ISSN 1069-6563

O'Neill, H.J. & Brubaker, K.L. (2003). Method and Apparatus for the Gas Phase Decontamination of Chemical and Biological Agents. *United States Patent*, US 6,630,105 B1, filed (September 2000), issued (October 2003)

Pan, G.X.; Spencer, L. & Leary, G.J. (1999). Reactivity of ferulic acid and its derivatives toward hydrogen peroxide and peracetic acid. *Journal of Agricultural and Food Chemistry*, Vol. 47, No. 8, (August 1999), pp. 3325-3331, ISSN 0021-8561

Peral, J.; Doménech, X. & Ollis, D. F. (1997). Heterogeneous Photocatalysis for Purification, Decontamination and Deodorization of Air. *Journal of Chemical Technology and Biotechnology*, Vol. 70, No. 2, (October 1997), pp. 117–140, ISSN 1097-4660

Pereira, V.J.; Weinberg, H.S.; Linden, K.G. & Singer, P.C. (2007). UV Degradation Kinetics and Modeling of Pharmaceutical Compounds in Laboratory Grade and Surface Water via Direct and Indirect Photolysis at 254 nm. *Environmental Science & Technology*, Vol. 41, No. 5, (March 2007), pp. 1682-1688, ISSN 0013-936X

Parisi, A.N. & Young, W.E. (1991). Sterilization with Ethylene Oxide and Other Gases, In: *Disinfection, Sterilization, and Preservation* (fourth edition), S.S. Block, (Ed.), 580-595, Lea & Febiger, ISBN 978-081-2113-64-8 London, Philadelphia

Prousek, J. (1996). Advanced oxidation processes for water treatment. Chemical processes. *Chemicke Listy*, Vol. 90, No. 4, (April 1996), pp. 229-237

Rauf, M. A. & Ashraf, S. S. (2009). Fundamental principles and application of heterogeneous photocatalytic degradation of dyes in solution. *Chemical Engineering Journal*,Vol. 151, No. 1-3, (August 2009), pp. 10-18, ISSN 1385-8947

Ray M. B. (2000). Photodegradation of the Volatile Organic Compounds in the Gas Phase: A Review. *Developments in Chemical Engineering and Mineral Processing*, Vol. 8, No. 5-6, (n.d. 2000), pp. 405–439, ISSN 0969-1855

Reich, R.R. & Caputo, R.A. (2004). Vapor-Phase Hydrogen Peroxide Resistance of Environmental Isolates. *Pharmaceutical Technology*, (August 2004), pp. 50 -58, ISSN 1543-2521

Roberts, S.; Khammo, N.; McDonnell, G. & Sewell, G.J. (2006). Studies on the decontamination of surfaces exposed to cytotoxic drugs in chemotherapy workstations. *Journal of Oncology. Pharmacy Practice*, Vol. 12, No. 2, (June 2006), pp. 95-104, ISSN 1477-092X

Rogers, J.V.; Sabourin, C.L.K.; Choi, Y.W.; Richter, W.R.; Rudnicki, D.C.; Riggs, K.B.; Taylor, M.L. & Chang J. (2005). Decontamination assessment of Bacillus anthracis, Bacillus subtilis, and Geobacillus stearothermophilus spores on indoor surfaces using a hydrogen peroxide gas generator. *Journal of Applied Microbiology*, Vol. 99, No. 4, (October 2005), pp. 739-748, ISSN 1365-2672

Russell, A.D. (1990). Bacterial Spores and Chemical Sporicidal Agents. *Clinical Microbiology Reviews*, Vol. 3, No. 2, (April 1990), pp. 99-119, ISSN 1098-6618

Russell A.D. (1991). Principles of Antimicrobial Activity, In: *Disinfection, Sterilization, and Preservation* (fourth edition), S.S. Block, (Ed.), 27-58, Lea & Febiger, ISBN 978-081-2113-64-8 London, Philadelphia

Rutala, W.A. & Weber, D.J. (1999). Disinfection of endoscopes: review of new chemical sterilants used for high-level disinfection. *Infection Control & Hospital Epidemiology*, Vol. 20, No.1, (January 1999), pp. 69-76, ISSN 1559-6834

Sagripanti, J.L. & Bonifacino, A. (1996). Comparative Sporicidal Effects of Liquid Chemical Agents. *Applied and Environmental Microbiology*, Vol. 62, No. 2, (February 1996), pp. 545-551, ISSN 0099-2240

Sapers, G.M.; Walker, P.N.; Sites, J.E.; Annous B.A. & Eblen D.R. (2003). Vapor-phase Decontamination of Apples Inoculated with *Escherichia coli. Journal of Food Science*, Vol. 68, No. 3, (April 2003), pp. 1003–1007, ISSN 1750-3841

Scatchard, G.; Kavanagh, G.M. & Ticknor, L.B. (1952). Vapor-Liquid Equilibrium. VIII. Hydrogen Peroxide – Water Mixtures. *Journal of the American Chemical Society*, Vol. 74, No. 15, (August 1952), pp. 3715–3720, ISSN 1520-5126

Schmaus, G.; Schierl, R. & Funck, S. (2002). Monitoring surface contamination by antineoplastic drugs using gas chromatography–mass spectrometry and voltammetry. *American Journal of Health-System Pharmacy*, Vol. 59, No. 10, (May 2002), pp. 956-961, ISSN 1079-2082

Shaffstall, R.M.; Garner, R.P.; Bishop, J.; Cameron-Landis, L.; Eddington, D.L.; Hau, G.; Spera, S.; Mielnik, T. & Thomas, J.A. (April 2006). Vaporized Hydrogen Peroxide (VHP®) Decontamination of a Section of a Boeing 747 Cabin. In: *DOT/FAA/AM-06/10 Federal Aviation Administration (FAA)*, Oklahoma City, 25. 7. 2011, Available from: www.faa.gov/library/reports/medical/oamtechreports/index.cfm

Sharma, D. (2005). Bhopal: 20 years on. *The Lancet*, Vol. 365; No. 9454, (January 2005), pp. 111-112, ISSN 140-6736

Sheth, V.B. & Upchurch, D.C. (1996). System for Detecting the Presence of Liquid in a Vapor Phase Sterilization System. *United States Patent*, US 5,482,683, filed (March 1994), issued (January 1996)

Simmons, G.F.; Smilanick, J.L.; John, S. & Margosan, D.A. (1997). Reduction of Microbial Populations on Prunes by Vapor-Phase Hydrogen Peroxide. *Journal of Food Protection*, Vol. 60, No. 2, (February 1997), pp. 188-191, ISSN 0362-028X

Švrček, J. (2010). The Research and Development of Decontamination Process Utilizing Vapor Phase Hydrogen Peroxide. PhD diss., Institute of Chemical Technology, Prague

Tschirch, J.; Bahnemann, D.; Wark, M. & Rathouský, J. (2008). A comparative study into the photocatalytic properties of thin mesoporous layers of TiO_2 with controlled mesoporosity. *Journal of photochemistry and photobiology. A, Chemistry*, Vol. 194, No. 2-3, (February 2008), pp. 181-188, ISSN 1010-6030

Unger-Bimczok, B.; Kottke, V.; Hertel, Ch. & Rauschnabel, J. (2008). The Influence of Humidity, Hydrogen Peroxide Concentration, and Condensation on the Inactivation of Geobacillus stearothermophilus Spores with Hydrogen Peroxide Vapor. *Journal of Pharmaceutical Innovation*, Vol. 3, No. 2, (February 2008), pp. 123-133, ISSN 1939-8042

Unger, B.; Rauschnabel, U.; Düthorn, B.; Kottke, V.; Hertel, Ch. & Rauschnabel, J. (2007). Suitability of different construction materials for use in aseptic processing environments decontaminated with gaseous hydrogen peroxide. *The PDA Journal of Pharmaceutical Science and Technology*, Vol. 61, No. 4, (July-August 2007), pp. 255-275, ISSN 1079-7440

U.S. Environmental Protection Agency. (2005). *Compilation of Available Data on Building Decontamination Alternatives*, EPA/600/R-05/036, Cincinnati, Ohio

U.S. Environmental Protection Agency. (2008). *Effects of Vapor-Based Decontamination Systems on Selected Building Interior Materials: Vaporized Hydrogen Peroxide*, EPA/600/ R-08/074, Cincinnati, Ohio

U.S. Peroxide. (2009). Physical Properties of Hydrogen Peroxide, In: *Hydrogen Peroxide Technical Library.*, U.S. Peroxide, LLC, Atlanta, 27. 7. 2011, Available from: http://www.h2o2.com/technical-library/physical-chemical-properties/physical-properties/default.aspx?pid=20&name=Physical-Properties

Vassal, S.; Favennec, L. Ballet, J.-J. & Brasseur, P. (1998). Hydrogen peroxide gas plasma sterilization is effective against Cryptosporidium parvum oocysts. *American journal of infection control*, Vol. 26, No. 2, (April 1998), pp. 136-138, ISSN 1527-3296

von Woedtke, Th.; Haese, K.; Heinze, J.; Oloff, Ch.; Stieber, M. & Jülich W.-D. (2004). Sporicidal efficacy of hydrogen peroxide aerosol. *Die Pharmazie*, Vol. 59, No. 3, (March 2004), pp. 207-211, ISSN 0031-7144

Wagner, G.W.; Sorrick, D.C.; Procell, L.R.; Brickhouse, M.D.; Mcvey, I.F. & Schwartz, L.I. (2007). Decontamination of VX, GD, and HD on a Surface Using Modified Vaporized Hydrogen Peroxide. *Langmuir*, Vol. 23, No. 3, (January 2007), pp. 1178-1186, ISSN 1520-5827

Wagner, G.W. & Yang, Y.-C. (2001). Universal Decontaminating Solution for Chemical Warfare Agents. *United States Patent*, US 6,245,957 B1, filed (September 1999), issued (June 2001)

Watkins, W.B. (2006). Decontamination Apparatus and Methods. *United States Patent*, US 7,145,052 B1, filed (September 2004), issued (December 2006)

Watling, D.; Ryle, C.; Parks, M. & Christopher M. (2002). Theoretical Analysis of the Condensation of Hydrogen Peroxide Gas and Water Vapour as Used in Surface Decontamination. *PDA Journal of Pharmaceutical Science and Technology*, Vol. 56, No.6, (November/December 2002), pp. 291-299, ISSN 1948-2124

Wold A. (1993). Photocatalytic properties of titanium dioxide (TiO2). *Chemistry of Materials*, Vol. 5, No.3, (March 1993), pp. 280–283, ISSN 1520-5002

Xu Y. (2001). Comparative studies of the $Fe^{3+/2+}$–UV, H_2O_2–UV, TiO_2–UV/vis systems for the decolorization of a textile dye X-3B in water. *Chemosphere*, Vol. 43, No. 8, (June 2001), pp. 1103-1107, ISSN 0045-6535

Zhao, J.; Chen, Ch. & Ma, W. (2005). Photocatalytic Degradation of Organic Pollutants Under Visible Light Irradiation. *Topics in Catalysis*, Vol. 35, No. 3-4, (July 2005) pp. 269-278, ISSN 10225528

Alternative Treatment of Recalcitrant Organic Contaminants by a Combination of Biosorption, Biological Oxidation and Advanced Oxidation Technologies

Roberto Candal, Marta Litter, Lucas Guz, Elsa López Loveira, Alejandro Senn and Gustavo Curutchet

Centro de Estudios Ambientales: Escuela de Ciencia y Tecnología e Instituto de Investigaciones e Ingeniería Ambiental, Universidad de San Martín, Campus Miguelete, Gerencia Química, Comisión Nacional de Energía Atómica, San Martín, Prov. de Buenos Aires, Argentina

1. Introduction

Industrial activity produces increasing amounts of effluent. Dyes and surfactants are typical examples of such pollutants, and both are present in various industries such as textiles, which are widely distributed in South-American, Indo-Asian and African countries. These industries are an important source of resources in the 3rd World Countries and also a source of pollution. Many compounds present in them are recalcitrant and therefore persistent in the environment, causing deleterious effects on the ecosystem (Padmavathy et al. 2003; Stolz et al. 2001). These compounds are usually found in very low concentrations and large volumes of effluent, characteristics that make very difficult its treatment by conventional means. Conventional techniques for removal of recalcitrant pollutants from wastewaters are based on the use of activated carbon, ionic exchange resins, chemical precipitation or membrane filtration. However, these technologies are usually not very convenient due to the large volumes to be treated and the low concentration of pollutants, excessive use of chemicals, accumulation of concentrated sludge and disposal problems, high cost of operation and maintenance of plant, sensitivity to other components of the liquid effluent (Balaji and Matsunaga 2002, Chen and Lin 2001). Traditional and alternative biological processes have received increasing interest owing to their cost, effectiveness, ability to produce less sludge and environmental benignity (Chen et al 2003; Volesky 2007) but some organic industrial and agricultural pollutants are recalcitrant to biological treatments . Several common compounds such as dyes, surfactants and pesticides, among others, are biorecalcitrant and can produce microbial death or other problems in water treatment plants. The biological treatment of liquid effluents containing this type of pollutants involves a previous separation step, or the isolation and use of specialized strains that can resist and degrade these toxic contaminants (Padmavathy et al. 2003).

In the case of effluents with these features, it is possible to use different combinations of processes to ensure proper treatment. Adsorption on low-cost substances, mainly biomass (biosorption), and combinations of special biological treatments (resistant strains in biofilm bioreactors) (Curutchet et al. 2001, Palmer and Sternberg 1999) with advanced oxidation technologies (such as photocatalysis, photo-Fenton, etc.) can be selected among the most versatile and economical treatments. Biosorption does not require expensive facilities. Advanced oxidation technologies partly mineralize the pollutants, using solar light as a source of energy and medium price oxidants, reducing the amount of contaminated sludges.

Biosorption is a process that utilizes various natural materials of biological origin, including bacteria, fungi, yeast, algae, etc. (Volesky 2007; Vijayaraghavan and Yun 2008). These biosorbents have heavy metal and organics sequestering properties, and can decrease the soluble concentration of these contaminants.

The use of biosorbents is an ideal alternative method for treating high volumes of wastewater with low concentrations of heavy metals (Volesky 2007) or persistent organic compounds (Wang et al. 2007, Patel & Suresh 2008). Compared with conventional treatment methods, biosorption has the following advantages (Volesky 2007): high efficiency and selectivity for absorbing contaminants in low concentrations, energy-saving, broad operational range of pH and temperature and, in some cases, easy recycling of the biosorbent.

However, biosorption generates huge quantities of sludge that should be treated or disposed to avoid secondary pollution (Stolz 2001). Most of the works in the literature study only the process of adsorption and do not mention the fate of the adsorbent loaded with contaminants and converted into a hazardous waste.

Knowledge of the mechanisms of degradation of the adsorbed contaminants (particularly in the case of dyes adsorbed on biomass, or biomass associated with matrixes found in natural environments such as clays and sediments) is very important to understand the processes involved in its fate in natural environments and to develop remediation alternatives for polluted water. In the case of organic contaminants, the adsorption enables fast and efficient decontamination of water (natural or residual), and the possibility of further degradation in solid phase allows to regenerate the adsorbent for further use or its decontamination to avoid expensive disposal processes.

Composting is known to stabilize biosolids (Tandy et al., 2009), transforming compounds to a high fraction of humic and fulvic acids and carbon dioxide. Therefore, composting biosolids is a good alternative for biowaste disposal (Ho et al 2010).

Adsorption of contaminants on metal oxides with photocatalytic activity is another way to remove pollutants by a fast adsorption followed by slow degradation under solar or UVA illumination. There are several metal oxides with high adsorptive capacity and photocatalytic activity. Most metal oxides display the property to catalyze the oxidation of organic (and inorganic) compounds under illumination with light with appropriate wavelength (energy higher than the metal oxide band-gap) (Litter 1999, Candal et al. 2004). The activation of a photocatalytic oxide involves the promotion of an electron from the valence band to the conduction band. Both species migrate to the surface of the catalyst where can recombine or react with adsorbed species like organics or water. The reaction of a hole with water produces highly oxidant OH• radicals, which also react with organic or inorganic species adsorbed and/or dissolved in water. TiO_2 is one of the most well known photocatalyst, and it can be used in water treatment in areas with high solar light illumination and in several commercial products. TiO_2 displays photocatalytic activity under UVA illumination. Iron oxide and oxohydroxide also present photocatalytic activity and their band gap lies in the visible region, which is appropriate for the use of solar light. Unfortunately its oxidant ability is lower, it is easy photo-corroded and have low chemical resistance (Lackoffand et al; 2002) However, it was recently reported the degradation of pesticides and dyes mediated by different iron oxides (Vittoria Pinna et al., 2007; Chien-Tsung et al, 2007; C. Pulgarin et al, Langmuir, 1995, Gilbert et al, 2007). Besides, colloidal iron oxides are present in natural waters displaying an important adsorption capacity which, combine with its photocatalytic activity, may play an important role in the final fate of contaminants in water.

On the other hand, adsorption of contaminants on biomass (for example, bacteria) and/or oxides with photocatalytic activity are processes that occur in natural environments when a polluted effluent reach a course of water.

This article describes fundamental and applied experiments that may contribute to the development of alternative processes for wastewater treatment and to a better understanding of the mechanisms involved in the fate and transformation of dyes and surfactants in natural environments such as streams.

2. Research methods

2.1 Culture media and microbiological methods

Appropriate culture media were used to isolate bacteria from consortia in reactors and from the José León Suarez channel an affluent to the Reconquista river in the neighborhood of Buenos Aires. The isolation medium had the following composition: 5.0 g L^{-1} glucose, 3.0 g L^{-1} peptone, 0.5 g L^{-1} $(NH_4)_2SO_4$, 0.5 g L^{-1} K_2HPO_4, 0.1 g L^{-1} $MgSO_4$, 0.010 g L^{-1} $CaCl_2$, 1.7% agar. The same medium without agar was used as liquid medium for growing the bacteria in the reactors. 100 mg L^{-1} of BKC was added as selective pressure in order to obtain apropiate consortia.

The culture medium used to feed the reactors (reactor medium) during the experiments had a different composition to avoid the formation of micelles, which interfere in the determination of BKC by HPLC (see below). The composition of such media was: 5.0 g L^{-1} glucose, 2.0 g L^{-1} $(NH_4)_2SO_4$, 5.0 g L^{-1} K_2HPO_4, 0.1 g L^{-1} $MgSO_4$, 0.01 g L^{-1} $CaCl_2$. In some experiments, the glucose concentration of the media was changed.

Biomass was measured by dry weight, turbidimetry, and plate counting, as indicated in each experiment.

2.2 Isolation of microorganisms

Native strains were isolated from contaminated streams with noticeable presence of dyes and tensioactives and a strong autodepurative capacity. These streams are affluents of Reconquista River, one of the most polluted rivers near Buenos Aires City (Argentina).

Other microbial consortia coming from conventional wastewater treatment plants were subjected to selective pressure (high concentrations of colorants or surfactants) in a packed bed bioreactor to select resistant organisms. The reactor was operated for six months, and cultivable microorganisms in the supernatant were isolated in solid media. The different strains obtained were characterized by biochemical analysis. The characterization of the genera was made by the API 20-E and 20 NE biochemical tests (Biomeriaux), catalase, oxidase and Gram test.

The isolation was carry out in agar plates with PCA medium. Strains isolated from the JLS channel were denominated StA, StB StC StD. Strains isolated from bioreactors were denominated SbA, SbB, SbC, SbD SbE.

2.3 Biosorption experiments
2.3.1 Biosorption dynamics and equilibria

Biosorption experiments were performed in batch systems with a constant amount of biomass of different individual and in consortium isolates and different concentrations of crystal violet (dye) or benzalkonium chloride (BKC) (surfactant). Experiments were conducted to determine the dynamics of the process and the equilibrium (isothermal) conditions for different model compounds. At different incubation times, the biomass was separated by centrifugation (5000 g) and the remaining dye or surfactant concentration was determined in supernatant. Crystal violet was measured by visible spectrophotometry at 590 nm and BKC was determined by HPLC (Ding et al, 2004).

The data were fitted by the Langmuir isotherm. Experiments with living biomass (active uptake) and died biomass (passive adsorption) were performed. For crystal violet, the effect of pH on the adsorption was studied.

2.4 Treatment of the liquid (effluent) and solid (sludges) obtained by adsorption of dyes. Liquid batch and solid phase dye degradation
2.4.1 Bacterial growth in liquid medium in the presence of dye

To study the effect of crystal violet on the growth of the isolated strains, an experiment of growth kinetics was carry out in agitated flasks. The same base culture medium was used with the addition of different dye concentrations. Optical density (OD) at 410 nm and dye concentration for spectrophotometry at 590 nm after centrifugation were measured. All the strains isolated from the JLS channel grew to the concentration of 0.023 at the same speed as the control without dye.

2.4.2 Solid phase dye degradation

Solid phase degradation experiments were carried out in plastic vessels of 50 ml using StB and StC strains. Biomass and sediment-adsorbed biomass were generated in agitated flasks with and without 1% w/v of previously desiccated sediment. Cultures in stationary phase with 2 g L^{-1} biomass were incubated with a know amount of dye. After equilibrium was reached, biomass was separated by centrifugation at 5000 g and the dye concentration in the supernatant determined by spectrophotometry. The dye-charged biomass was washed and mix with 3 g of desiccated aquatic plants (*Salvinia sp.*) isolated of the same environment than bacterial strains and cultivated in a greenhouse in dechlorinated tap water. This mixture was placed in the containers and composted for different times. Remaining dye in the mixtures was measured by extraction of the dye with 10 g L^{-1} Sodium dodecyl sulphonate (SDS) and spectrophotometric quantification.

2.5 Batch and continuous BKC biodegradation coupled with TiO$_2$-photocatalysis

Experiments were performed in agitated vessels and in a continuous packed bed reactor (CPBR). Both types of reactors were inoculated with bacteria obtained from a conventional wastewater treatment plant from which several subcultures with 150 mg L^{-1} BKC as selective pressure were made. The CPBR was operated in continuous mode at 50 mL h^{-1} feeding rate. The reactor was formed by two cylindrical vessels 60 cm long and 5 cm internal diameter, with a packed bed of glass chips (-5 +3.5 mesh). One of the reactors was fed from the bottom with the liquid media. The outlet of this reactor was the feeding media of the second reactor, which was fed from the upper part. The first cylinder was oxygenated by bubbling air from the bottom, in the opposite direction to the feeding liquid. The second cylinder was not oxygenated, but the liquid was always in contact with air as it moved towards the outlet, placed in the bottom part of the cylinder. In regular experiments, the reactor was fed with the liquid medium described above (2500 mg L^{-1} Chemical Oxygen Demand (COD), mainly glucose) and the corresponding amount of BKC. Samples of the liquid were taken at the inlet, at the outlet and at a point between the two cylinders.

The aqueous effluent obtained after microbiological treatment was submitted to photocatalytic treatment. The photocatalytic BKC oxidation was performed in a 0.6 L recirculating batch system consisting of an annular glass reactor (415-mm length, 32-mm internal diameter), a peristaltic pump (APEMA BS6, 50 W), and a thermostatted (298 K) reservoir. The TiO_2 suspension (1 g L^{-1}) containing BKC was recirculated at 1 L min^{-1} from the reservoir through the photoreactor. The total volume of the circulating mixture was 450 mL, of which 100 mL were inside the photoreactor. Air was bubbled in the reservoir at 0.2 L min^{-1} all throughout the irradiation time. The illumination source was a black light tubular UV lamp (Philips TLD/08, 15 W, 350 nm $< \lambda <$ 410 nm, maximum emission at 366 nm) installed inside the annular reactor. A photon flux of 7.4 µeinstein s^{-1} L^{-1} was determined by actinometry with ferrioxalate, assuming a 366 nm monochromatic light.

TiO_2 (Aeroxide® TiO_2 P25) was incorporated to the effluent in order to obtain a 1.0 g L^{-1} suspension. The system was ultrasonicated for 1 minute and recirculated in the reactor for 30 minutes to reach the adsorption equilibrium (as determined in preliminary experiments). Periodically, samples were taken from the suspension and filtered through a 0.2 µm cellulose acetate membrane before analysis or for biological treatment.

Alternatively, aqueous solutions containing only BKC (200-50 mg L^{-1}) were treated by photocatalysis during different periods in the reactor described before. The obtained solutions were mixed with a carbon rich media and submitted to biological treatment

3. Results

3.1 Isolation and characterization of cultivable bacteria strains

Five strains (StA, StB, StC, StD and StE) were isolated from samples of water from the José León Suarez (JLS) channel, a tributary of Reconquista river. This river, considered the second most contaminated river of Argentina, receives an important load of pollutants of domestic and industrial origin. The track of the basin in which the channel is located practically combines all the elements typical of a hyper-degraded area: informal occupation of the plain of flooding, high population density, extreme poverty, clandestine industrial and sewage discharges and presence of the biggest sanitary fill of the metropolitan region.

Strains were selected by morphological differences of the colonies in agar plates with PCA medium. Biochemical tests were carried out for metabolic characterization. StB strain shows correspondence within *Enterobacter cloacae*. StC Strain was identified as belonging to the *Pseudomonas* genus. The other three strains were Gram negative rod, catalase + and oxidase. Biochemical characterization did not allow genera identification of these bacteria.

Presence of *Escherichia coli* and other fecal Enterobacteria was detected by isolation in DEV and EMB (Merck) media, showing fecal contamination of the stream.

Another six strains were isolated from bioreactors working in our laboratory by several months with selective pressure (BKC 100 mg L^{-1}). Twenty days were needed to form a biofilm of BKC-resistant bacteria on the glass support. After 30 days, the film had homogeneously covered all the support inside the column, as shown in Figure 1. Six strains could be isolated from the biofilm (strains SbA to SbF).

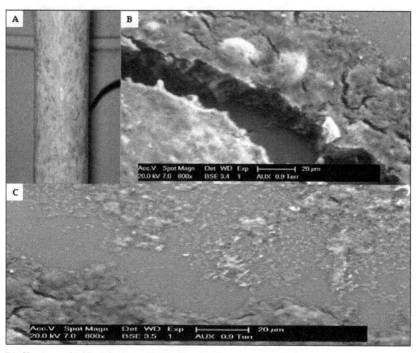

Fig. 1. *Biofilm in packed bed reactor.* A) Biofilm in the support. B) Scanning electronic microscopy of the biofilm, where the thickness of biofilm is shown. C) Scanning electronic microscopy of the biofilm after 30 days from inoculation of reactor.

Biochemical tests allowed the identification of two BKC-resistant bacterial genera (*Pseudomonas sp.* and *Saccarococcus sp.*). This agrees with literature reports of degradation of quaternary ammonium compounds (QACs) by microorganisms present in activated sludges, where the majority of microorganisms able to utilize QACs as the carbon and energy source (QACs degraders) were classified as *Pseudomonas sp., Xanthomonas sp.* and *Aeromonas sp.* (Zhang 2011, Ismail 2010).

3.2 Biosorption experiments
3.2.1 Crystal violet experiments

3.2.1.1 Adsorption dynamics

Figure 2 shows crystal violet concentration in solution vs. time. For the 5 strains (StA to StE), one hour was necessary to reach the equilibrium.

Experiments in the presence of an energy source (glucose) show the same pattern. This suggests a strong and fast physicochemical adsorption and no significant contribution of active uptake mechanisms (at least in the time range studied).

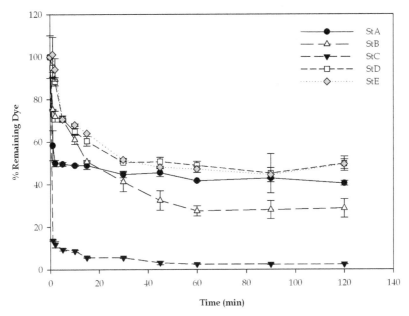

Fig. 2. Percentage of dye remaining in solution vs. time.

Sorption kinetics was fitted with an equation corresponding to a pseudo-second order model (Equation 1).

$$\frac{t}{q} = \frac{1}{k_2 q_{eq}^2} + \frac{1}{q_{eq}} t$$

Equation 1. Pseudo-second order model of biosorption dynamics.

where t is the time, q_{eq} is the specific adsorption reached at equilibrium and k_2 is the second-order rate constant.
Values of k_2 and correlation coefficients are shown in Table 1.

	R^2	k_2 (g s^{-1} mg^{-1})
A	0.9988	0.020
B	0.9943	0.006
C	0.9999	0.084
D	0.9955	0.009
E	0.9895	0.006

Table 1. Dynamic constants and correlation coefficients for the pseudo-second order model of biosorption dynamics.

The pseudo-second order model is based in chemisorption. A monolayer of adsorbate is formed on the sorbent surface by ionic interaction. The dynamic equilibrium is reached when all the bonding sites are saturated (Ho et al. 1999). Based on the previous model, the results obtained in this work indicate that the adsorption of crystal violet on the bacteria is due to electrostatic interaction between the dye and the wall of the cells.

3.2.1.2 Adsorption of crystal violet by the isolated strains. Effect of pH

Figure 3 shows the variations of specific adsorption (Q_{eq}) with the initial pH of the solution for 50 mg L^{-1} initial concentration of the dye. It is seen that Q_{eq} does not change significantly in a wide range of pH (from 5 to 8). From pH 3 to 5-6, Q_{eq} rises drastically. pH values below 3 or above 8 lead to changes in color or solubility of the dye, and were not studied.

The results agree with a typical behavior for biosorption of cationic species. Changes in the protonation of the active sorption groups in biomass (carboxyl, hydroxyl, amino, etc.) lead to a minor specific adsorption (Volesky 2007).

The maximum Q_{eq} value was reached in the typical range of pH found in natural waters and most of the industrial effluents containing dyes (Chen 2003, Akar 2010).

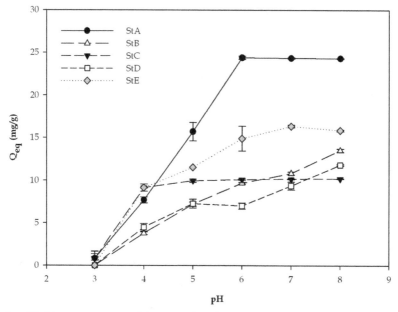

Fig. 3. Specific biosorption vs. pH for crystal violet adsorbed in strains isolated from the JLS channel.

3.2.1.3 Crystal violet adsorption isotherms

Figure 4 shows the adsorption isotherms for strains StA to StE. As shown in Figure 2, the strains show Q_{eq} between 100 and 300 mg L^{-1} and different profiles.

Linearization of the isotherms of Figure 4 was attempted by a Langmuir model. This was made only to get a phenomenological description of the plots and for calculation of specific adsorption Q^0. The respective extracted adsorption parameters are presented in Table 2. Strain B shows a sigmoidal shape and was fitted with a Hill model.

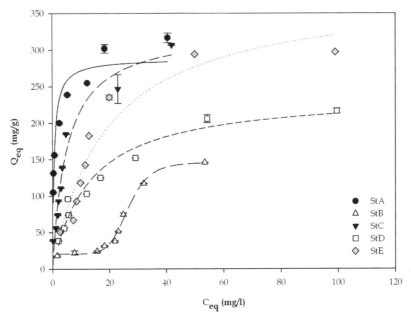

Fig. 4. Adsorption isotherms of crystal violet on the five isolated strains. The initial crystal violet concentration was in the 10-200 mg L^{-1} range.

	R^2	Q^0 (mg g^{-1})	K$_L$ (mg L^{-1})
A	0.9027	288 ± 15	0.59 ± 0,15
B	0.9927	125 ± 6	26.7 ± 0,6
C	0.9573	328 ± 22	5.1 ± 0,9
D	0.9564	240 ± 15	13 ± 2
E	0.9215	381 ± 40	19 ± 5

Table 2. Constants and correlation coefficients for the Langmuir and Hill biosorption models.

The strains isolated in the JLS channel show higher Q_{max} values than those reported in the literature for cationic dyes adsorbed on other strains (Chu and Chen 2002, Fu and Virarahavan 2000). These high adsorption capacities (close to 30% in mass) suggest the importance the biosorption processes in the natural attenuation of the contaminants discharged to the stream.

3.2.2 BKC experiments

Kinetic studies indicated that BKC adsorption on the biomass in the studied range is very fast. After 20 min, the saturation equilibrium was reached in all cases. Due to these results, adsorption isotherms were evaluated after 30 min incubation. Figure 5 shows the isotherms obtained by adsorption of BKC on the six isolated strains (strains SbA to SbF).

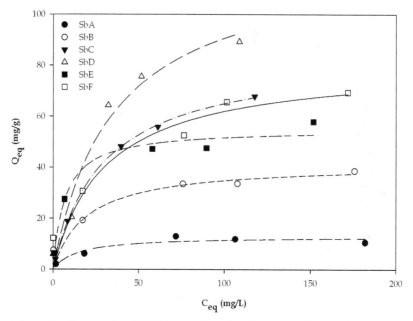

Fig. 5. Adsorption isotherms of BKC on the six isolated strains. The initial BKC concentration was in the 7-209 mg L^{-1} range.

In a similar way as in the crystal violet adsorption experiments, linearization of the isotherms of Figure 5 was attempted by a Langmuir model (only for phenomenological description of the plots and for calculation of Q^0). The respective extracted adsorption parameters are presented in Table 3.

	R^2	Q^0 (mg g^{-1})	K_L (mg L^{-1})
A	0,977	11	9,3
B	0,987	37	7,9
C	0,999	85	31
D	0,875	104	21
E	0,985	56	10
F	0,971	71	14

Table 3. Adsorption parameters obtained from the linearization of the adsorption isotherms applying the Langmuir model.

With the data obtained from the adsorption isotherms extracted for each strain, it can be estimated that the BKC sorption capacity on this consortium is in the range 11-110 mg BKC per gram of dry biomass. In further experiments with the consortium, an average value of 50 mg BKC g^{-1} biomass was taken, in agreement with previous data found for BKC adsorption on activated sludges.

3.3 Treatment of sludges from adsorption processes. Adsorption-composting coupled experiments with crystal violet
3.3.1 Solid phase dye degradation

Figure 6 shows the percentage of degradation of adsorbed dye on biomass (strains StB and StC) and biomass-sediment.

StB and StC show 40 and 30% of dye degradation in only 12 days. In the systems with biomass associated to sediment, the degradation rate is delayed. This behavior may be due to the distribution of adsorbed dye between the biomass and the abiotic surfaces (clays, humic substances, etc.). In systems with biomass and sediments, the highest proportion of sediment could favor adsorption on the abiotic component. Under these conditions, the population of microorganisms that catalyze the degradation of this fraction of the dye would be less than the present when the dye is directly adsorbed on biomass.

Although experiments with higher incubation times are needed to understand the precise involved mechanisms, the degradation rates observed in the case of C and B strains are high enough to show that this degradation process in solid phase has a high potential for the treatment of wastes produced in the remediation of contaminated sludge and produced by biosorption of dyes.

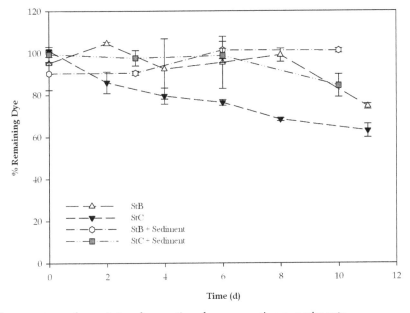

Fig. 6. percentage of remaining dye vs. time for composting experiments.

3.4 Biological-photocatalytic coupled treatment of surfactants

Biodegradation experiments were conducted in shaken flasks and a packed bed reactor with a consortium of microorganisms adapted to grow as a biofilm in the presence of BKC. Operational variables were studied to optimize the process, mainly the presence and concentration of organic matter other than the studied recalcitrant contaminants, such as is usually found in real effluents.

The photocatalytic degradation of benzyl alkyl ammonium was studied by Hidaka et al. (2002) and in our laboratories. Degradation of these compounds by oxidative photocatalysis is possible, but mineralization is slow. However, photocatalysis is an interesting alternative to reduce the BKC concentration entering biological plants, to avoid the biomass death provoked by the presence of relatively high BKC concentrations. In this case, the role of the oxidized byproducts on the performance of the biological reactor should be investigated. Also, the presence of an extra source of carbon should be considered in order to feed the growing biomass. Some studies in batch and supported reactors were performed in this work.

Alternatively, when a relatively low BKC concentration is present in the effluent, biological treatment followed by photocatalysis may be useful to completely remove BKC from the treated water. In this case, the effect of organic matter (other than BKC) on the photocatalytic reactor should be tested.

3.4.1 Batch experiments

In the photocatalytic experiments, BKC concentration was rapidly reduced from 100 to 5 mg L^{-1} after 150 min irradiation (97% total degradation). However, TOC was reduced only 21% (from 213 mg L^{-1} to 168 mg L^{-1}) in the same period, indicating a low mineralization degree with several byproducts generated during the treatment. BKC decay followed an excellent pseudo-first order kinetics (R^2 = 1), with k = 5 × 10^{-4} s^{-1}.

When the pure and the photocatalyzed BKC solutions (with added nutrients but no another carbon source) were inoculated and incubated, few changes in BKC concentration were observed, but the suspended biomass decreased during the first 50 h incubation time, until a minimum value was reached. The viable bacterial population determined by plate counting also decreased dramatically during the whole incubation period (not shown). These results indicate that the cells died as a consequence of the pollutant and/or the low level of biodegradable organic matter (carbon and electron source), unable to sustain the microorganisms.

When the experiments were repeated with the incorporation of a carbon source (glucose in the reactor medium), the results changed notably. BKC concentration decreased approximately 22% for pure BKC at the end of experiment (230 h). The patterns in BKC concentration and bacterial growth (data not shown) suggest that BKC is adsorbed by the growing biomass. The average sorption capacity of the biomass in system with pure BKC (30 mg BKC L^{-1}) calculated from data of isotherms is enough to produce the observed reduction in BKC concentration by biosorption. When BKC and its oxidation byproducts are present, BKC concentration also decreased, although the amount eliminated in 230 hours was 75% of the initial amount. Unlike the previous case, the profile of the curves (data not shown) suggests a different mechanism for BKC elimination, e.g., a combination of biosorption and biodegradation.

3.4.2 Biodegradation of pure and photocatalyzed BKC solutions in a packed bed reactor operated in continuous mode (CPBR)

3.4.2.1 Medium optimization in CPBR experiments

The batch experiments demonstrated that the presence of an energy and carbon source other than BKC is necessary to maintain the biomass alive and available for adsorption and biodegradation of BKC. The optimum concentration of this carbon source was found to be around 2500 mg L^{-1} COD (data not shown).

Alternative Treatment of Recalcitrant Organic Contaminants by a Combination of Biosorption, Biological Oxidation and
Advanced Oxidation Technologies

223

3.4.2.2 Coupled CPBR-photocatalytic treatment

3.4.2.2.1 Experiments with 100 mg L-1 BKC

The biological system was fed with a solution containing culture plus 102 mg BKC L^{-1}, and the experiment was run during 110 h. Samples were taken during 5 days and analyzed for COD and BKC. Figure 7 shows BKC and COD concentration at inlet and outlet at different times of the CPBR operation. The total BKC elimination rate was calculated. BKC concentration in the R1 outlet was near 50% of the R1 inlet concentration. The BKC elimination rate remained almost constant, about 6.5 mg L^{-1} h^{-1}, which means that in this reactor about 30% of the incoming BKC was biodegraded. There is a slight tendency of the rate to decrease at longer times, possibly because of deterioration of the biofilm. In R2, total BKC elimination rate decreased with the operation time, due to the decrease of the biodegradation rate, which may be a consequence of the low concentration of the carbon source entering R2 and leads to the degradation of biomass. Both reactors reduce BKC concentration to below 40 mg L^{-1}.

Figure 7a shows a substantial reduction in COD. COD consumption rate in R1 was in the range 263-206 mg L^{-1}h^{-1}, and in R2 19-63 mg L^{-1} h^{-1}. The lower COD consumption rate in R2 would be explained in terms of the Monod model (Zhang 2011) due to the low COD concentration (carbon and energy source) at R2 inlet. Continuous operation of R1 leads to diminish slightly the efficiency of the reactor, but R2 was able to assimilate all the COD. The COD degradation rate agrees again with the Monod model. The decrease of COD degradation rate in R1 led to an increase of COD concentration in R2 inlet, which led to an increase of COD degradation rate in this reactor.

A 500 mL sample of the R2 outlet was collected after the first 36 h of biological treatment ([BKC] = 38 mg L^{-1}), and submitted to a photocatalytic treatment. The BKC concentration decreased to 27.6 mg L^{-1} (28%) after TiO$_2$ incorporation and, after 2 h of HP treatment, more than 99% BKC was eliminated. Again an excellent pseudo-first order decay (not shown, R^2 = 0.94, k = 6 × 10^{-4} s^{-1}) was obtained. These results demonstrate the feasibility of using photocatalysis as a post-biological treatment to eliminate the recalcitrant pollutant. As observed, the biological treatment reduced the organic charge present in solution from 2222 to 246 mg L^{-1} , concentrations that do not hinder the activity of the photocatalyst (at least with the nutrients used in this work); further photocatalytic treatment eliminates the remaining toxic contaminant.

3.4.2.2.2 Experiments at 180 mg L-1 BKC

A new set of experiments were run by feeding the CPBR with a solution containing 2500 mg COD L^{-1} and 180 mg BKC L^{-1}. COD decreased notably at the outlet of both R1 and R2. BKC concentration also decreased at the beginning of the experiment, but after 2 days of continuous working, the concentration rose dramatically (not shown). These results would indicate that the biofilm collapses after 2 days and that BKC is released to the environment. R2 adsorbs part of the BKC but, after 5 days, the concentration in the solution rose until values close to the inlet concentration.

These results indicate that effluents containing up to 100 mg L^{-1} of BKC and biodegradable organic matter can be purified by a biological reactor followed by photocatalytic treatment. The bioreactor eliminates most of the organics and the photocatalytic treatment eliminates the remaining BKC. In contrast, higher BKC concentrations (about 180 mg L^{-1}) degrade the biofilm and avoid this type of coupling.

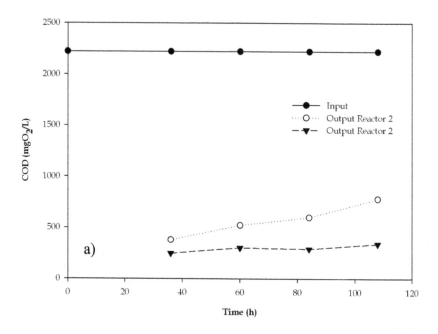

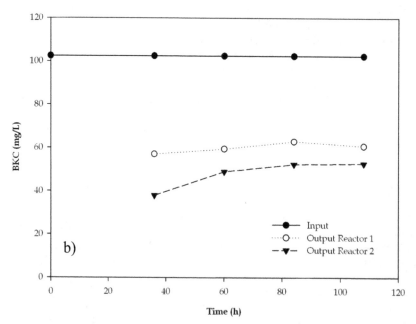

Fig. 7. Biological BKC treatment before the photocatalytic treatment: a) COD vs. time; b) BKC concentration vs. time

Alternative Treatment of Recalcitrant Organic Contaminants by a Combination of Biosorption, Biological Oxidation and
Advanced Oxidation Technologies

225

3.4.2.3 Coupled photocatalytic-CPBR treatment

A BKC sample previously treated by HP with an initial concentration of 81 mg L^{-1} was diluted
with culture medium, leading to a final solution with 48 mg BKC L^{-1} and 2050 mg COD L^{-1},
and this solution was submitted to the biological treatment. Results are shown in Figure 8.

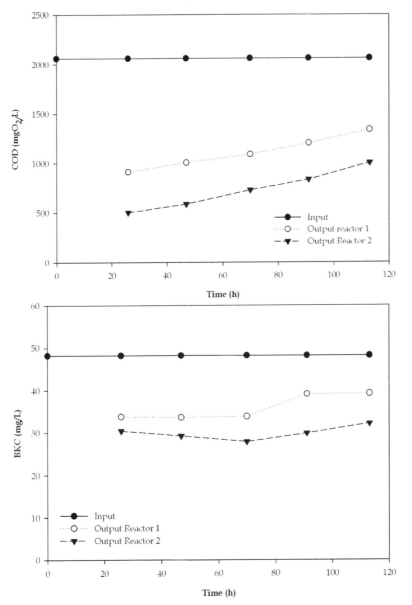

Fig. 8. *Biological treatment after photocatalysis:* A) DQO vs. time; B) BKC concentration vs. time.

The BKC removal rate in R1 was constant around 2 mg L^{-1} h^{-1} during the first 70 h; then it decreased to 1.3 mg L^{-1} h^{-1}, probably by degradation of the biofilm. These values are also considerably lower than 6.5 mg L^{-1} h^{-1}, value found in the previous biological-photocatalytic experiment; this, as observed also for COD removal, strongly suggests that incomplete degradation products are more toxic to the biofilm than pure BKC. In R2, the total removal rate increased BKC from 0.48 to 1.0 mg L^{-1} h^{-1}, resulting lower than the rate in the previous experiment. The COD degradation rate in R1 decreased 37% throughout the experiment. Compared to the previous experiment in which the initial rate was 263 mg L^{-1} h^{-1} and its decrease in time was 23%, it can be assumed that (below 100 mg L^{-1} BKC) the presence of partially degraded products produced during photocatalysis have a partial inhibitory effect on biomass, stronger than that of BKC. R2 shows a relatively constant COD degradation rate (60-48 mg L^{-1} h^{-1}), similar to that obtained in R2 in the previous experiment and lower than the one shown by R1. This fact is consistent with the Monod model, according to the input substrate concentrations (R1 output). In R2, the concentration of inhibitory compounds (BKC and intermediates) is lower than those in R1, and the biofilm activity did not decrease over time.

4. Conclusions

The results of this study show that coupling processes based on different technologies such as adsorption, biodegradation and photocatalysis are potentially useful in both, the understanding of the dynamics and fate of pollutants in superficial waters such as streams and rivers and for develop alternatives for effluent treatment.

The adsorption of dyes on biomass of native bacterial strains is a very fast process that occurs in a wide range of pH with very high specific adsorption (near 30 % of the dry weight). In natural waters, biomass is commonly growing as biofilms on sediment surfaces. This fact could lead to the immobilization of the contaminant. The dye-charged biomass alone or associated with sediments shows dye degradation capacity in composting processes.

With respect to potential effluent treatment processes, the isolated native strains show interesting capacities: StA strain shows very high affinity for crystal violet and could be used to treat large volumes of diluted effluents, and StE strain shows very high specific adsorption, and could be used to treat smaller volumes of concentrated effluents.

From the technological point of view, adsorption-based processes could be used to reach a very fast separation of the contaminants from large volumes of water. Later use of advanced oxidation technologies could be used to the final polish of the effluent, while composting of the contaminated biomass appears to be an excellent alternative to the treatment of the formed sludge, avoiding secondary contamination or high disposal costs.

With regard to the treatment of BKC, the process can be improved if the samples are submitted to a coupled treatment of a photocatalytic treatment combined with a biological system. For this purpose, different configurations of coupled photocatalytic-biological reactors can be adopted, depending mainly on the BKC concentration and the total organic load.

For BKC concentrations up to 100 mg L^{-1}, the CPBR-HP treatment configuration shows some advantages. In the biological reactor, 50% BKC is degraded and this drastically reduces the COD of the effluent. The subsequent photochemical treatment leads to total BKC removal without losing efficiency by oxidizing other readily degradable compounds present in the matrix.

Alternative Treatment of Recalcitrant Organic Contaminants by a Combination of Biosorption, Biological Oxidation and
Advanced Oxidation Technologies

227

In the case of BKC concentrations below 100 mg L^{-1}, the HP pretreatment does not work properly, because the toxicity of the photodegradation byproducts on the biofilm is higher than that of pure BKC.

In the case of higher BKC concentrations (180 mg L^{-1}), the bacterial biofilm is not able to be sustained over the time; therefore, it is necessary to perform a previous HP pretreatment to reduce BKC concentration.

In spite that the biomass activity decreases with time, the deleterious effect of BKC and byproducts on biofilm activity is less important compared with that of an effluent containing a very high charge of BKC and directly submitted to the biological treatment. As shown, if BKC concentration is too high (for example 180 mg L^{-1} or more), the biomass is strongly affected and BKC is released to solution. However, if the highly concentrated BKC solution is first photocatalytically treated, the biosystem can support the effluent containing the remaining BKC and its oxidation byproducts.

5. Acknowledgements

This work was performed as part of Agencia Nacional de Promoción Científica y Tecnológica PICT-512, PAE 22257, CONICET PIP 11220090100079, and Universidad de San Martín, UNSAM SA08/011.

6. References

Akar,T. Sema Celik, Sibel Tunali Akar (2010).Biosorption performance of surface modified biomass obtained from *Pyracantha coccinea* for the decolorization of dye contaminated solutions. *Chemical Engineering Journal*, 160, 2 (466-472)

Balaji, T.; Matsunaga, H. 2002. Adsorption characteristics of As(III) and As(V) with titanium dioxide loaded Amberlite XAD-7 resin. *Anal. Sci.* 2002;18:1345–1349.

Candal, R., S.A. Bilmes, M.A. Blesa 2004 "Semiconductores con Actividad Fotocatalítica"; en *Eliminación de Contaminantes por Fotocatálisis Heterogénea*, capítulo 4, M.A. Blesa, B. Sánchez Cabrero Ed. CIEMAT, ISBN 84 7834 489 6.

Chen, C. Pei-Ssu Wu, Ying-Chien Chung. (2009) Coupled biological and photo-Fenton pretreatment system for the removal of di-(2-ethylhexyl) pftalate (DEHP) from water. *Bioresource Technology*, 100, 19, (4531-4534)

Chen, J.; Lin, M. S. 2001. Equilibrium and kinetic of metal ion adsorption onto a commercial. H-type granular activated carbon: experimental and modeling studies. *Water Res.* 2001; 35:2385–2394.

Chen, K. J.Y.Wu, D.J. Liou, S.C.J. Hwang, Decolorization of the textile dyes by newly isolated bacterial strains, *J. Biotechnol.* 101 (2003) 57–68.

Curutchet, G. E. Donati, C. Oliver, C. Pogliani, M.R. Viera, Development of *Thiobacillus* biofilms for metal recovery, in *Methods in Enzymology*. 2001. p. 171–186.

Ding, X. S. Moub, S. Zhaoa, Analysis of benzyldimethyldodecylammonium bromide in chemical disinfectants by liquid chromatography and capillary electrophoresis, *J. Chromat.* A 1039 (2004) 209–213.

Dixit, A. A.J. Tirpude, A.K. Mungray, M. Chakraborty, Degradation of 2,4-DCP by sequential biological–advanced oxidation process using UASB and UV/TiO$_2$/H$_2$O$_2$, *Desalination* 272 (2011) 265–269.

El-Moselhy, M. (2009). Photo-degradation of acid red 44 using Al and Fe modified silicates. *Journal of Hazardous Materials*, 169, 3, (498-508)

García, M.T., I. Ribosa, T. Guindulain, J. Sánchez-Leal, and J. Vives-Rego, Fate and effect of monoalkyl quaternary ammonium surfactants in the aquatic environment. *Environ. Pollut.*, 2000. 111(1): (169-175).

Hidaka, H. J. Zhao, E. Pelizzetti, N. Serpone, 1992 Photodegradation of surfactants. 8. Comparison of photocatalytic processes between anionic sodium dodecylbenzenesulfonate and cationic benzyldodecyldlmethylammonlum chloride on the TiO2 surface, *J. Phys. Chem.* 96 (1992) 2226–2230.

Ho, C., S.T. Yuan a, S.H. Jien b, Z.Y. Hseu 2010. Elucidating the process of co-composting of biosolids and spent activated clay- *Bioresource Technology* 101 (2010) 8280–8286

Ismail, Z. U. Tezel, S. G. Pavlostathis, Sorption of quaternary ammonium compounds to municipal sludge, *Water Res.* 44 (2010) 2303 – 23.

K. Vijayaraghavan, Yeoung-Sang Yun. (2008) Bacterial biosorbents and biosorption. *Biotechnology Advances*, 26, 3, (266-291)

Lackoffand M., R. Niessner; 2002. Atmospheric Aerosols, and Soil Particles; *Environ. Sci. Technol.* 36, (2002), 5342-5347.

Litter, M.I., Heterogeneous Photocatalysis. (1999) Transition metal ions in photocatalytic systems. *Appl. Catal. B: Environ.*, 23, (89-114).

Muñoz, I. José Peral, José Antonio Ayllón, Sixto Malato, Paula Passarinho, Xavier Domènech (2006) Life cycle assessment of a coupled solar photocatalytic-biological process for wastewater treatment. *Water Research*, 40, 19, (3533-3540)

Padmavathy, S., Sandhya, S., Swaminathan, K., Subrahmanyam, Y.V., Chakrabarti, T.,Kaul, S.N., 2003. Microaerophilic-aerobic sequential batch reactor for treatment of azo dyes containing simulated wastewater. *Process Biochemistry*, Volume 40, Issue 2, February 2005, Pages 885-890

Palmer,J and C. Sternberg, Modern microscopy in biofilm research: confocal microscopy and other approaches. *Curr. Opin. Biotechnol.*, 10 (1999) 263–268.

Patel,R Suresh, S. 2008. Kinetic and equilibrium studies on the biosortion of reactive black 5 dye by *Aspergillus foetidus. Bioresource Technology*, Vol 99, No 1, pp 51-58

Pinna V., M Zema, C Gessa, and A Pusino; 2007. Structural Elucidation of Phototransformation Products of Azimsulfuron in Water; *J. Agric. Food Chem* (2007) 55, 6659-6663

Pulgarín, C. J. Kiwi, (1995) Iron Oxide-Mediated Degradation, Photodegradation, and Biodegradation of Aminophenols; *Langmuir* (1995) 11, 519-526.

Stolz, A., 2001. Basic and applied aspects in the microbial degradation of azo dyes. *Applied Microbiology and Biotechnology* 56, 69e80.

Tandy, S., Healey, J.R., Nason, M.A., Williamson, J.C., Jones, D.L., 2009. Heavy metal fractionation during the co-composting of biosolids, deinking paper fibre and green waste. Bioresour. Technol. 100, 4220–4226.

Volesky, B. (2007) *Biosorption and me.* Water Research, 41 (4017-4029)

Wang, Y. Yang Mu, Quan-Bao Zhao, Han-Qing Yu (2006). Isotherms, kinetics and thermodynamics of dye biosorption by anaerobic sludge. *Separation and Purification Technology*, 50, 1, (1-7)

Wang; C. 2007 Photocatalytic activity of nanoparticle gold/iron oxide aerogels for azo dye degradation; *Journal of Non-Crystalline Solids* 353 (2007) 1126–1133.

Zhang, C. U. Tezel, K. Li, D. Liu, R. Ren, J. Du, S.G. Pavlostathis, Evaluation and modeling of benzalkonium chloride inhibition and biodegradation in activated sludge, *Water Res.* 45 (2011) 1238-1246.

Permissions

The contributors of this book come from diverse backgrounds, making this book a truly international effort. This book will bring forth new frontiers with its revolutionizing research information and detailed analysis of the nascent developments around the world.

We would like to thank Dr. Tomasz Puzyn and M.Sc. Eng Aleksandra Mostrag-Szlichtyng, for lending their expertise to make the book truly unique. They have played a crucial role in the development of this book. Without their invaluable contribution this book wouldn't have been possible. They have made vital efforts to compile up to date information on the varied aspects of this subject to make this book a valuable addition to the collection of many professionals and students.

This book was conceptualized with the vision of imparting up-to-date information and advanced data in this field. To ensure the same, a matchless editorial board was set up. Every individual on the board went through rigorous rounds of assessment to prove their worth. After which they invested a large part of their time researching and compiling the most relevant data for our readers. Conferences and sessions were held from time to time between the editorial board and the contributing authors to present the data in the most comprehensible form. The editorial team has worked tirelessly to provide valuable and valid information to help people across the globe.

Every chapter published in this book has been scrutinized by our experts. Their significance has been extensively debated. The topics covered herein carry significant findings which will fuel the growth of the discipline. They may even be implemented as practical applications or may be referred to as a beginning point for another development. Chapters in this book were first published by InTech; hereby published with permission under the Creative Commons Attribution License or equivalent.

The editorial board has been involved in producing this book since its inception. They have spent rigorous hours researching and exploring the diverse topics which have resulted in the successful publishing of this book. They have passed on their knowledge of decades through this book. To expedite this challenging task, the publisher supported the team at every step. A small team of assistant editors was also appointed to further simplify the editing procedure and attain best results for the readers.

Our editorial team has been hand-picked from every corner of the world. Their multi-ethnicity adds dynamic inputs to the discussions which result in innovative outcomes. These outcomes are then further discussed with the researchers and contributors who give their valuable feedback and opinion regarding the same. The feedback is then collaborated with the researches and they are edited in a comprehensive manner to aid the understanding of the subject.

Apart from the editorial board, the designing team has also invested a significant amount of their time in understanding the subject and creating the most relevant covers. They scrutinized every image to scout for the most suitable representation of the subject and create an appropriate cover for the book.

The publishing team has been involved in this book since its early stages. They were actively engaged in every process, be it collecting the data, connecting with the contributors or procuring relevant information. The team has been an ardent support to the editorial, designing and production team. Their endless efforts to recruit the best for this project, has resulted in the accomplishment of this book. They are a veteran in the field of academics and their pool of knowledge is as vast as their experience in printing. Their expertise and guidance has proved useful at every step. Their uncompromising quality standards have made this book an exceptional effort. Their encouragement from time to time has been an inspiration for everyone.

The publisher and the editorial board hope that this book will prove to be a valuable piece of knowledge for researchers, students, practitioners and scholars across the globe.

List of Contributors

Amilcar Machulek Jr. and Fabio Gozzi
Universidade Federal de Mato Grosso do Sul, Departamento de Química – UFMS, Brazil

Volnir O. Silva, Leidi C. Friedrich and Frank H. Quina
Universidade de São Paulo, Instituto de Química and NAP-PhotoTech – USP, Brazil

José E. F. Moraes
Universidade Federal de São Paulo, Escola Paulista de Engenharia – UNIFESP, Brazil

Cláudia Telles Benatti
Faculdade Ingá – UNINGÁ, Brazil

Célia Regina Granhen Tavares
Universidade Estadual de Maringá – UEM, Brazil

Malik Mohibbul Haque and Mohammad Muneer
Department of Chemistry, Aligarh Muslim University, India

Detlef Bahnemann
Institut fuer Technische Chemie, Leibniz Universität Hannover, Germany

Fanxiu Li
College of Chemical & Environmental Engineering, Yangtze University, Jingzhou, Hubei, China

Junko Hara
Institute for Geo-resources and Environment, National Institute of Advanced Industrial Science and Technology, Japan

Ruixia Wei and Shuguo Zhao
Hebei Polytechnic University, Tangshan, Hebei, China

Songsak Klamklang
Technology Center, SCG Chemicals Co., Ltd., Siam Cement Group (SCG), Bangkok, Thailand

Hugues Vergnes
Laboratoire de Génie Chimique, UMR CNRS 5503, BP 84234, INP-ENSIACET, France

Kejvalee Pruksathorn and Somsak Damronglerd
Department of Chemical Technology, Faculty of Science, Chulalongkorn University Bangkok, Thailand

A. G. Chmielewski
Institute of Nuclear Chemistry and Technology, Warsaw, Poland
University of Technology, Warsaw, Poland

Yongxia Sun
Institute of Nuclear Chemistry and Technology, Warsaw, Poland

Petr Kačer, Jiří Švrček, Kamila Syslová and Jiří Václavík
Institute of Chemical Technology, Prague, Czech Republic

Dušan Pavlík, Jaroslav Červený and Marek Kuzma
Institute of Microbiology, Prague, Czech Republic

Roberto Candal, Marta Litter, Lucas Guz, Elsa López Loveira, Alejandro Senn and Gustavo Curutchet
Centro de Estudios Ambientales: Escuela de Ciencia y Tecnología e Instituto de Investigaciones e Ingeniería Ambiental, Universidad de San Martín, Campus Miguelete, Gerencia Química, Comisión Nacional de Energía Atómica, San Martín, Prov. de Buenos Aires, Argentina

Printed in the USA
CPSIA information can be obtained
at www.ICGtesting.com
JSHW011425221024
72173JS00004B/674